FOOD AND AGRICULTURAL POLICY

ECONOMICS AND POLITICS

McGraw-Hill Series in Agricultural Economics

CONSULTING EDITOR

Peter J. Barry, *University of Illinois*

Halcrow, Spitze, and Allen-Smith: Food and Agricultural Policy: Economics and Politics
Kay and Edwards: Farm Management
Looney and Uchtmann: Agricultural Law: Principles and Cases
Norton and Alwang: Introduction to Economics of Agricultural Development
Seitz, Nelson, and Halcrow: Economics of Resources, Agriculture, and Food

FOOD AND AGRICULTURAL POLICY

ECONOMICS AND POLITICS

SECOND EDITION

Harold G. Halcrow
Department of Agricultural Economics
University of Illinois, Urbana-Champaign

Robert G. F. Spitze
Department of Agricultural Economics
University of Illinois, Urbana-Champaign

Joyce E. Allen-Smith
Department of Agricultural Economics
University of Illinois, Urbana-Champaign

McGRAW-HILL, INC.
New York St. Louis San Francisco Auckland Bogotá
Caracas Lisbon London Madrid Mexico City Milan
Montreal New Delhi San Juan Singapore Sydney Tokyo Toronto

This book was set in Times Roman by Ruttle, Shaw & Wetherill, Inc.
The editors were Anne C. Duffy and John M. Morriss;
the production supervisor was Kathryn Porzio.
The cover was designed by David Walters.
Project supervision was done by Tage Publishing Service, Inc.
R. R. Donnelley & Sons Company was printer and binder.

FOOD AND AGRICULTURAL POLICY

Economics and Politics

This book is printed on acid-free paper.

1 2 3 4 5 6 7 8 9 0 DOC DOC 9 0 9 8 7 6 5 4 3

ISBN 0-07-025800-7

Library of Congress Cataloging-in-Publication Data

Halcrow, Harold G.
Food and agricultural policy: economics and politics / Harold G. Halcrow, Robert G. F. Spitze, Joyce E. Allen-Smith.—2nd ed.
p. cm.
Rev. ed. of: Agricultural policy analysis / Harold G. Halcrow. 1984.
Includes index.
ISBN 0-07-025800-7 (alk. paper)
1. Agriculture and state—United States. 2. Agriculture and state. 3. Food industry and trade—Government policy—United States. I. Spitze, R. G. F. II. Allen-Smith, Joyce E. III. Halcrow, Harold G. Agricultural policy analysis. IV. Title.
HD1761.H3513 1994 93-17345
338. 1′873—dc20

ABOUT THE AUTHORS

HAROLD G. HALCROW earned his Ph.D. in Economics from the University of Chicago, where he was a Farm Foundation Fellow. He received his B.S. from North Dakota State University and his M.S. from Montana State University. Currently, he is Professor Emeritus of Agricultural Economics at the University of Illinois, Urbana-Champaign, where he has taught and conducted research since 1957, and also served as head of the Department of Agricultural Economics for 13 years. Prior to this he served on the faculties of Montana State University and the University of Connecticut. He has also served as Visiting Professor at Stanford and the University of California, Berkeley. Dr. Halcrow is author or coauthor of eight books, numerous research publications, journal articles and extension reports, and editor of two books. He served as book review editor and then editor of the *Journal of Farm Economics,* predecessor to the *American Journal of Agricultural Economics.* He is a Fellow of the American Agricultural Economics Association (AAEA), and has been cited five times by the AAEA and the former American Farm Economic Association for professional excellence in research or communication. He served for 20 years as the AAEA representative on the Board of Directors of the National Bureau of Economic Research, New York, and has worked as consultant to the U.S. Department of Commerce, the U.S. Department of Agriculture, and the State of Illinois Economic Technical Advisory Committee.

ROBERT G. F. SPITZE is Professor of Agricultural Economics at the University of Illinois, Urbana-Champaign, with a Ph.D. from the University of Wisconsin, Madison. He has previously served on the faculties at the University of Wisconsin at Madison, University of Tennessee at Knoxville, and as Visiting Professor at the University of London Wye College and USDA Economic Research Service. Dr. Spitze has been recognized by the American Agricultural Economics Association (AAEA) with its awards for Distinguished Teaching, Policy Contribution, Quality of Communication for Publication (Honorable Mention), and International Association of Agricultural Economists (IAAE) Travel Grant. He is author of numerous professional articles and bulletins published by the AAEA, IAAE, regional associations, universities, and private media. Dr. Spitze has edited two policy books, (1) *Agricultural and Food Price and Income Policy: Alternative Directions for the United States and the Implications for Research* and (2) *Agricultural and Food Policy: Issues and Alternatives for the 1990s,* and was founding editor of *Policy Research Notes.* He has been a leader in professional research committees focusing the attention of the profession and policy-making public on the issues, knowledge, and choices relevant to the 5-year cycles of agricultural and food policy development in the United States.

JOYCE E. ALLEN-SMITH is Associate Professor of Agricultural Economics at the University of Illinois, Urbana-Champaign. She received her B.S. in Economics from North Carolina A & T State University, and her Ph.D. in Agricultural Economics from the University of Illinois, Urbana-Champaign. She has held the following positions: economist with the Economic Research Service, U.S. Department of Agriculture, where she was the project leader for food assistance policy research; senior research associate at the Joint Center for Political and Economic Studies, Washington, D.C.; and has been a Kellogg Fellow at the National Center for Food and Agricultural Policy, Resources for the Future. Dr. Allen-Smith has published numerous articles, bulletins, and chapters in books focusing on food policy, poverty, and human resource development. She has received several Administrator's Special Merit Awards from the Economic Research Service for outstanding contributions in these areas. Dr. Allen-Smith is an active member of the American Agricultural Economics Association (AAEA), having served as chair of the Committee on Opportunities and Status of Blacks in Agricultural Economics, on the Steering Committee of the Committee on Women in Agricultural Economics, on the advisory council of *Choices,* and on various other committees including the award committee for Distinguished Policy Contribution.

CONTENTS

PREFACE

Food and Agricultural Policy: Economics and Politics, Second Edition, is an entirely new textbook focusing on public policies and alternatives of the twentieth century and for the emerging issues of the twenty-first. The book encompasses the agricultural producing and the food consuming sectors. It recognizes that the foundations of public policy rest on both the general theory of economics and the theory and practice of politics, or political science.

Food and agricultural policy is growing both in scope and importance. Hence, *Food and Agricultural Policy: Economics and Politics* concerns the actions that individuals take through their government to develop the food and agricultural sector, stabilize the incomes of producers and the food supply of consumers, facilitate agricultural trade, protect and improve the environment, assure the quality and safety of food and drinking water, and encourage development of the rural economy. New policies must be developed to achieve production that is sustainable, and that is capable of meeting rapid growth in the world population while minimizing the adverse effects on the environment. Now is an exciting time to study these issues and their policy options.

In addition to analyzing national agricultural price and income policy, the authors identify and analyze important factors in the development and growth of agriculture and the food economy. Natural resources policy is highlighted in terms of actions required to sustain increasing productivity and growth. Problems of improving food quality and safety are recognized. Significant effects of monetary policy on agriculture and rural communities are examined and linked to agricultural credit and the nation's fiscal policies. Human capital policy, which involves research, education, and training, is studied as an integral part of food and agricultural policy. The final chapter, ''Looking forward to the Twenty-First Century,'' brings together some of the problems, uncertainties, possibilities, and opportunities for food and agricultural policy makers in the years and decades ahead.

This book has been designed and written to be used as the text for the first general course in food and agricultural policy at either the advanced undergraduate or beginning graduate level in colleges and universities. A solid foundation in economic principles and theory is important and assumed. Other helpful subjects include farm management or production economics, agricultural prices and marketing, some of the physical and biological sciences relating to agriculture, political science, food science, sociology, and geography.

This project began as major revision of the award-winning *Agricultural Policy Analysis* (McGraw-Hill, 1984). As the work progressed, it became increasingly evident to the authors that this is indeed a new, forward-looking study. During this period, several distinguished policy researchers and teachers made important contributions with detailed and very helpful reviews. These individuals are: Andrew P. Barkley, Kansas State University; Paul W. Barkley, Washington State University; Peter J. Barry, University of Illinois; Josef M. Broder, The University of Georgia; John Campbell, Oklahoma State University; Neilson C. Conklin, Arizona State University; Michael Dicks, Oklahoma State University; A. L. Frederick, University of Nebraska, Lincoln; James Kendrick, University of Nebraska, Lincoln; Marshall A. Martin, Purdue University; Daryll E. Ray, University of Tennessee; Tim Wallace, University of California, Berkeley; and Robert Weaver, Pennsylvania State University.

Many others have also made important contributions. The authors were encouraged throughout by the enthusiastic support of Anne Duffy, agricultural editor at McGraw-Hill. The readability of the book was improved by the editing of Sheila A. Ryan, editorial consultant, Champaign, Illinois. The most indispensable help in production of the manuscript was the highly efficient and excellent secretarial work provided by Barbara Bohor, Stephanie Spaulding, Phyllis Blackford, Kathy Abbott, Ann Marie Berner, Pat Roosevelt, and Elaine Ebeling, members of the agricultural economics staff at the University of Illinois. The authors have received generous support and encouragement from David L. Chicoine, department head. In addition, colleagues at the University of Illinois and students have contributed through discussions, questions, and comments.

With all of this excellent help and advice, there should be few errors of omission or commission. But to the extent that there are, the authors accept responsibility.

Harold G. Halcrow
Robert G. F. Spitze
Joyce E. Allen-Smith

FOOD AND AGRICULTURAL POLICY

ECONOMICS AND POLITICS

PART **ONE**

CONTENT AND SCOPE OF FOOD AND AGRICULTURAL POLICY

CHAPTER 1

WHAT IS FOOD AND AGRICULTURAL POLICY?

Throughout the world, the relationships between government and agriculture are important determinants of how the food and agricultural economy will be organized, develop and grow, prosper, and relate to other parts of the economy. These relationships define the scope and substance of food and agricultural policy.

IN THIS CHAPTER

By studying Chapter 1, you will learn how to answer the following questions:

1. What is the scope and substance of food and agricultural policy?
2. What are the determinants of such policy?
3. What are some of the important problems and issues for food and agricultural policy?

DEFINING POLICY

What is food and agricultural policy? What is its range or scope? To answer these questions, we must first determine what is meant by policy. Then we can analyze the scope of public food and agricultural policy in an economic context, explain why we have such policy, and understand the importance of studying it.

A Definition of Policy

Policy is defined as a deliberate course of action, as contrasted with a haphazard or capricious type of activity, chosen and followed by a public body, private firm, family, or individual. Policy generally implies wisdom or prudence in managing affairs, based on a definitive plan or program created through a process of thought and reason. This does not imply perfect knowledge or infallibility. Policy is a universal activity of individuals, families, firms, political movements, organized groups, governments, and agencies of government.

Purpose of Policy

Existence of a policy signals that there was or is some problem for which someone considered alternative responses and then chose a course of action that seemed desirable in view of existing knowledge and feelings as well as foreseeable options. Neither societies nor individuals have to ponder whether problems or policy issues will arise. They occur inevitably in the course of people living, an economy developing, and a nation surviving. Difficulties arise. Established courses of action are found wanting. Different values emerge. New knowledge changes our perceptions. Conflicts appear among individuals and groups. For instance, the waste from one person's agricultural production may become pollutants for other people as well as the producer. An existing policy may become untenable so that a new policy must be forged.

Policy is formed by examining newly perceived problems, by exploring alternative choices and their expected results, and by choosing a new, more satisfying course of action. A policy is selected because the anticipated consequences or ends, along with the associated costs or means, are valued as superior to other alternative actions examined. Hence, the purpose of policy is to solve a specific problem by acting to achieve desired results.

Distinguishing Public Policy

Distinguishing public policy from private policy rests on how the policy is made rather than whom it affects. Public policy is made by individuals in their roles as citizens; in groups and organizations with a political objective or goal; and in participatory governments at local, state, or national levels. By contrast, private policy is made by individuals, families, firms, and policy makers of authoritarian governments. Through actions of government, policy affects everyone, but the decisions are determined by distinctively different processes in participatory and authoritarian political systems. A study of policy does not cover the full discipline of political science, but it involves the political systems on which governments are based. On this basis, we draw sharp distinctions among different systems for making policy.

In a democracy, such as the United States, numerous individuals and interest groups are involved in making policy, as contrasted with an authoritarian government that is run by one individual or a single group, such as a military dictatorship or a one-party state. Hence, in the United States, making public policy involves many compromises by different individuals and interest groups, represented by various organizations, political parties, and governmental agencies. Public policy is an action based on the quest for desirable results, as citizens act through government, to solve a problem experienced by many individuals. Even though public policy is designed to be in the public interest, it does not follow that it will be perceived as being in everyone's private interest.

Defining Public Food and Agricultural Policy

Public food and agricultural policy (hereafter food and agricultural policy) may be defined as a course of public action taken by participatory governments directed primarily, but not exclusively, toward the problems of the farm, rural communities, food consumption, and agribusiness sectors of society. Broadly speaking, this policy involves the full range of public decisions that influence what products individuals and firms decide to produce, how they will be produced, and for whom they will be produced. Figure 1-1 provides a schematic of the definition of the inclusive concept of policies.

Food and agricultural policy applies mainly to two broad sets of markets: the agricultural input markets, through which the resources and commodities used in farming (and ranching) are made available for production; and the agricultural product markets, through which food products are processed and marketed for consumption at home and abroad. Policy in the input markets applies to the use of land and other natural resources, agricultural credit, industrial products used by agriculture, and the human resources employed in all sectors of agriculture. Policy in the product markets applies to quantities to be marketed, prices, producer incomes, food distribution, and trade. Product-market policy involves issues such as laissez-faire markets, price supports, production and marketing controls, target prices, grain reserves, international trade in farm food products, trade agreements, export subsidies, tariff and nontariff barriers to trade, food subsidies, and food inspections.

REASONS FOR PUBLIC FOOD AND AGRICULTURAL POLICY

We have food and agricultural policy for the same reason that we have democratic governments. It is a necessary part of public order and is designed to improve public welfare when private actions result in unacceptable conflict. In all societies, problems arise regarding the production, distribution, and safety of food and the economic well-being of farm families, rural nonfarm families, and rural communities, but not all societies decide their governmental policy the same way.

Over the first 200 years of U.S. history, from the founding of the nation through the second decade of this century, food and agricultural policy was dominated by what may be called *developmental policy*—governmental policy directed at developing and supporting family farms and the inputs of the total agricultural sector. This policy shaped the economic structure of agriculture, and the conditions under which food would be produced and the system would develop and grow. The Morrill Act of 1862, which initiated the land-grant college system, was one in a long series of acts that provided public support for agricultural research and education. This policy provided the foundation for the scientific and technological revolution that has transformed traditional agriculture into its modern commercial status.

The increase in output of this modern industrial system has forced down prices of food and other agricultural products relative to the prices of most inputs used in production. Historically, returns to farm labor have been unstable and periodically depressed. These conditions contributed to policy makers enacting *price and income policy*—governmental policy directed primarily at supporting agricultural commodity

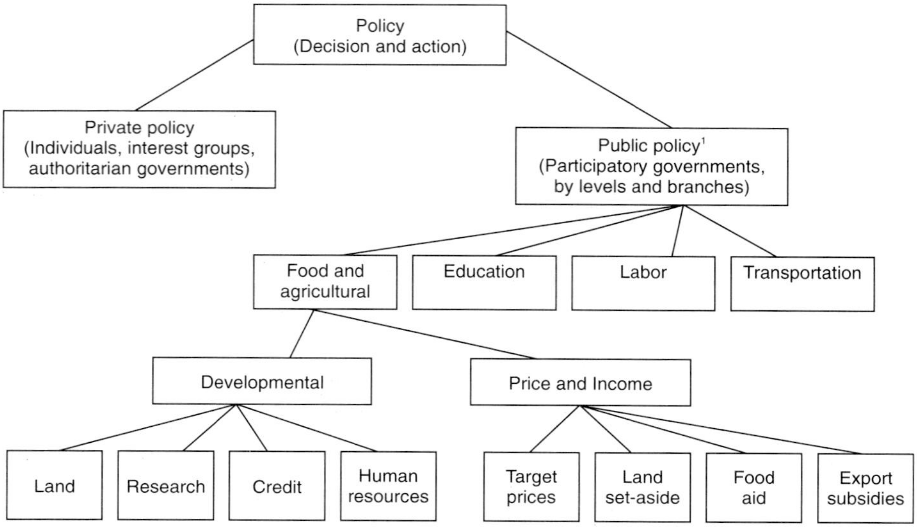

[1]The public policies that are cited are representative examples.

FIGURE 1-1
A schematic classification of policy, emphasizing food and agricultural policy.

prices and the incomes of farmers. Although the farm population has declined, primarily as a result of economic forces, there has remained considerable political pressure from both farm and agribusiness interests for government to continue to aid farmers.

So the farm sector, which traditionally has been as close as any sector to the model of free enterprise under pure competition, involves public policy for at least four reasons. Some of these reasons are characterized in economics as market failures. First, increasing the productivity of agriculture is essential for meeting the growing world demand for food, and government is widely and deeply involved in helping almost all sectors of agriculture to increase their productivity. Second, economic stability is important, too, and this has involved some governmental policy. Third, government policy greatly influences income distribution between the farm and nonfarm sectors as well as among consumers. As income distribution is often not satisfactory and poverty rates are relatively high among some population groups in rural society, there is a continuing interest in policy. Fourth, some private actions by farmers may result in excessive soil erosion, damage to the environment, and contamination of food and drinking water. The detrimental effects of such actions necessitate a public response because the major costs are widely shared throughout the economy.

EVOLUTION OF PUBLIC FOOD AND AGRICULTURAL POLICY

The evolution of food and agricultural policy is characterized by continual revision and expansion, rather than by abrupt, dramatic shifts. In North America, for example, for many centuries before Europeans settled on the eastern shores of this continent, most

Native American tribes had customs dealing with hunting, fishing, and territorial priorities. Sometimes when agreements were broken, warfare ensued. Then in the late 1700s, with the founding of the United States, new policies were fashioned for property rights, land distribution, and ownership. Agricultural policies for the entire nation were structured.

Since then, food and agricultural policy development can be characterized as an evolution having at least three important characteristics: (1) an evolution of policy content, (2) more comprehensiveness of problem areas, and (3) an expansion in the number and diversity of interest groups participating directly in policy development. Seldom has there been a complete reversal of policy or even major redirection, except in times of crisis. Rather, change has been gradual and incremental. The policy was never concerned just with farm production, and it is even less so now. Food and agricultural policy encompasses many areas including conservation, trade, food aid, credit, and rural development. Nor was it ever just the concern of farm groups, and certainly it is not now. Others, including consumers, rural residents, and taxpayers, have a stake in policy. In addition, the very terminology that identifies it has changed.

A schematic, historical perspective of both the changing terminology and the problem areas is presented in Figure 1-2. In the earliest years, *farm policy* was usually the term applied to policy that was directed at issues of farmers and the rural society. In the twentieth century, the term gradually changed to *agricultural policy*. Finally, in recent decades, a commonly used term is *agricultural and food policy*, or *food and agricultural policy*.

The categories of policy listed in Figure 1-2, however, barely hint at the broad and sweeping drama of the policy story. The early developmental policies involved laws and rules for land distribution; public education and research; the building of rural roads, canals, and railroads; development of banking and credit to serve rural needs; maintenance of public law and order for a rural society; the beginning of irrigation and regulation of marketing facilities; and the beginning of a conservation policy for minerals, forests, and related natural resources. These early public policies substantially shaped agriculture's distinctiveness and contributed greatly to its productivity.

Since the beginning of the 1920s, price and income policy for food and agriculture has been driven by a changing set of policy circumstances, by accelerating rates of change, by the advances of new technology, and by a broadening of domestic and world markets. The United States and nearly all other countries have faced new sets of problems that have challenged each nation's policy-making process. The goal of food and agricultural policy study and analysis is to learn how to approach the subject, using the various disciplines that are involved in policy making. The chief ones are economics and politics (or the political economy), and this is the emphasis of this book.

Importance of Studying Public Food and Agricultural Policy

Because of the importance of food and agricultural policy in influencing our lives, self-interest alone may be sufficient reason to make studying it worthwhile. The public choices affect our daily lives and well-being through their effects on the availability, price, and quality of food; the safety of our drinking water; the cleanliness of our lakes and rivers; the incomes of farmers; and the state of the rural economy. Our response

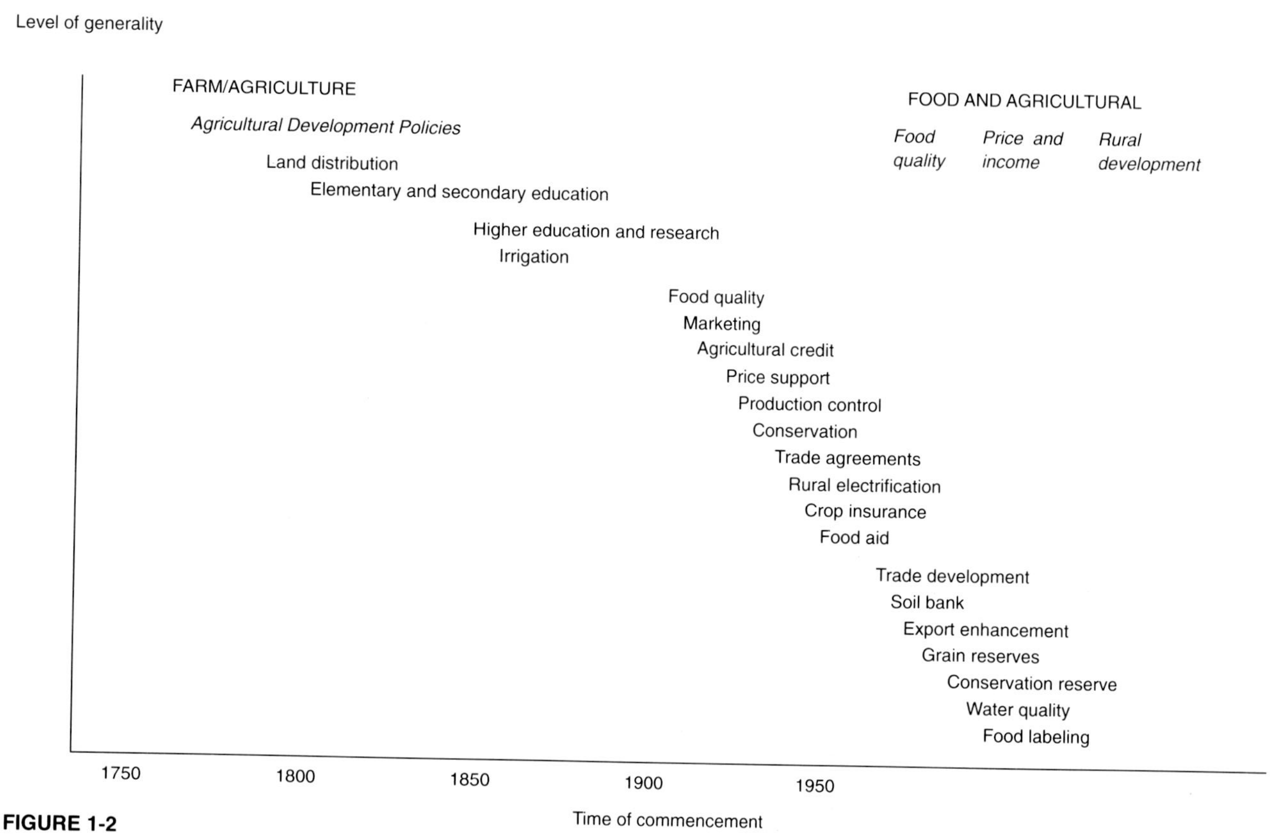

FIGURE 1-2
Evolution of food and agricultural policy in the United States

to these choices will affect our prosperity, health, and happiness. Also, study of food and agricultural policy is—or should be—worthwhile because of our current and future roles in the political processes of our democratic system. One way or another, through these roles we each help determine what choices are made.

Analyzing the effects of current public policy and alternative policy responses is the essence of policy study. You can learn how to conduct the analyses by studying food and agricultural policy, and this will benefit you throughout life.

BELIEFS, VALUES, AND PUBLIC POLICY

All recommendations of policy involve judgments of value concerning what shall or shall not be done, and why. These judgments are rooted in beliefs about what is good, or thought to be good and hence of value, and involve judgment as to what should or should not be done to try to achieve that which is deemed good. Such beliefs differ from those dealing simply with what is true in an empirical sense, to which no particular values may be attached.

Beliefs and Values Underlying Policy

People often share the same beliefs about what is good and what is not, but they may act in different ways because of differences in the strength of their beliefs or in the relative importance they attach to their beliefs. Myths as well as factual understanding affect beliefs. Basic to the democratic principles, there is no single enduring correct answer to the question of what values and weights shall be accepted and consequently no single correct concept of the social benefits and costs of a particular policy or program. This means that all public policy is subject to comment and criticism from one standpoint or another or is subject to valuation in terms of individual and group preferences. Markets provide measures of certain economic costs encountered by firms, but there are other costs that are external to a firm, such as pollutants or soil erosion, and only a well-designed public policy may be able to deal with these costs effectively.

The list of important values that underlie food and agricultural policy decisions is long and includes: national security, public safety, freedom, progress, standard of living, stability, equality, individual rights, free speech and press, free assembly and religion, justice, opportunity, family farm, fair prices, parity, productivity, and various creeds such as those of democracy, private or public enterprise, and self-integrity. Governmental involvement in food and agricultural policy is also related to concerns in the total society about achieving adequate productivity to meet growing food needs, about attaining and maintaining economic stability in agriculture, and about more satisfactory income distribution among farm and nonfarm families.

Distinguishing Policy from Beliefs, Values, and Goals

One of the most difficult concepts is how to distinguish policy from values, beliefs, or goals. Policy is affected by beliefs and values, yet it is critically different. Values are generalized attributes of a state of mind or of an object, while policies are particular representations of values. High income may have exceptional value to producers. Food security has value to consumers. Neither of these goals or values, however, may be adequate justification for a particular policy placing controls within the food system.

Similarly, high income or food security may be viewed as an end for which a means is sought. However, to separate such an end from the means to achieve it, either in theory or in practice, can lead to unwanted consequences and flawed policy making. A rational policy cannot be chosen as desirable without assessing the costs of achieving it. Also, it is not possible to rationally choose a policy means without assessing the likely consequences that may result from the policy. For instance, adverse effects must be considered along with a particular suggested or targeted end.

A policy, as a managed and rationally chosen course of action, implies an evaluation of the benefits as well as the costs of the action itself. For example, to ensure that its food supply is completely safe, a country might be wise to choose a policy of having all meat inspected for evidence of disease. Yet with the same general values and ends, a society might choose a quite different policy, or even no policy, after considering alternative procedures, costs, and the likely results. Similarly, a farm producer might choose to diversify among several enterprises with a policy directed at income maintenance and risk aversion, even though alternative enterprise schemes would be expected to earn a higher average but more variable income. Members of a society may espouse the value of equality while rationally choosing policies that result in a skewed income distribution.

Beliefs, values, and goals contribute to policy, along with many other factors, but they are not policy. Policy embodies both means and ends, jointly identified, evaluated, and chosen.

PROBLEMS FOR PUBLIC FOOD AND AGRICULTURAL POLICY

Food and agricultural policy deals with issues and problems that are of concern to the public and to their elected and appointed representatives. Policies arise from problems—real, persistent, troublesome, hurting problems—experienced by private decision makers and public officials in the course of their living, producing, and consuming or administering public programs. Policy makers do not wake up in the morning wondering what policy they will make that day. Rather, they are preoccupied with assessing which of the myriad of problems merit public attention and which, if any, should remain in the arena of private policy.

Public food and agricultural policies focus on problems in the production, marketing, trade, distribution, and consumption of food and fiber; the level and stability of farm income; human resource adjustment; and the development of rural communities. Problems and policies related to food and agriculture have been prominent throughout the history of the United States, from issues concerning the ownership and use of farmland during colonial times to the application of agricultural chemicals today. The current list of problems for policy is long and includes:

- maintaining adequate agricultural prices and incomes
- facilitating world agricultural trade
- assuring sufficiency and quality of food
- using land and water resources wisely

- financing the agricultural sector
- improving the productivity and mobility of human capital
- developing the rural economy
- meeting the challenges of emerging public issues

This book focuses on the major problems experienced in the food and agricultural sector and alternative policies to deal with the problems. While the book uses examples primarily from the United States, their implications are not limited to one country. Policies of any nation can be understood only in the environment of economic, social, and political forces and the resulting policies of other countries, which increasingly are interrelated worldwide.

LOOKING AHEAD

As we look to the future, we must realize that the success of food and agricultural policy is tied to three interlocking goals:

1 achieving a prosperous, dynamic, and growing food and agricultural system
2 solving the major problems experienced by producers, processors, distributors, and consumers within the food and agricultural sector
3 developing the system of agricultural production and marketing so that it is sustainable for future generations

In the next two chapters, we focus on the participants in the policy-making process and how public policy is made. Chapter 4 builds the foundation for understanding policy by examining the U.S. economic situation for food and agricultural policy. Chapter 5 describes the world economic situation as it affects food and agricultural policy. Chapters 6 through 11 deal with price and income policy, international trade policy, and food aid policy, all of which were initially designed to support and stabilize farm and food prices and producers' incomes. Major policies relating to agricultural resource use and conservation, food quality and safety, farm credit, and human capital are covered in Chapters 12 through 15. In the final chapter we project some of the more specific policy alternatives for the twenty-first century.

TERMS AND CONCEPTS

Policy	Developmental policy
Public policy	Price and income policy
Private policy	Policy evolution
Food and agricultural policy	Beliefs, values, goals
Input markets policy	Means and ends of policy
Product markets policy	Policy problems

QUESTIONS AND STUDY SUGGESTIONS

1 What is policy? Public policy? Food and agricultural policy?
2 Why do we have public food and agricultural policy?

3 What have been some of the main developments during the evolution of U.S. public food and agricultural policy?
4 What is a policy-related belief? A value? Explain how these concepts help to demonstrate why people may differ on the acceptability of a proposed policy.
5 What have been the most important problems for food and agricultural policy? Give reasons for your answer.
6 What do you believe will be the most important problems for food and agricultural policy in the next decade? In the next half century?
7 When a public policy is adopted, what effect does it have on the discretions of people to take actions and achieve their wants?

CHAPTER **2**

WHO MAKES FOOD AND AGRICULTURAL POLICY?

Policies prevail wherever societies are found throughout the world and are as old as the history of the world. Public food and agricultural policies evolve as governments respond to the political will of citizens and attempt to ensure that the most basic needs of their people are met. The purpose of this chapter is to examine who makes these policies, what roles governmental institutions play in policy making in the United States, and which interest groups help to shape public policy.

IN THIS CHAPTER

By studying Chapter 2, you will learn how to answer the following questions:

1. Who are the participants in public policy making?
2. How do the various branches of government contribute to public policy making?
3. How do private interest groups participate in making food and agricultural policy?

PARTICIPANTS IN PUBLIC POLICY MAKING

If policies are decisions and their resulting actions, as found in Chapter 1, who makes or determines them? *Policies, whether private or public, are made by individuals, by interest groups, and by governments.*

Individuals

Individuals make policy in the normal course of living, in pursuing their careers, in functioning as members of groups with similar interests, or in participating in the decision-making processes of a representative government. Problems occur and choices are made in each of these activities to achieve desired results. A student decides to take a particular course; a farmer decides to buy additional land; a consumer decides on a consumption pattern; or an investor decides to purchase stock in a corporation.

In each case, individuals as policy makers confront a problem and adopt a policy that can be identified and studied. However, because of the uniqueness of each decision maker and the complexity of internal and external factors determining the choice, identifying just why the policy was chosen often eludes us. Similarly, even though some of the results of each policy can be identified and analyzed, other results unique to certain individuals or groups will elude investigation. Regardless of difficulties inherent in analyzing the rationale for individual private policy making, it seems clear that the decision grew out of a felt need to resolve a problem in such a way as to achieve some anticipated outcome.

Individual participation in public policy making is entirely voluntary, and hence citizens may choose to ignore policy development, thereby accepting out of ignorance or indifference the choices of others even in matters vital to their welfare. An individual who does participate in the public policy process plays one or more of the following distinct roles: (1) citizen; (2) member of an interest group such as corn growers; (3) leader of an occupational organization, an interest group, or a private business such as a farm chemical distributor; (4) staff member of a nonprofit foundation or center such as a consumer federation; (5) public researcher or educator such as an agricultural economics professor; (6) appointed or elected public official such as a member of Congress, and (7) an administrator or a staff member for an official or public agency such as a federal conservation service.

Because individuals may play more than one role—for example, citizen, economist, staff member for a partisan official—they are constantly challenged to properly discharge their distinct private and public responsibilities. In these various policy roles, individuals vote, evaluate policies and alternatives, discuss policy issues, personally contact other citizens or decision makers, organize interest groups, design policy proposals, lobby policy makers, or serve as policy makers.

Interest Groups

A public policy-making interest group may be defined as an organization of individuals (or families) who are drawn together primarily for some focused and shared purpose that has emerged as a public issue beyond the realm of private decision making. Such groups might be organized by those who produce identical products, pursue the same occupation, enjoy similar hobbies, are concerned about the same community issue, identify with one political party, or belong to a particular religious denomination. Most interest groups never become involved in the public policy process, but simply pursue their private interests for various economic or social purposes as a corporation, civic club, church, labor union, professional association, or literary society.

However, increasingly in a socially complex, economically interdependent nation, many groups become actively involved in some public issues. Usually they do so because they think that public action can help with unresolved problems in their private endeavors or because they are concerned about a policy being proposed by another interest group. For example, a group of canoeists may advocate governmental action to preserve a naturally flowing river or a group of consumers may advocate a food-reserve program. Furthermore, individual and group decision making may be interrelated. That is, a livestock farmer following private practices to control a cattle disease may advocate a similar policy position for a cattlemen's interest group, which eventually leads to a public policy mandating such management practices. Individuals working jointly in a public policy interest group may be actively pursuing conflicting public goals in other endeavors, may be working toward a shared objective in more than one group, or may have no other association.

Governments

Governments make policy to resolve problems, but all governments do not make policies in the same manner. Many forms and processes of governmental decision making have existed, do exist, or even could exist in the world.

To understand how choices are made, it is important to recognize that around the world at least two distinct forms of governmental policy making exist at opposite poles: (1) a representative form of government, which is designed to encompass the maximum participant input into policy making (also referred to as participatory or democratic decision making); and (2) an authoritarian form of government, which is designed to maximize the input of a singular political leader or group, as the main source of decisions (also referred to as dictatorial decision making). As with all dichotomous classifications, numerous gradations exist between these polar governmental forms.

The representative political system rests on the fundamental premise that diverse values should be allowed among the citizenry, sometimes referred to as "freedom." The authoritarian system rests on the primacy of the singular decision source. The first emphasizes the quality of the composite policies of individuals, their interest groups, and their chosen policy makers. The second emphasizes the quality of the policy sources, be it a person or a small monolithic group, an ultimate authoritative figure, a doctrine, or accepted theory. One focuses on the changing values of the participants, the other on the enduring transcendence of the authority. One exalts pluralism and diversity of policy possibilities, and the other the exclusiveness of the accepted policy. The quality of one is directly proportional to its ability to encompass many informed interests. The quality of the other is proportional to purity of the value source uncontaminated by ill-formed, extraneous influences. One represents many different interests, while the other represents a singular, or at most a few, accepted interests. *Compromise* among the diverse interests is the hallmark of the representative system, while *persuaded or forced acceptance* of the choice of the authority is the hallmark of the authoritarian system.

These forms of governmental policy making are differentiated on the basis of who makes the policy, or the process of decision making, rather than on whom the policy affects. In a participatory system, the government is the composite of the citizens choosing to participate, the interest groups, and the elected or appointed final decision makers. For U.S. public food and agricultural policy, the latter include the president, the appointed officials of the U.S. Department of Agriculture (hereafter USDA) and other related cabinet departments, and the houses of Congress.

Where are public policies to be found? Governments use many different processes at different levels to make their public policies. That variability expresses the search for alternative avenues to represent more effectively the diverse values and political power of the citzen participants. In the United States, as in most other countries, there are national public policies. However, policies also exist at local, state, district, and provincial political and geographical levels. Regardless of level, the forms and processes are similar. Only the extensiveness of the participation and the effects differ. For example, a nation can have a public policy of target prices for corn, along with voluntary set-aside of acreage for producers. A state can have a public policy for inspecting meat-processing plants to ensure food quality. A county or city can have a public policy of zoning for orderly and compatible land uses. In all cases of public policies, only experience can finally test whether the results serve a recognized public purpose.

Public policies are found not only among different levels of government, but also among different branches of each level. Governmental policy is often equated with action by a legislative body such as a county board, a state legislative assembly, a congress, or a parliament. Indeed, these bodies originate many public policies, although they are not the only sources. In the unique policy processes of the United States, public policy has additional origins. The executives, such as a president, governor, mayor, or county executive, must often orginate legislative actions and give final approval to them. Executives must also have authority to make innumerable discretionary policy choices. For instance, when an executive under authority of law decides to ban imports of fruits as a means of disease control or to license the use of certain agricultural chemicals, public policies are shaped as much as if legislation had been adopted. Furthermore, a public policy is shaped when a judge or jury reaches a verdict about the legitimacy of a land tax, the use of stream water for crop irrigation, or regulation of a futures market. One must examine these different governmental decision processes at the different levels, both to understand the effect of existing public policy and to influence policy development.

Interrelations Between Political and Economic Systems

If public policies are developed by governmental processes, then what are the relationships among political and economic systems? All nations have a political system to determine how governmental policy is formed and an economic system to determine how the production and consumption processes are organized. Even though these two systems are interrelated, they have distinct functions, they can be conceptualized separately, and they exist in many different combinations. Once the political system is formed and a government is in control of the nation, the government adopts the economic process or system. Both the political and the economic systems change over time. Hence the kind and extent of a nation's policy is partly determined by the nation's evolving political and economic systems. A schematic diagram of the general types of both systems is presented in Figure 2-1.

The three types of economies in each of the three political systems are simply illustrative of the many combinations that can be defined in greater detail by the disciplines of political science and economics. Further elaborations are unnecessary for our purposes, however. As with all classifications, these are abstractions from the

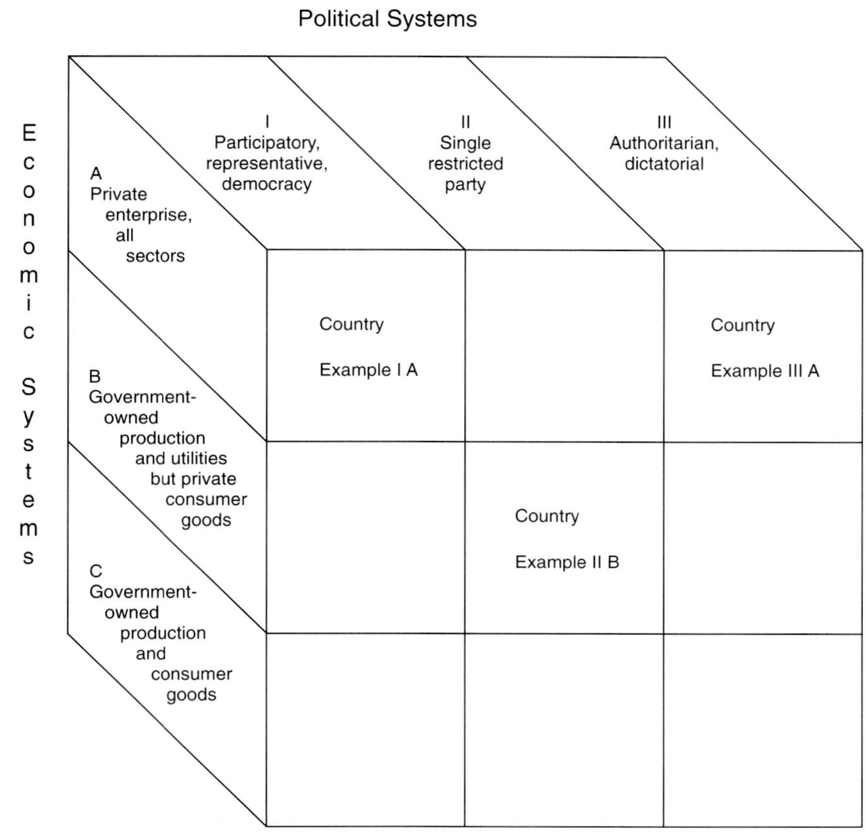

FIGURE 2-1
A matrix of the political and economic systems relevant to public policy.

complexity of the world of experience, where a particular case may not fall neatly within any single classification.

In Figure 2-1, let us designate country I A, for example, as a Type I participatory political system with a Type A private enterprise economic system. This probably best describes the United States and many other representative-type nations; yet the United States is still characterized by considerable public ownership of some production goods, such as utilities, and even some consumption goods, such as national parks. Furthermore, its political processes, even though tending increasingly toward the enfranchisement of all of its people, still fall short of the levels of participation exhibited in some European nations.

It is perhaps too simplistic to equate the Type I political system with the Type A economic system, or Type III political system with Type C. It should be remembered that some countries, such as pre-World War II Japan and Germany, with highly authoritarian governments and capitalist economies, could be represented best by country Type III A. The Soviet Union and most Eastern European countries between World

War II and 1990 could be characterized by country Type II B, even though most housing was also government owned. Every possible combination of country examples could exist in theory and has existed or does exist among the 204 countries and geographic units around the world. Hence agricultural and food policy is developed by many different types of governments, and even when the policy is public, it can take many forms.

STRUCTURE OF U.S. PUBLIC POLICY MAKING

In the United States, public policy is made by both private and public decision makers and groups. The purpose of the policy process is to give citizens the chance to express their preferences when governments at the various levels are developing ways to solve problems experienced by private decison makers. Food and agricultural policies that emerge as the ultimate compromises from this process concern production, distribution, and consumption of agricultural products and the well-being of rural people and their communities.

Schematic of Public Policy Making

Unique in the U.S. participatory political process is the balanced, shared authority among the executive, legislative, and judicial branches of government, each with a distinctive function and all constrained by a very slowly changing federal constitution (see Figure 2-2). In contrast, most other participatory political systems around the world have some form of a parliamentary system in which the executive and legislative functions are collapsed into a single, final decision-making body. These systems also commonly have a judicial arm, but in many cases it is not guided by a constitution. Sometimes they have a second-tier legislative body, such as the United Kingdom House of Lords, but usually without substantive authority. The U.S. political structure has continued to evolve slowly over two centuries, thus generally strengthening its representativeness. For example, voting rights were extended to non–land-owning citizens, African Americans, women, and younger voters; direct election of the president was provided; and periodic reapportionment of the congressional House of Representatives and other legislative bodies across the nation was mandated. The structure of the public policy-making process, with a focus on food and agricultural policy, is presented graphically in Figure 2-3. Primary public participants in policy making are the Congress (legislative), president (executive), and Supreme Court (judicial).

Legislative Participants

Within the legislative branch, the primary participants are the members of the House of Representatives (435 members apportioned among the states on the basis of population) and the Senate (100 members with two per state, irrespective of population). Members of Congress introduce bills, which then must be approved, usually by majority vote in subcommittee, followed by the full committee, and eventually separately by the House and Senate (hereafter, the chambers). At each level, the bill may be substantially altered as compromises are made. Because differences are almost inevitable, reconciliation must come from a conference committee of the two chambers. Approval of that

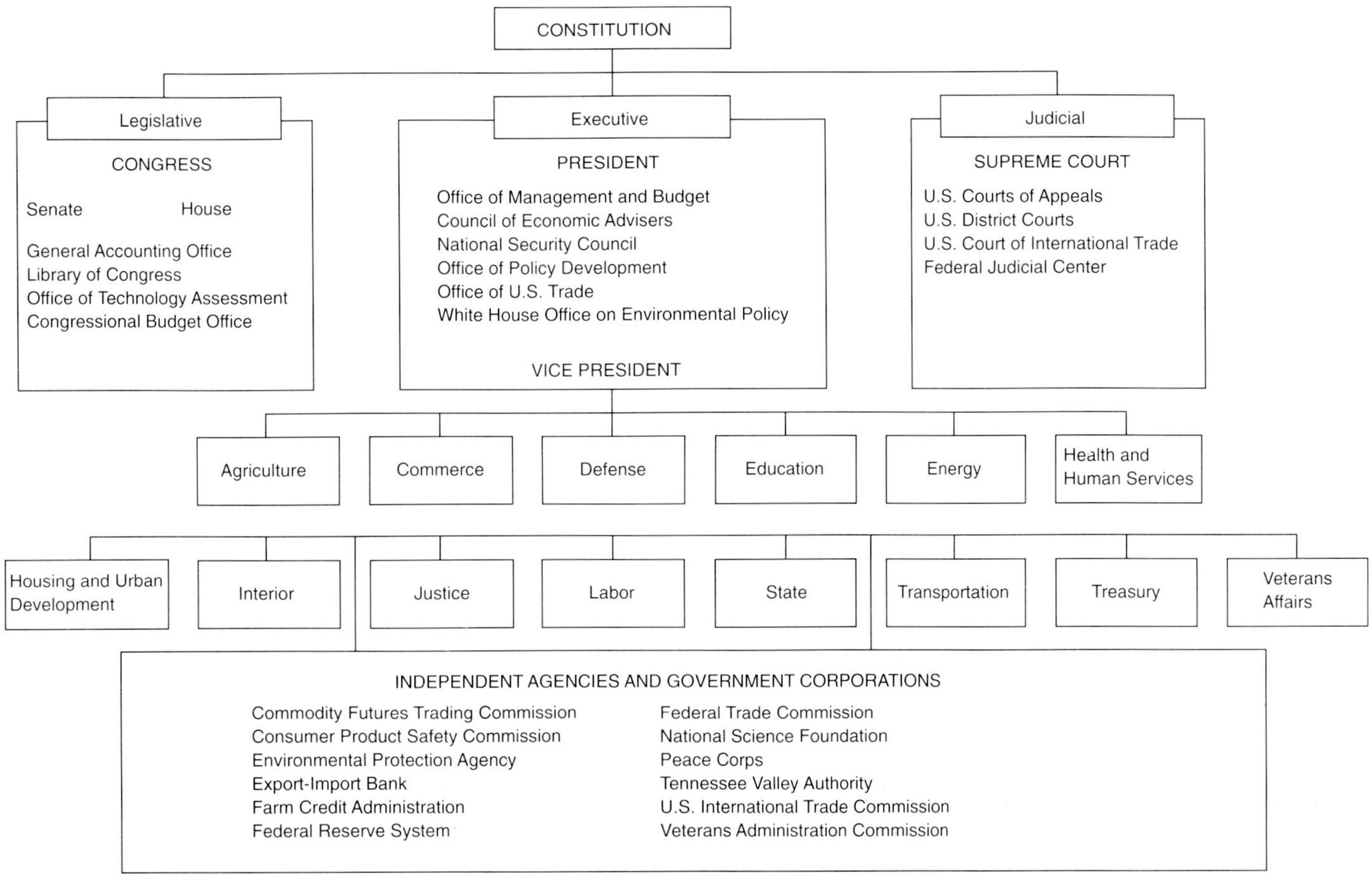

FIGURE 2-2
Structural organization of the government of the United States including selected agencies. *Source:* Adapted from *Washington Information Directory, 1988–89,* by Congressional Quarterly Books, Washington.

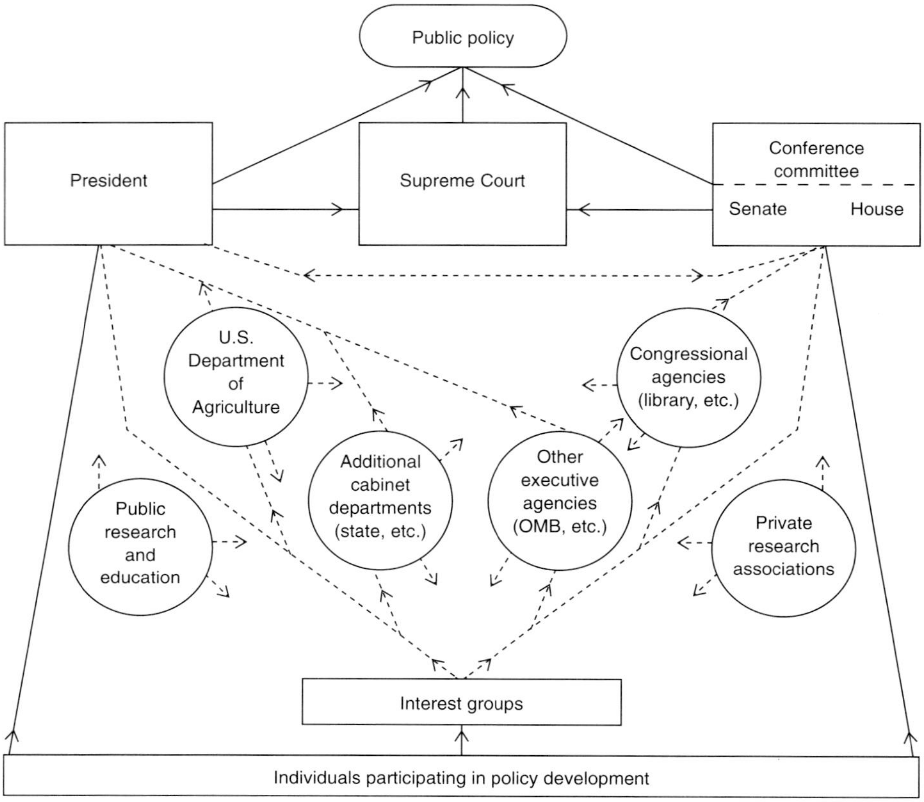

FIGURE 2-3
Schematic of public policy making in the United States. Solid lines denote direct voting; dotted lines denote indirect influence.

congressional action by the president—or passage by a two thirds vote over a veto—is finally necessary before a bill becomes law and a policy is created.

Legislative participants with special influence include the leadership of the two chambers and the committee chairs, who are members of the majority party of that chamber. The proportion of party members on each committee is determined by the overall party split in that chamber. Also, somewhat unique to the U.S. system is the prominence of open public testimony and hearings that precede debate and voting among the members. At every phase of the process, private input flows to the cabinet officials and members of Congress through formal and informal channels. In turn, policy research findings and educational materials are distributed at each participant level—hearings, subcommittees, committees, and executive agencies, as well as directly to the president and members of Congress. Ultimately, however, public policies are the decisions, forged through compromise, of elected public policy makers whose judgments arise from their economic, noneconomic, technical, and common sense values and knowledge.

Executive Participants

Within the executive branch, agriculture is the only private industry identified by its commodities that has its own cabinet-level department. This attests to agriculture's recognized economic importance in the economy, its dominance in the early history of the nation, and its prominence in the political structure of the nation. Increasingly, however, other departments have had vital input into public food and agricultural policy and its implementation. Examples are the Department of State in all trade matters and foreign food aid; the Department of the Interior in natural resource issues; the Department of Health and Human Services in food quality and safety policies; the Department of the Treasury in credit questions; and independent agencies such as the Office of Management and Budget and the Environmental Protection Agency. Also, the Farm Credit System serves as an important example of a federally sponsored agency that is indirectly controlled through elected and appointed officials acting under federal law. Within the Department of Agriculture, which is pivotal in policy development as a part of the Executive Branch, a multitude of agencies provides input and the means of implementation. Examples are the Economic Research Service, Commodity Credit Corporation, Agricultural Stabilization and Conservation Service, Soil Conservation Service, Farmers Home Administration, Forest Service, Agricultural Marketing Service, and Food and Nutrition Service.

Judicial Role

Participants in the judicial branch issue judgments as to the constitutional legality of any challenged policies arising from legislative and executive decisions. Judicial participants include the nine appointed, lifetime members of the Supreme Court, similar members of each state Supreme Court, and numerous federal, state, and local court justices. Judicial decisions shape public policy by establishing precedent for policy makers, through rulings on constitutionality of policies, and through interpretations of the changing values of society. For example, in 1936 the Supreme Court invalidated principal parts of the 1933 Agricultural Adjustment Act (AAA). In that decision, it nullified the power of the secretary of agriculture to levy processing taxes and to enter into acreage reduction contracts with farmers. Within two years, as both the membership of the Court and public perception changed, a revised AAA of 1938 was adopted, reinstating some of these same provisions, which were later found to be constitutional.

STRUCTURE OF PRIVATE PARTICIPATION IN PUBLIC POLICY

Roles of Interest Groups

Although interest groups serve individuals as they pursue many private goals, such as occupational, social, and recreational goals, one of the most important is facilitating their participation in public policy making. Individuals cannot contact every policy maker or attend every public hearing, but they can be represented through interest groups. Interest groups have emerged in abundance to provide an outlet, an identification, and a collective presence in matters of common interest. For example, members may be drawn together because all of them produce dairy products, or because all are concerned about the purity of their childrens' food, or because all enjoy undisturbed natural beauty. A relatively free political, social, and economic system both permits and fosters such private organizational activity.

Interest groups facilitate participation in public policy making through direct professional contact (lobbying), presentations of policy positions at hearings, coalition building with other groups, dissemination of policy positions via the media, joint funding of various activities, access to information relevant to policy concerns, and facilitation of contacts with other members through meetings.

Interest groups emerged in the original colonies to express concerns about the tobacco trade, taxation, and private property, and they flourished after the founding of the nation. In 1835, the widely traveled French writer, Alexis de Tocqueville, observed that

> In America the liberty of association for political purposes is unbounded. . . . There are no countries in which associations are more needed. . . . Americans of all ages, all conditions, all dispositions, constantly form associations. They have not only commercial and manufacturing companies, in which all take part, but associations of a thousand other kinds—religious, moral, serious, futile, general or restricted, enormous, or diminutive. . . . If it be proposed to advance some truth, or to foster some feeling by the encouragement of a great example, they form a society . . . [they] seem to regard it as the only means they have of acting.[1]

In the early 1800s, about half of the U.S. population lived on farms or worked with closely related businesses, and many of the earliest organizations reflected agricultural interests. In the mid 1800s, the Civil War brought new problems in its wake, such as unstable banking, periodic collapses of the monetary system, and low farm product prices. Widespread protests, known as the Populist Revolt, ushered in many protest groups that flourished for more than 30 years. One of the most notable was the American Society of Equity, founded in 1902, which spawned many marketing cooperatives in the Midwest, Great Plains, and tobacco-growing states before it also helped lay the foundation for a farmer–labor political party. These agrarian-based groups figured in many elections well into the twentieth century.

When an expanded use of Political Action Committees (PACs) was authorized in 1974, many interest groups, including agricultural groups, began using them to represent their members' interests in policy development. PACs have facilitated member contributions directly to political candidates, and PAC funds and influence have grown rapidly. In particular, PACs have been prominent for dairy and rural electric cooperatives and have begun to permeate general farm organizations.

Central to a participatory system, interest groups allow each citizen to contribute to policy making, even though that input may be somewhat modified in the process by the accommodations of differences within the group. The interest groups that are increasingly involved in food and agricultural policy development fall into several broad categories: (1) general farmer, (2) farm commodity, (3) agricultural cooperative or noncooperative business, (4) consumer, and (5) general interest. These organizations provide educational and information services, often through trade publications, develop policy positions through conventions of delegates, directly influence policy and policy implementation through lobbying, interpret new programs during membership meetings, and indirectly influence policy decisions by making financial contributions or supporting candidates in elections. Organizations parallel in functions to the above exist for all other major industries, labor organizations, businesses, and professions.

General Farmer Organizations

A *general farmer organization* can be defined as an agricultural interest group that meets three criteria: (1) is recognized currently as an active farmer-oriented group; (2) represents farmers in many regions of the nation who produce different products; and (3) serves not only the policy interests of rural families, but also their economic and social needs. Four noteworthy organizations, each with a distinct origin, history, ideological orientation toward policy, and present status, have met these criteria: the National Grange, National Farmers Union, American Farm Bureau Federation, and National Farmers Organization.

National Grange The Grange is unique among agricultural organizations because it is the only such organization formed in the nineteenth century to survive into the twentieth century. Founded in 1867 as a fraternal order patterned after the Masonic Lodge, the Grange was primarily a social and quasi-religious organization with a major objective of helping overcome the adversity of agricultural people following the Civil War, and relieve the animosities between rural people of the North and South. Although it grew slowly at first, sharp declines in the prices of farm commodities, combined with a severe financial bank panic in 1873, provided incentives for rapid growth in Grange membership. By 1875, local chapters had been chartered across the nation in all but four states. Although membership peaked in the mid-1870s and then gradually fell to 1880, the Grange continued to be a potent political force and a vehicle for expressing rural discontent and reform.

The Grange movement initiated or embraced a wide spectrum of reforms. Grangers fought against high freight rates on agricultural commodities and farm supplies, monopolies in transportation and marketing facilities, and what they regarded as restrictive and unfair banking practices. The Grange movement became identified with political candidates who were anti-industry, antibank, and even antigovernment. The antigovernment wing involved Grange support of freer trade in general, but a tariff policy favorable to agriculture, economy in government, low taxes, antitrust laws, and civil service reform. Grange support for credit and banking reform continued through the early years of this century. The Grange had an important influence in writing and politically supporting the Federal Farm Loan Act of 1916 and the eventual creation of the Farm Credit System. Perhaps of most significance, from the beginning of farm price and income policy making in the 1920s until recent years, the Grange has actively promoted its views on farm price and income policies, generally favoring a two-price plan distinctly shaped for each major export commodity. This plan would generate higher prices in domestic markets than in foreign markets, and be combined with some form of production or marketing controls.

The Grange has its headquarters near the White House in Washington, D.C., and it often serves as a conciliatory force among conflicting groups. Probably the second largest general farmer organization, the Grange is most active in the Northeast and in the northern Midwest.

National Farmers Union When this organization was founded in Texas in 1902, it was known as the Farmers Educational and Cooperative Union of America. It began as a service organization for the family farm, emphasizing a business-type approach to problems of buying farm supplies and selling agricultural commodities. It even attempted to manage the pricing of farm products by gaining control of production by farmers' collective action, but this proved more difficult in farming than in the urban business world. From its beginning in the early 1900s, the Union established cooperative enterprises of many kinds, particularly in the South—cotton gins, warehouses, grain elevators, fertilizer-buying pools, petroleum cooperatives, and occasionally a bank, hotel, or livery stable.

In the 1930s and 1940s, the Union moved into the plains states. In these states, its growth and membership stability were secured by a system of grain elevators headed by the Farmers Union Grain Terminal Association with headquarters in St. Paul, Minnesota. This association continued to expand, with country elevators in more than a dozen states and large terminal elevators on the Great Lakes, the Mississippi, and the West Coast. In addition, Union-sponsored cooperatives became important in petroleum, livestock, and dairy operations and in the insurance field.

The Union has sought to be the chief representative of the smaller family farm. To implement this claim, the Union became an ardent supporter of parity (explained in Chapter 6) as the basis for a policy of high price supports and compulsory production control, along with strong government programs to achieve their goals. Because of its strong grain-marketing affiliation, the Union is particularly active in the wheat regions. Perhaps slightly smaller than the Grange, the Union has national headquarters in Denver and Washington, D.C., where it can urge government to take a stronger role in agricultural affairs.

American Farm Bureau Federation This organization was started by the Chamber of Commerce in a New York county in 1911 to support specialists to teach better farming methods. It then spread through other county organizations called farm bureaus, which provided the base for state organizations. In 1919 these groups formed the national American Farm Bureau Federation, which moved quickly into national policy. To serve farmers, the Farm Bureau established cooperatives, primarily to purchase farm and home supplies, including insurance. Eventually these groups expanded into county, state, regional, and national cooperatives, serving both the purchasing and the marketing needs of their members. The diversity of this movement and its business success helped to account for the rapid growth and stability of the Farm Bureau.

During this period, financial support for the county educational programs came from many sources, among them county governments, railroads, bankers, and such businesses as Sears, Roebuck and Co., International Harvester, and the Chicago Board of Trade. Through the Smith–Lever Act of 1914, the federal government established a system of county extension agents, with matching state and county funds to be administered through the land-grant universities. The Farm Bureaus offered a ready-made local organization to garner financial support. As a result, close ties soon developed among the Farm Bureau, the Cooperative Extension Service, and the land-grant universities. These ties led to the rapid growth of strong extension programs, but also posed a dilemma when a separation was mandated in the 1950s in order to clearly distinguish the public (Extension Service) from the private (Farm Bureau) institutions. Even though earlier general farmer organizations had to some extent preempted the field of agricultural marketing cooperatives, the Farm Bureau's entry into purchasing and insurance cooperatives proved timely as farming became more commercialized and more dependent on purchased inputs. The Farm Bureau spread rapidly throughout the nation as a general farmer organization and as a sponsor of statewide agricultural supply services.

Because of its unique origin and emphasis on agricultural business, the Farm Bureau developed a relatively conservative enterprise philosophy, quite different from the protest movements that launched most other such organizations.[2] In the 1920s, a pronounced split developed between the Farm Bureau and the others about the type of organization and policy that could best serve American agriculture. Farm Bureau leaders favored the ongoing involvement of the federal government in agricultural price and income policy in the 1930s. Subsequently, however, through its national policy positions and leaders, the Farm Bureau has consistently favored reduced direct governmental intervention and the "market system" approach.[3] It usually advocates what it calls "temporary price and income support" until the market forces are able to function properly. Thus it has generally approved the evolving voluntary production controls, minimal price supports, compensatory payments, and the more aggressive export promotion policies embodied in recent food and agricultural acts.

Active throughout the country, the Farm Bureau has its national headquarters in Park Ridge, Illinois, in the heart of the Midwest where it dominates, and strong lobbying offices in Washington, D.C. Its membership of more than two million farmers, and associate members for some of its services, far exceeds any other general farmer organization, thus helping the Farm Bureau to wield recognized influence in policy development.

National Farmers Organization (NFO)[4] The NFO is the newest addition to the interest groups that meet the criteria of a general farmer organization, but it is still struggling to survive. Founded in the 1950s in the heart of the Cornbelt, it grew out of the livestock industry's severe post-World War II economic crisis. Farmers' dissatisfaction with the level of prices received and perceptions of inadequacies in federal price support programs provided fertile ground for the formation of a new direct-action organization. In 1955, the NFO built its program on the idea that farmers, through concerted action to hold designated commodities off the market, could have an important influence in getting higher prices—an idea that was first proposed by the American Society of Equity and the Farmers' Holiday Association of the 1930s.

The NFO began by organizing meetings with dissatisfied farmers and having them sign binding marketing contracts agreeing to hold certain commodities off the market when signaled to do so. The NFO did not reach its peak of activity until the 1960s. On August 28, 1962, an estimated 20,000 people attended a mass NFO meeting in Des Moines, Iowa, where the crowd vigorously approved an "all-out holding action" to start at once. When this action was initiated, various other general farmer organizations and commodity groups strongly protested. Acts of violence, either by NFO supporters or by others, aroused fears and public opposition. Some public officials denounced the developments, and lawsuits were filed.

The effects of the action on livestock prices were not clear, so the NFO shifted its efforts toward marketing contracts to control marketing among its members, and toward more traditional lobbying for ongoing price and income policies. Even though NFO members tend to spurn governmental interference, opting for private bargaining power instead, they nevertheless have embraced public price and income support until their primary objectives could be achieved. Headquartered at Ames, Iowa, in the center of the Midwest where it claims its greatest support, the NFO continues as the smallest of the general farmer organizations, still pursuing a somewhat different economic agenda from other groups.

Recent Movements Other general farmer organization efforts have been attempted, usually in a protest mode, but almost all have been short-lived. One such effort that is worthy of note is the American Agricultural Movement (AAM), which arose out of dramatic marches and tractor drives to Washington, D.C., in 1977 and 1978, to protest the worsening economic conditions that befell farming after the earlier prosperous export-driven years of the mid-1970s. Although the AAM has experienced internal conflicts and has not established a recognized national headquarters with the usual staff services, its leaders and rank-and-file members are strongly committed and aggressively pursue their policy objectives in the various national arenas of decision making.

Commodity Organizations[5]

Farm commodity interest groups appeared early in the history of commercial agriculture, particularly around export products such as tobacco and cotton. In the late 1900s, these groups spread to include essentially every commodity of any market importance. They have been greatly aided by check-off promotion programs that have sprung up to capture both compulsory and voluntary producer contributions. Such programs are usually sanctioned by votes of producers, as provided by law. These groups have been rooted in a recognition of the limited inelastic markets for most primary farm products, in the explosion of mass-media advertising, and in attempts to develop increased export potentials. The two-tier commodity organizations that are common involve state groups and federations at the national level. Some prominent commodity groups are the National Association of Wheat Growers, National Cattlemen's Association, National Corn Growers Association, National Pork Producers Council, and National Milk Producers Federation.

In general, the crop and dairy producer groups are intent on strengthening their commodity price and income programs, while most livestock, poultry, and soybean organizations are more lukewarm toward government intervention. The stances of these latter groups are associated with the less inelastic nature of demand for their products, producers' dependency on farm products as a major-cost input, and the infeasibility of the commonly used intervention techniques. The rise of commodity groups has tended to diminish participation in the general farmer organizations, which in turn have had to respond to the needs through commodity subsections.

Commodity groups are usually supported by modest, voluntary membership dues, except where compulsory check-offs may be involved. In return, the groups provide members with current information through their own trade publications, a ready avenue of contact with policy makers and market representatives, and vigilant lobbyists on duty constantly in state capitals and in Washington, D.C. Commodity group headquarters are often located in the heart of their producing area, but always with an adjunct office in the nation's capital. Some groups have focused on national policies that might help their members. Other groups have attempted to form coalitions with farm or nonfarm groups having mutual interests. However, commodity groups usually have concentrated on policy matters that are central to their interests, which they have pursued independently and diligently.

Agricultural Cooperatives and Other Agribusinesses

Agricultural businesses and their interest groups, whether farmer–producer cooperatives or stockholder corporations, have moved steadily into food and agricultural policy development. Most of them have tended to stay aloof during controversial policy debates, emphasizing instead their traditional production, marketing, and consumer-relation concerns. However, some have begun to participate actively in policy development as they have felt some of the effects of policy decisions, such as massive land retirement programs on fertilizer use, high price supports on exports, or food quality regulations on processing techniques. The profits and economic survival of many of these businesses depend on buying or selling a product to farmers or food consumers, who are the ultimate focus of most food and agricultural policies.

Examples of these agriculture-related groups are the National Rural Electric Cooperatives, National Cotton Council, Farmland Industries, Rice Millers Federation, National Council of Farm Cooperatives, National Broiler Council, and American Bakers Association. The U.S. Tobacco Institute, which represents the entire tobacco industry, has a strong interest in promoting the U.S. tobacco program and increasing tobacco exports. Some others have broader interests in market development or trade, particularly the American Meat Institute, National Food Processors, American Frozen Food Institute, National Frozen Food Association, and National Soybean Processors Association. The National American Wholesale Grocers Association is a distributor group. The Food Marketing Institute represents retail interests. The Farm Credit Council is interested in policy affecting credit cooperatives, and the United Farm Workers is interested in migrant labor legislation.

Many large corporations also invest time and financial resources in food and agricultural policy. Among them, Cargill, a large grain trading firm, generally favors freedom from governmental intervention in its grain-trading activities, but Archer-Daniels-Midland, as the largest producer of corn sweeteners, benefits from government intervention to keep prices high for sugar, and it also strongly favors ethanol subsidies. Both Mars and Pizza Hut want lower prices for their dairy and sugar inputs, and they have been corporate leaders of the coalitions opposing dairy and sugar price-support programs. Pioneer Hybrid Seeds works in a leadership capacity for the American Seed Trade Association. Monsanto has done the same through the National Agricultural Chemicals Association.

Agricultural businesses serve their members or stockholders in policy development by publishing membership newsletters, sponsoring meetings, lobbying with members of Congress and other policy makers, and contributing to political candidates and parties. Sometimes, when the trade organizations try to represent the policy interests of both producers and marketing firms, they face internal conflicts, and many therefore lobby differently for various programs within the overall policy. However, with the continued growth and globalization of the food and agricultural industry, the farmer cooperatives and stockholder corporations will continue to participate actively in future food and agricultural policy.

Consumer Organizations

Consumer-oriented organizations—whether focused specifically on food or more generally focused—have existed for decades. Recently, however, with new concerns about food quality and safety, their prevalence and participation in policy making have increased. Even though their primary attention is on food processing and distributing, they also actively pursue policy measures in food production and trade that affect food prices, availability, and quality.

Examples of these organizations are the Consumer Federation of America, Public Voice for Food and Health Policy, Community Nutrition Institute, Food Research and Action Committee, and Consumers Union. Because many of the consumer organizations are coalitions or federations of other groups and businesses, or because their leaders are concerned about food issues, they participate in a wide range of activities such as providing testimony, lobbying government officials, and issuing reports or media releases on current topics for general consumer use. They also garner financial support for major studies about hunger, food consumption, quality, and safety issues.

Consumer organizations may join other groups around some policy issues of mutual interest. For example, when the Food, Agriculture, Conservation, and Trade Act of 1990 was in the early stages of development, a coalition of 11 national consumer and environmental groups jointly issued several high-quality publications setting forth its agenda. The coalition offered many provisions and was recognized as influential in the decision-making process. Such organizations are likely to continue as major participants in developing food and agricultural policy because of public concerns about food adequacy, nutrition, quality, and safety.

General Interest Groups

General interest organizations have grown rapidly because of the open system of policy development, the expansion of the mass media, the explosion of technological knowledge, and the growth of an aging and better educated citizenry. These groups are much more varied in composition and agenda than were the previous ones, but their distinguishing characteristic is a focus on some policy issue perceived as affecting most if not all citizens. Thus they engage in many activities, from direct lobbying to mass membership drives, and from publishing newsstand materials to supporting volunteer election campaign networks.

Examples of general interest organizations involved in food and agricultural policy are Common Cause, Public Citizen, League of Women Voters, American Farmland Trust, Bread for the World, Center for Science in the Public Interest, and National Catholic Rural Life Conference. The intensity of concerns of individuals and the extent of their participation in general interest organizations is most likely to be related to changing food situations and to economic and social conditions at home and abroad. Although the conditions and leaders will change, many citizen groups will probably continue to participate in the making of food and agricultural policy.

LOOKING AHEAD

Understanding the experience of past food and agricultural policy and the participants in the public policy process provides an indispensable foundation for policy study. Policy is an ongoing endeavor to be developed around new problems and issues that continually confront citizens and policy makers. In the next chapter, you will learn about the process by which public policy is made, determined, and implemented, with particular reference to food and agricultural policy of the United States.

ENDNOTES

1 Alexis de Tocqueville, *Democracy in America,* Schocken Books, New York, 1961, 4th printing 1970, vol. I, p. 191; vol. II, pp. 128–129.
2 John L. Shover, *Cornbelt Rebellion: The Farmers' Holiday Association,* University of Illinois Press, Urbana, 1965, p. 107.
3 Christiana McFadyen Campbell, *The Farm Bureau and the New Deal: A Study in Making of National Policy, 1933–40,* University of Illinois Press, Urbana, 1962, pp. 44–67.
4 This section draws from George Brandsberg, *The Two Sides of the NFO's Battle,* Iowa State University Press, Ames, 1964.
5 Excellent coverage of these commodity groups can be found in H. D. Guither, *The Food Lobbyists,* Lexington Books, Lexington, MA, 1980.

TERMS AND CONCEPTS

Executive	General farmer organizations
Legislative	Commodity groups
Congress	Agribusinesses
Judicial	Consumer organizations
Interest groups	General interest groups

QUESTIONS AND STUDY SUGGESTIONS

1 Why does the type of political system affect the role of individuals and interest groups in policy development?
2 What is the overall structure of public policy making in the United States? Who are its primary public and private decision makers?
3 How do interest groups play a unique role in public policy development?
4 What is a general farmer organization? What is the background of these organizations in the United States?

5 In what ways have the roles of the other interest groups changed in their policy involvement in recent years?
6 Why has the activity of interest groups increased in the United States over the years?
7 What do you see as the place of organizations in the development of future food and agricultural policy?
8 How do the general farm organizations differ in respect to their support for specific price and income policies for agriculture?
9 Why have certain commodities, such as wheat, tobacco, cotton, and dairy products played such a prominent role in food and agricultural policy?
10 It is sometimes suggested that the major farm organizations should be united in their policy proposals for the good of agriculture. What prevents them from doing so?
11 Distinguish between issues in which the major farm organizations appear to be united and those on which division seems to persist.

CHAPTER 3

HOW FOOD AND AGRICULTURAL POLICY IS MADE: THE POLICY PROCESS

Public policy is made by a complex, continuous process that can be studied and understood just as we learn about the processes of an economic market system. Both private and public policies shape the food and agricultural system of the United States, yet these two important types of policies are made by different processes and for different purposes. The purpose of this chapter is to identify and characterize the public policy process of participatory governments, with particular emphasis on that process in the United States. We will examine how food and agricultural policy is made generally, and then study the development of the Food, Agriculture, Conservation, and Trade Act of 1990 (hereafter referred to as the 1990 Act).

IN THIS CHAPTER

By studying Chapter 3, you will learn how to answer the following questions:

1 What are the stages in the process of developing public policy?
2 What are the primary determinants of public policy?
3 What are the usual steps in public policy making, and were they evident in the making of the 1990 Act?
4 How is public policy implemented?

STAGES OF POLICY DEVELOPMENT

In a participatory government, such as exists in the United States, what are the typical stages for developing public food and agricultural policy? We suggest a conceptual model having four sequential stages: (1) identifying a problem directly affecting the food and agricultural economy, (2) increasing public awareness of the problem, (3) developing alternative policy proposals and predicting likely consequences, and (4) reaching public decisions and actions including many adaptations in implementations.

Identifying Problems

Public policy making begins when an individual or an interest group becomes sufficiently concerned about a problem or problems to reach beyond private action to develop political action. The perception is the critical element of the problem, whether "real" or "symptomatic." In policy making, a problem is defined as a situation having two or more alternative solutions; identifying a problem therefore begins with an estimate or appraisal of a situation. Activity to make such an estimate or to begin the process of public policy making can be triggered by an adverse experience, new knowledge about a situation, an unexpected turn of events from an existing policy otherwise viewed as satisfactory, a divergence of reality from some perceived goal, or a combination of events. For example, in the early 1950s, U.S. farmers put increasing political pressure on the federal government to sell their surplus commodities for foreign currency as well as for U.S. dollars. This was the beginning of a policy that has been carried out chiefly under the Agricultural Development and Trade Assistance Act of 1954, known more widely as Public Law (P.L.) 480. Since 1954, under this policy $1 to $1.5 billion of U.S. agricultural commodities have been exported annually.

Increasing Public Awareness

The concerns of one person or a few individuals must spread to the public because a few unhappy citizens alone cannot bring about a public policy. Public awareness normally involves enlisting the support of at least one major interest group or organization and more than just one national leader. This support sharpens the identity of the problem and generates ideas that are proposed as solutions. For example, it took more than two decades of discussion, private fund raising, and private experimentation before Congress seriously considered for public policy a national cooperative (federal, state, and local) educational extension service for agriculture and home economics. Since 1914, this service has functioned as a widely respected model of educational innovation and an integral part of the unique U.S. land-grant university system.

Developing Alternative Proposals

Policy making reaches a critical crossroad when public discussion has generated identifiable alternative proposed policies with predictable consequences. Proposals are refined, interest groups coalesce around the various options, and objective analysis and knowledge are at a premium. Educators, researchers, and informed leaders play a critical role in identifying useful alternatives and their probable consequences. One alternative is to opt for no public policy, leaving the problem entirely in the private policy sphere. For example, many schemes for helping low-income households were considered and analyzed through experience and research before the national Food Stamp Program was established to provide food assistance to such households in all counties.

Reaching Public Decisions and Action

The final stage is reached when the participatory government makes a public decision and takes action. Until a compromise is negotiated among policy makers, often at the eleventh hour, neither a policy, its ends and means, nor its likely specific consequences will be known. Once this stage is reached, a policy becomes more certain, and generally some of the consequences will be defined more clearly. The 1990 Act, for example, became law and set policy only after more than two years of extensive discussions throughout the nation, months of debate, compromising and decision making by both houses of Congress, days of intense negotiations in a joint congressional conference committee, and innumerable compromises between the Congress and the Administration. Then, during the five-year life of the Act, there were many adaptations in implementation.

DETERMINANTS OF PUBLIC POLICY

During these stages of development, what can be distilled as the determinants of public food and agricultural policy? An adequate theory incorporates the most important factors, even though it is recognized that measurable empirical representation for some of the variables is not easily estimated. The determinants could be expressed as the diagram in Figure 3-1 or, alternatively, in a relationship of factors influencing an independent variable or policy outcome as follows:

where $PP^{FA} = f(P, ES, K, VB, I)$

PP^{FA} = public policies for food and agriculture
P = present policies
ES = economic and social conditions, particularly current trends directly relevant to the food and agricultural sector
K = knowledge that participating individuals or groups have about alternative policies for the public problem at issue and about the likely consequences of each
VB = values and beliefs of the participating individuals and groups
I = influence levels of private interest groups affected by the alternative policies

Thus, food and agricultural policy, as an economic, social and political activity, is determined by several factors that can be identified, studied, and analyzed.

Present Policy Existing policy is probably the most important determinant of future public policy. In a participatory political system, decision-making builds on or adapts any policy already in effect rather than starting from a point of no policy. Current issues and problems are perceived within the context of present policy. After all, policy makers who forged compromises from conflicting private interests to achieve current policy are not inclined to forget or ignore their common ground. Thus, the Food Security Act of 1985 was the logical starting gate for policy for the 1990s. As characteristic of

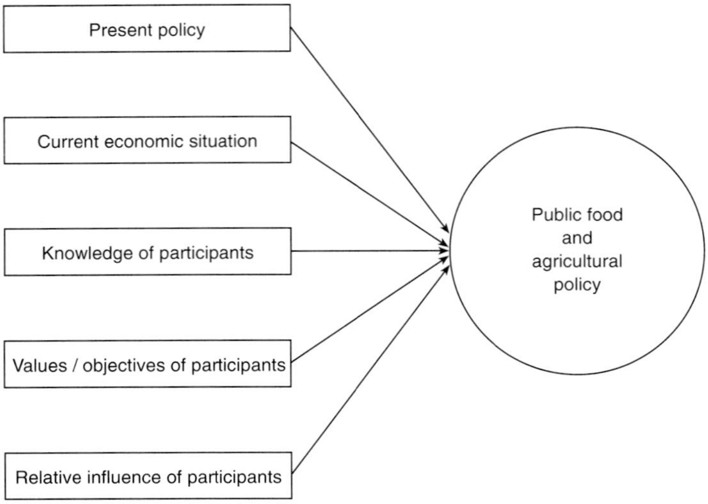

FIGURE 3-1
Determinants of public policy.

public policy, 19 of the 20 major food and agricultural acts over the past half century have been based on previous policy.

Economic and Social Situation The current economic and social situation, like current policy, is a major determinant of future policy. Policy makers, as decision makers, interpret their definitions of problems from the current situation, and anticipate the actions most likely to improve the situation. Furthermore, current trends can best predict future directions of the relevant economic and social variables. For example, if inflation had not grown to double digits in the late 1970s, the Agriculture and Food Act of 1981 would probably not have contained roughly six percent escalations in target prices. Similarly, if 40 percent of the agricultural export market had not been lost in the early 1980s, the 1985 Act would probably not have included its unprecedented bundle of export subsidy schemes. Thus, the substantial economic recovery in the food and agricultural sector from the farm crisis of the early 1980s, and the heightened concerns about reducing the national budget deficit, substantially affected the provisions of the 1990 Act.

Participants' Knowledge The knowledge, understanding, and awareness of all participants—citizens and leaders alike—about the economic situation, the policy experience, and future alternatives relevant to the issues on the agenda, help to determine future policy. This implies that incomplete or inaccurate information will affect policy right along with reliable, thorough knowledge. Guesses, speculation, and ideological exhortation are fair game in a free participatory society, but may be of limited value for relevant, rational policy making. Public research and educational endeavors, such as from land-grant universities and USDA, make a more valuable contribution to policy.

In past policy decisions, the public needed to know the likely effects of production control, a trade embargo, and a lowering of commodity price support levels. Now the public needs dependable information about the relationship between cropping systems and groundwater quality and about the use of chemicals and food safety, among other issues. Leaders, researchers, and educators must not only search for information about the policy issues at hand, but acknowledge the absence of reliable information.

Participants' Values and Objectives Policy is also determined by society's underlying, enduring values and objectives. These include such important perceptions as the role of government, agriculture's place in the nation's economy, the fairness of the economic system, the level of farmer income, food safety, environmental quality, and the status of nonrenewable resources. Although private decision makers may be guided by a single specific policy objective, public policy makers must find a workable consensus and eventual compromise from a myriad of competing private objectives and policy proposals.

Only when a compromise is struck can the objectives and priorities of the public policies be clearly perceived. Of course, general objectives, such as adequate incomes for farmers and a secure food supply, can be readily discerned from decades of food and agricultural policies, but these generalizations give little guidance for many of the necessary future policy choices. Thus, relatively enduring values and objectives might be gleaned from past policies, but it is more important to note that those values now held by citizens and interest groups will help shape future policy decisions.

Political Influence of Individual and Interest Group Participants In a participatory political system, no one dictates policy. Elected and appointed leaders, professionals, interest groups, and others all help determine it, but just as in economic decision making, the effect of each is somewhat proportional to the relative political influence in the process. This influence is affected by the participant's economic power, knowledge, and organizational adeptness, as well as by how power is distributed via the structure of the political system. Thus, future policies will be affected by the structural composition of Congress, the strategies selected by the Executive, and by the relative influence exerted by producer and consumer interest groups.

STRUCTURE OF POLICY MAKING

The process of governmental policy making as it is accomplished in participatory and in authoritarian political systems is as different as private and public decision making. Even though a commonality exists among countries with participatory governments, differing structures for their public policy making are clearly evident around the world. The most prominent contrasts are found between the parliamentary systems, which predominate around the world, and the divided legislative, executive, and judicial systems, which are symbolized by the federal, state, and local governments of the United States.

Different political parties and divided authorities within governments characterize policy making in the United States. This structure of government necessitates negoti-

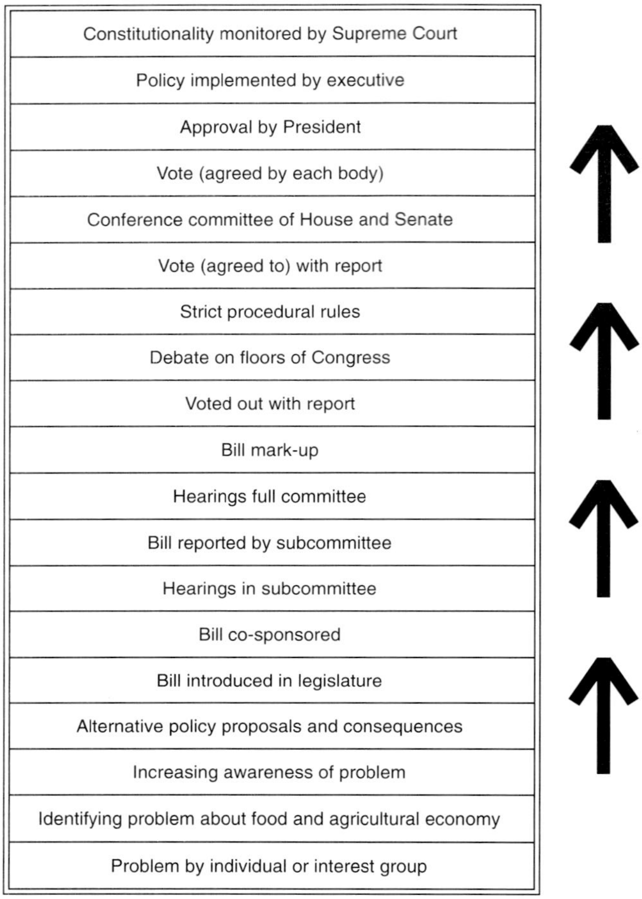

FIGURE 3-2
Steps in making public food and agricultural policy in the United States. Public policy making starts at the bottom with problems and works to the top.

ation and compromise at all of its levels of policy making and even between its executive and legislative branches. In contrast, within the parliamentary system, many primary compromises are struck at the time the government is "formed."

Steps in Public Policy Making

If we focus on the structure of public policy making in the United States, what are the important steps in the process? These customary steps are depicted in Figure 3-2 in a process that, in reality, is cyclical and continuous. New problems emerge as a public policy is formed and implemented and the process commences again.

The first step of the public policy making process is to assess problems experienced by individuals or interest groups. This will help to determine whether a public response is likely to be successful or not. If the problem is persistent, pervasive, and identifiable enough, it may survive though the additional stages of policy making. These include increasing public awareness and developing alternative policy proposals. Eventually, a step may be reached involving introduction of a bill for consideration by the legislature. In the United States, this could occur in either or both of the houses of Congress.

For example, suppose that erosion of farmland is perceived as a serious public problem. Farmers and urban environmental interest groups may believe that soil productivity is being reduced. They may learn that reservoirs are being filled with silt, that ground water is becoming polluted, that wildlife and water sports are being jeopardized by the polluted waterways, and that flooding is being exacerbated downstream.

Usually, more than one bill is introduced for any policy problem. However, in order to survive in the process, at least one bill must garner support from politically influential cosponsors. This process of clarifying the nature and severity of a public problem and of refining proposals into viable bills, commonly involves public hearings at which concerned individuals or representatives for interest groups may appear. Such hearings usually occur first at a subcommittee level followed by a full committee. For example, the U.S. House of Representatives Committee on Agriculture, through which all food and agricultural bills must pass, functions through six subcommittees each of which has jurisdiction over a particular commodity (for example: Livestock; General Commodities), or programs vital to the nation, or consumers (for example: Environment, Credit, and Rural Development; Foreign Agriculture and Hunger). The full committee comprises about 45 members of Congress, divided roughly proportional to the sizes of the majority and minority parties represented. Subcommittees are usually organized similarly, and are composed of about 20 members each.

A mark-up session signals that a policy proposal (bill) has broad and deep enough support among participants across the society to test actual policy makers votes. The marking-up process literally means modifying the meaning and purpose of policy through publicly recorded votes by the committee on the language and provisions in the bill. Decision makers must weigh potential benefits from a policy choice against costs, including the economic concept of transaction costs, or those involved with information, contracts, and policing. If the language of a bill is supported by a majority of members, it is voted out of that body to reach the next important step. This is the debating and voting on the open floor of the House or Senate. The bill is accompanied by a voluminous report setting forth the policy process, rationale, and decisions up to that point.

At the critical step of floor debate, the previous decisions of an issue-friendly committee must yield to the scrutiny and judgment of other members of Congress. Strict procedural rules are followed during the decision making, particularly in the larger House of Representatives, in recognition of scarcity of time and competing public concerns. In addition, as policy proposals move toward consideration by a house of Congress, the executive branch (president) usually signals its concerns, interpretations, and preferences through its cadre of special staff. Because of these many hurdles, only a few public policy issues ever survive all of the policy steps leading to approval by a house of Congress. Even then, it would not only have been altered but, in order to achieve adequate support, it would often have been expanded beyond the original issues.

Often, a parallel process in the other house of Congress yields a somewhat similar bill. If so, compromises must still be found in a conference committee representing the parent committees for the bills from each house. Even this conference compromise must be accepted by each house. Also, the president's position on the bills before the conference committee influences the resolution of the issues among the three public decision-making bodies: the House, Senate, and president. If the president disapproves and vetoes a bill, it requires a two-thirds vote of each house to successfully override. Even though the president signs a bill, or a veto is overridden, the policy making process is incomplete if the U.S. Supreme Court declares the new law to be unconstitutional. If it is constitutional, the policy must still be implemented by "action agencies" of federal, state, or local governments. The development of the 1990 Act is a good example of this public policy-making process. Furthermore, the expanding of the 10-year national Conservation Reserve Program in that Act exemplifies the public response to the concerns traced earlier in this chapter of farmers and environmental interest groups about the private use of erodible farmland.

Policy Process of the 1990 Act

The depth and breadth of knowledge of citizens, political leaders, and policy makers who participated in developing the 1990 Act was very extensive. For the previous two years, public and private information channels contained an abundance of research and educational reports. Information bearing directly on the 1990 policy development included a national survey of farmers' preferences on selected current policy issues,[1] extensive background information published by the U.S. Department of Agriculture,[2] analytical studies of alternative policy options,[3] and critiques by panels of researchers and educators compiled as books.[4, 5] Also, many major farmer, agribusiness, and consumer organizations issued their own proposals.

The House Committee on Agriculture began hearings June 29, 1989, developed subcommittee mark-up bills on various titles during the following year, and reported its comprehensive decisions for House action on July 3, 1990. The committee's report included a 1,006-page legal format accompanied by a 914-page description.[6] That proposal was debated on the House floor for five days, and this phase ended with passage of House Bill 3950.[7]

In general, the bill continued most provisions of the 1985 Act, but froze target prices, reduced loan rates more gradually, and allowed more flexibility in planting. Also, it gave increased attention to food safety, water quality, rural development, forestry, sustainable agriculture, and research support. Again, social, health, ethical, and other noneconomic values emerged. The estimated cost for commodity programs was $55 billion, compared with the cost of $79 billion for the 1985 Act, projected during its development, both estimates covering five years. Actual costs invariably differ from those projected because of changing economic conditions during the succeeding years of implementation.

The Senate Committee on Agriculture, Nutrition, and Forestry began hearings on April 5, 1989, and reported its comprehensive bill for action June 6, 1990, in a 1,081-

page legal format accompanied by a 1,282-page description.[8] This bill was debated on the Senate floor for seven days, concluding with passage of Senate Bill 2830.[9] Similar to the House version, it expanded the marketing loan to more commodities, provided more flexibility in loan rates, revised the administration of P.L. 480, emphasized the protection of wetlands and water quality, and prohibited exports of chemicals that had been banned for domestic use. The estimated cost for the commodity programs for five years was $54 billion. Costs varied from those of the House to different provisions in the bill.

A legislative conference committee composed of 7 senators and 16 representatives was appointed to reconcile the differences between the two bills. The committee negotiated for eight days, with staff from the Administration always in attendance, and reported its compromise version October 16, 1990. This version had to be altered, however, when the long-debated deficit-reducing budget resolution called for more than $13 billion cuts in the estimated costs of the total bill, mostly in the farm commodity programs. Other programs, particularly the food aid programs, generally escaped this reduction.

The House then approved the final compromise on October 24 (vote 318 to 102) and the Senate on October 25 (vote 60 to 36); the president signed the 1990 Act on November 28, 1990.[10]

IMPLEMENTING PUBLIC POLICY[11]

Policy implies both decision and action. One without the other is either ideas absent of application or activity void of purpose. Similarly, public policy implies both the policy process and policy implementation. Implementation of any decision is a necessary and very important part of public policy, and can extend over a substantial period. Many policies at the state level are in essence implementations of national policy. Implementing policy may be as controversial as the legislative policy making process itself. Implementation may trigger further policy for clarification or modification in light of initial results. Public policy often provides considerable discretionary authority even to the degree of whether any action is to be taken. Through this discretion, the administering authority is also shaping further policy. During this time, the dynamics are similar to those in the policy-making process that involved legislative and executive approval. Thus, interested groups, including elected policy makers, continue to be attentive to policy implementation.

The U.S. Department of Agriculture (USDA) is the primary implementing body for food and agricultural policy in the United States. As indicated in Figure 3-3, each of the USDA undersecretaries and assistant secretaries is responsible for administering different programs, and hence implementing various policies. For example, the price and income policies are implemented mainly through several agencies of the USDA. The Agricultural Stabilization and Conservation Service (ASCS), with staff in some 2,800 counties, administers the farm price and income programs. The domestic food-aid programs, which usually absorb some 50 percent of the USDA budget, are administered through the Food and Nutrition Service. The Foreign Agricultural Service deals with agricultural export programs, including foreign food aid. The Soil Conservation

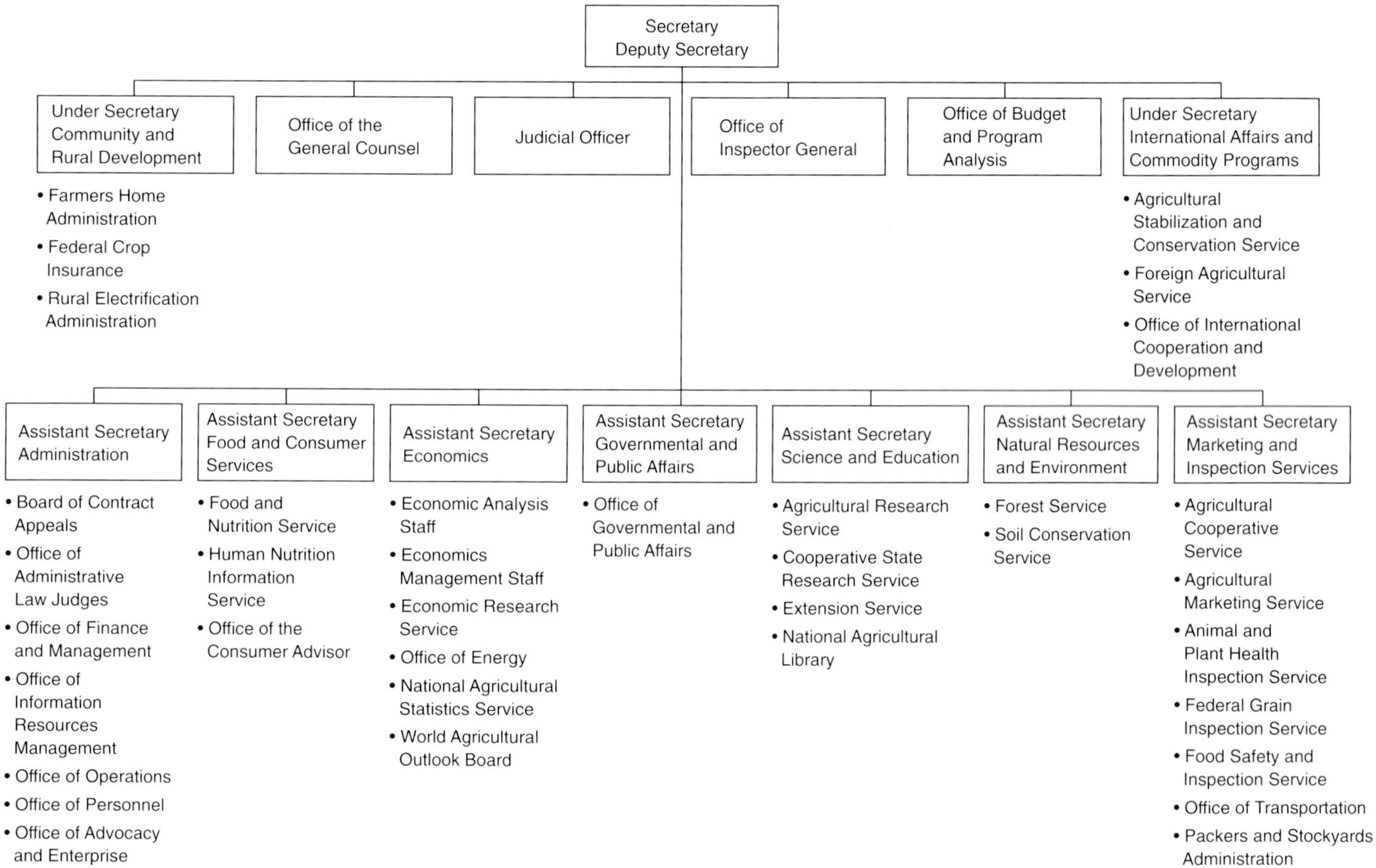

FIGURE 3-3
Organization chart of the U.S. Department of Agriculture.

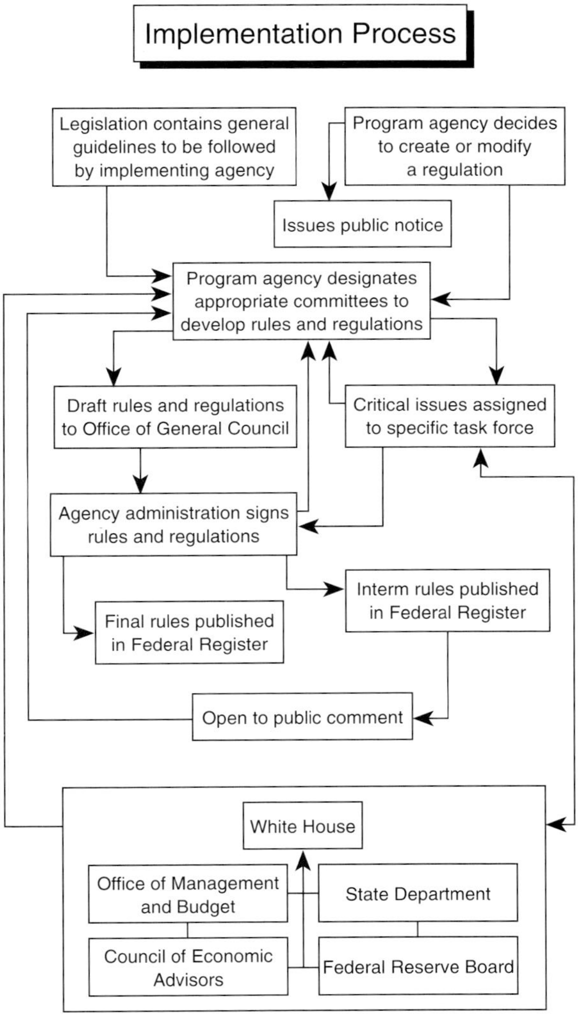

FIGURE 3-4
Implementation process for United States public policy.[12]

Service has staff in most counties to assist farmers with soil conservation. The Farmers Home Administration serves as a lender of last resort for some qualifying farmers, agribusinesses, rural homeowners, and governments that are unable to get credit from other sources. Finally, and perhaps most important, the federal government in cooperation with states and counties helps support agricultural research and education policies through agencies such as the Agricultural Research Service and the Cooperative Extension Service. As this implementation process proceeds, technical statutes might be interpreted, public comment invited, and rules promulgated for use at the USDA staff level, and in state and county offices (Figure 3-4).

In summary, this chapter illustrates the necessity of understanding how food and agricultural policy of a participatory government, such as the United States, is made, why we have public policies, how much a major policy costs, and what agencies implement them. For students, successful understanding also requires knowledge of the basic concepts of policy making, the political organizations and forces at work, and the indispensable theoretical and empirical economic tools that are involved.

LOOKING AHEAD

Major issues for the future continue to develop and broaden, some grow in urgency or magnitude, and many involve other nations. The next chapter takes the first step in discussing some of these issues by examining the national situation for food and agricultural policy. Following this, we consider certain aspects of the world situation as it affects this policy.

ENDNOTES

1 H. D. Guither, B. F. Jones, M. A. Martin, and R. G. F. Spitze, *U.S. Farmers' Preferences for Agricultural and Food Policy in the 1990s,* North Central Regional Research Publication 321, North Central Regional (NCR) Educational Materials Project, Ames, IA, 1989.

2 U.S. Department of Agriculture, *Agricultural Food Policy Review—U.S. Agricultural Policies in a Changing World,* Economic Research Service, Agricultural Economic Report No. 620, U.S. Government Printing Office, Washington, DC, November, 1989.

3 Food and Agricultural Policy Research Institute, *An Evaluation of Price Support Equilibration Options for the 1990 Farm Bill.* Center for Trade and Agricultural Policy, Ames, IA, April, 1990.

4 R. G. F. Spitze, ed., *Agricultural and Food Policy—Issues and Alternatives for the 1990s,* University of Illinois, Department of Agricultural Economics, Urbana, 1990.

5 C. S. Kramer, ed., *The Political Economy of U.S. Agriculture—Challenges for the 1990s,* National Center for Food and Agricultural Policy, Resources for the Future, Washington, DC, 1989.

6 U.S. Congress, House of Representatives, *Food and Agricultural Resources Act of 1990,* House Report 101–569, Part 1, U.S. Government Printing Office, Washington, DC, 1990.

7 U.S. Congress, *Congressional Record,* August 3, 1990, Part III, U.S. Government Printing Office, Washington, DC, 1990.

8 U.S. Congress, Senate, *Food, Agriculture, Conservation, and Trade Act of 1990,* Senate Report 101–357, U.S. Government Printing Office, Washington, DC, 1990.

9 U.S. Congress, *Congressional Record,* July 31, 1990, U.S. Government Printing Office, Washington, DC, 1990.

10 U.S. Congress, House of Representatives, *Food, Agriculture, Conservation, and Trade Act of 1990,* House Report 101–916, U.S. Government Printing Office, Washington, DC, 1990.

11 Credit for parts of this section are due Professor M. R. Dicks, Department of Agricultural Economics, Oklahoma State University.

12 This chart was also developed by M. R. Dicks.

TERMS AND CONCEPTS

Public policy process
Stages in the development of public policy
Congressional committees and subcommittees
Discretionary administering authority
Congressional committee hearing

Participatory governmental policy making
Public policy determinants
Steps in public policy making in the United States
Bill introduction
Bill mark-up
Congressional conference committee
Policy compromise
Policy implementation
Presidential veto

QUESTIONS AND STUDY SUGGESTIONS

1 Explain how and why policy involves both decision and action.
2 What is meant by *a policy problem*? How does it differ from a policy belief? A policy value? A policy goal?
3 Briefly characterize the process of policy development. What are the specific steps in policy making?
4 What are the determinants of public policy? How does economic analysis contribute to policy making?
5 How would you describe the differences in policy making between a private business corporation and a participatory government? Are these differences in respect to how the policy is made, or in respect to whom the policy affects?
6 In public policy making, are policy goals offered at any time during the process? Who establishes the goals?
7 List and explain the steps of the policy process in the United States.
8 Discuss these steps in the development of the Food, Agriculture, Conservation, and Trade Act of 1990.
9 How can policy implementation be viewed as an extension of the policy development process?
10 Identify the primary USDA agencies involved in implementing the farm price/income, food distribution, research, soil conservation, credit, and trade provisions of the 1990 Act.

PART TWO

ECONOMIC AND POLITICAL FOUNDATIONS

CHAPTER **4**

U.S. ECONOMIC SITUATION FOR FOOD AND AGRICULTURAL POLICY

The food and agricultural policy of a nation is largely determined by the changing economic situation in that country. The economic trends, adjustments, and resulting problems in each country are the sources of the public issues from which this policy may emerge. These problems are often associated with macroeconomic events and the national policies they precipitate. They are also rooted in the situation created by changes in farm production, consumer income, population, trade, and other market factors. Advancing technology precipitates many of these changes but does not automatically provide remedies for all of the problems.

The purpose of this chapter is to offer insights into these changes and their interrelationship with public policy through the use of the familiar economic framework of supply and demand. These changes, which are often driven by technological developments in supply, as well as by changes in demand, determine prices in a market economy. Such prices, along with costs, outputs, and off-farm income, determine producer incomes. Taken together, producer prices, consumer food supply, income of farm families, and the economic situation in rural communities help to shape much of food and agricultural policy.

IN THIS CHAPTER

By studying Chapter 4, you will learn how to answer the following questions:

1. What are the agricultural supply–demand determinants and balances?
2. What is the relevance of supply and demand elasticities to food and agricultural policy?
3. How have farm product prices and farm family incomes been affected by the national economic environment?
4. Why are future technologies important to food and agricultural policy?

SUPPLY CHARACTERISTICS

The economic tools of supply and demand are used to explain the levels and fluctuations of the quantities and prices of agricultural inputs and outputs. The analytical methods that are used in developing these explanations are the same for agricultural markets as for the other economic sectors. Many organizational and institutional characteristics of agricultural markets are distinctive for policy, however, and these must be analyzed and understood.

Definition and Determinants

Supply may be defined as a schedule or curve showing the various quantities of a resource, service, product, or commodity that individuals or firms produce and market under given conditions at each price in a range of possible prices during a specified period of time. In such a schedule, and with some factors fixed, the quantity supplied is a positive function of the price.

A very large number of supply schedules can be conceptualized for commodities in both agricultural input and output markets. These schedules exist for each and every commodity used in production or produced and sold in the agricultural commodity markets. For example, there are supply schedules for inputs such as fertilizer, farm machinery, pesticides, labor and management; for broad groups of commodities such as meat or grain; for individual commodities such as pork or beef; and for grades of a commodity such as No. 1 dark hard northern spring wheat or No. 2 yellow corn.

Various factors interact in the longer run to determine the supply schedules for agricultural commodities. These can be conceptualized in abbreviated, functional form:

$$Sq = f(P, H, L, C, T, W, A, PG)$$

where Sq = quantity of a commodity supplied by farmers and agribusiness firms
P = prices expected for the commodity
H = human capital consisting of labor and management
L = land cost and availability
C = capital inputs in addition to land and human capital
T = technology of production
W = weather and other natural forces
A = alternative commodity returns
PG = policies of government

Some of these factors are physical or biological, such as the weather or the growing conditions for raising crops or livestock. Some are social factors, and some require production knowledge from such fields as agronomy and animal science. Particularly critical for food and agricultural policy, however, are the following economic or technical factors: (1) the existing technology of production, (2) prices of various inputs or commodities used in production, (3) prices of alternative commodities that can be produced, (4) expectations concerning prices in both the input and output markets, and (5) policies such as taxes and subsidies.

A change in one of the basic supply determinants will cause the supply curve for a

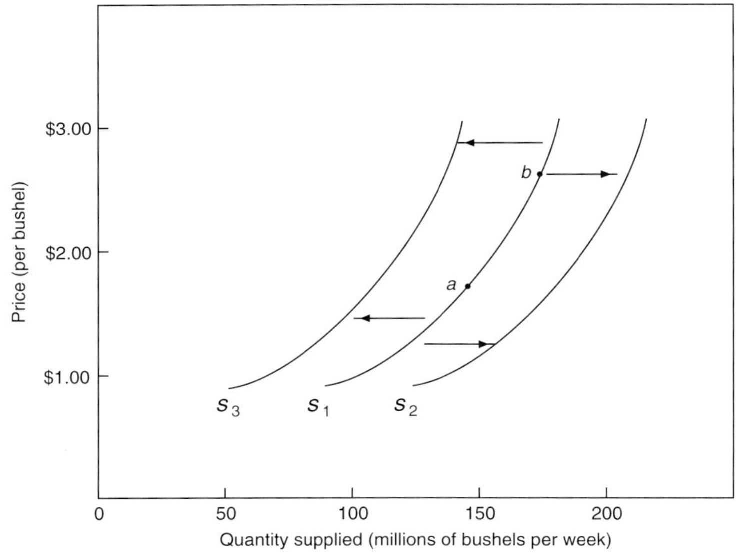

FIGURE 4-1
Supply of an agricultural commodity. Changes, for example, in the supply of corn, occur when there is a change in physical or biological conditions or in one or more of the other determinants of supply. An increase in supply is reflected by a shift to the right in the curve, as from S_1 to S_2. A decrease is reflected in a shift from S_1 to S_3. A change as from point *a* to *b* on S_1 is not a change in supply, but a change in the quantity supplied, which involves a change in the price of corn, all other things being equal. Agricultural supply curves are normally positive, or upward sloping, corresponding to the horizontal summation of that part of the marginal cost curves of individual firms above minimum average variable cost in the short run, or above minimum average (total) cost in the long run.

commodity to shift either to the right or to the left. A shift to the right (from S_1 to S_2 in Figure 4-1) signals an increase in supply, with producers willing to offer more of a commodity at each possible price. A shift to the left (from S_1 to S_3) signals a decrease, with producers offering less at each price.

Important Determinants of Agricultural Supply

Two determinants—technology and input prices—are the most important components of production costs. Increases in agricultural supply largely depend on advances in technology; for example, planting hybrid corn instead of open-pollinated varieties. Supply increases also result from reduction in the prices of inputs, as occurred with the growth of the fertilizer industry.

Changes in the price of one farm commodity can shift production of another commodity. A decline in the price of corn, for example, may cause farmers to shift to soybeans, thus increasing the supply of soybeans. One of the major problems in policy analysis is to determine the effects that a specific commodity program may have on other commodities.

Expectations about the future prices of a commodity affect a farmer's willingness to produce that commodity. These expectations affect both current and future supply curves. In the nineteenth century and the first quarter of the twentieth century, the dominant factors in increasing agricultural production were the increases in land, farm labor, and nonlabor capital. Since then, the dominant factors have been technological advances and declining real prices for inputs.

Production decisions of farmers are affected by governmental policies. Taxes and subsidies have complex effects on supply, depending on whether they affect marginal cost or marginal return. For example, an increase in a sales tax on purchased inputs will increase marginal cost and reduce supply (the curves shift to the left). An increase in the property tax does not change marginal cost or the supply curve in the short run, but it will increase average cost, and hence reduce property values, other things being equal.

A farm subsidy paid on current output, or a target price set above the average market price, will increase marginal revenue and the quantity supplied, but will not alter the supply curve itself. The land retirement program—which has been unique to the United States in seeking to control farm output—reduces the amount of land that a participating farmer may plant and also reduces supply.

Technology and Productivity

As a general result of continuing rapid advances in agricultural technology, farm output has increased per unit of input, and prices of farm outputs have declined compared with prices of farm inputs. The supply of farm commodities has increased, as significant increases in productivity have come from successfully applying new technologies that are either cost reducing or output increasing.

The farm sector interacts with two markets, one for inputs and the other for outputs, each of which has its own set of interacting supply and demand curves. Under advancing technology, if the increases in demand for outputs are larger than the increases in supply, output prices may increase relative to input prices in the short run. However, over the long run, given continuing increases in agricultural productivity and the present general market structure and performance, farm output prices must drop relative to input prices. This tends to depress incomes from farming and force adjustments in agriculture.

Since 1910, many U.S. farm families have prospered, but many others have failed, and many young people who have wanted to farm have not been able to do so because of inadequate capital or low net income prospects. Some members of farm families have obtained off-farm employment to supplement income from farming. For more than half a century, as advancing technology has displaced human labor, reduced entry and out-migration, rather than early retirement, have been the main factors in reducing the farm population. The average annual decline in the farm population was 3.8 percent in the 1950s, 4.6 percent in the 1960s, 2.0 percent in the 1970s, and about 2.7 percent in the 1980s. Linked to this dramatic adjustment, increasing productivity in farming, as measured in output per hour of labor, has consistently out-paced that of the nonfarm economy. To illustrate, since the 1950s, the annual increase in farm labor productivity

TABLE 4-1
INDEXES OF U.S. AGRICULTURAL INPUT, OUTPUT, AND PRODUCTIVITY (1977 = 100)

Year	Total input	Total output	Productivity
1940	97	50	52
1950	106	61	58
1960	99	76	76
1970	96	84	87
1980	103	104	101
1985	91	118	129
1990	88	119	135

Sources: Economic Research Service, USDA, *Agricultural Outlook* and *Outlook Charts* (various years), and *Economic Report of the President,* February 1992.

ranged from 2.7 percent to 6.7 percent per decade, while the annual increase in labor productivity of the nonfarm population ranged from 1.9 to 2.4 percent per decade.

The term *technology* generally implies a specific state of art and science that is used to transform a combination of inputs into outputs (goods or services). Modern farmers use combinations of land, human capital, and industrial inputs such as machinery and chemicals. Technological change occurs when there is a change in the relationship between inputs and outputs, either in the form or the mix of inputs, or the form and the mix of outputs. Some changes have been very important, such as the change from horses to tractors (1910–1940), the introduction of hybrid corn (1930–1950), and continuing improvements in seeds, pesticides, fertilizers, farm machinery, and other industrial inputs (mainly since the early 1930s).

Productivity refers to the ratio of output to input. An advance in productivity means an increase in output relative to input. In measuring *total productivity,* all inputs used in production—land, labor, buildings, equipment, energy, fertilizers, chemical pesticides, and others—are value-weighted to yield an index number of total input. An index of output is constructed by value-weighting all agricultural commodities at an appropriate point in the marketing chain. Also, the average annual rates of change can be computed for total input, total output, and productivity (Table 4-1). Since 1900, total U.S. farm output has increased at an average annual rate of +0.9 to 3.1 percent in selected periods, which must be balanced with increased demand to maintain market stability (Table 4-2).

The total productivity measure in Tables 4-1 and 4-2 does not capture all of the costs and benefits of technological change that are important. External factors include the effect of technological change on the environment (ground-water pollution from fertilizers or pesticides, for example), and external social costs such as the forced liquidation of family farms (farm family labor migrating off the farm), which are related to the farm technology and increasing returns to large scale. The labor released from farm work may be viewed as a net economic and social gain, however, because it increases the total labor supply for nonfarm production.

TABLE 4-2
AVERAGE ANNUAL RATES OF CHANGE (PERCENT PER YEAR) IN U.S. AGRICULTURAL INPUT, OUTPUT, AND PRODUCTIVITY, 1900–1989

	1900–1924	1925–1949	1950–1959	1960–1969	1970–1979	1980–1989
Total input	1.1	0.2	−0.5	−0.3	1.0	−1.9
Total output	0.9	1.6	2.4	1.3	3.1	1.0
Productivity	−0.2	1.3	2.6	1.6	2.1	2.8

Sources: Economic Research Service, USDA, *Agricultural Outlook and Outlook Charts* (various years), and *Economic Report of the President,* February 1992.

Partial productivity measures the ratio of output to one or more inputs. Some frequently used ratios refer to land, animals, or farm labor. Since the mid-1930s, for instance, in considering production per unit of land only, the national average of corn yields in the United States has increased from less than 40 bushels to more than 120 bushels per acre. Since 1950, the number of dairy cows has been reduced by more than one half, and dairy cows are producing more milk on one-third less feed. Since 1800, the farm labor required to produce 100 bushels of wheat has dropped from 373 hours to fewer than 9 hours. Since 1940, average farm labor productivity has increased by more than 10 times, and by 20 times or more on some more highly automated farms.

DEMAND CHARACTERISTICS

Definitions and Determinants

Demand may be defined as a schedule or curve showing the various quantities of a resource, service, commodity, or product that individuals or firms are willing and able to purchase at each price in a range of possible prices during a specific period of time. Viewed from the standpoint of quantity, demand indicates the various prices that will be received for various quantities offered for sale.

When these concepts are applied to agriculture, note that there are many demand curves. There is a demand for total agricultural output; for groups of commodities such as meats or vegetables; for individual commodities such as potatoes or fluid milk; and for specific grades, classes, or even brands of a given commodity. Also, there are several levels of demand. For instance, there is a *primary demand* at the *retail* or *consumer level,* from which the demands at the *wholesale* and the *farm levels* are derived. From these levels, there are demands by farm and agribusiness firms for the various inputs used in production, which are all derived from the demands for the agricultural commodities in wholesale and retail markets.

The relationship between price and quantity on a given demand curve is generally inverse, or negative, a characteristic of the law of demand. Again, as in the case of supply, a *change in demand* refers to a shift of the demand curve. The *quantity demanded* at a price refers to a point on a demand curve. As in supply, the quantity demanded can change when the price changes without the demand curve itself changing (Figure 4-2).

Determinants of the quantity demanded can be conceptualized in abbreviated, functional form:

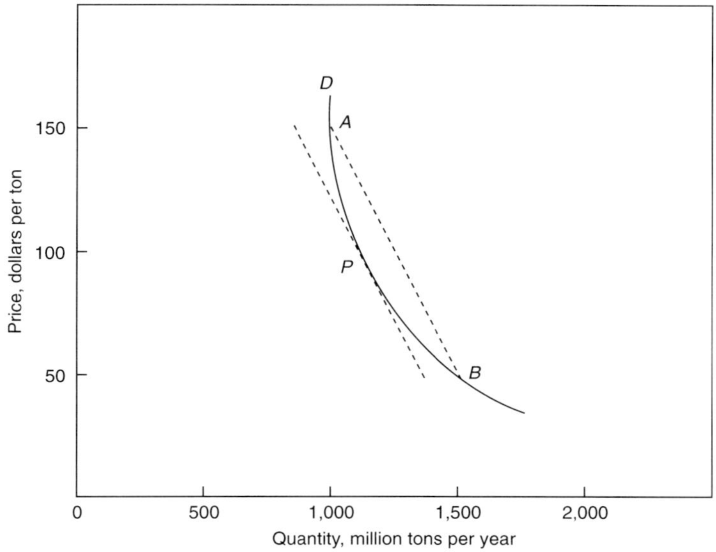

FIGURE 4-2
Demand curve and elasticity; world demand curve for grain (hypothetical). In this diagram, the line that is tangent to the demand curve at *P* is parallel to *AB*. Elasticity measured at point *P* is called *point elasticity*. Elasticity of the arc between *A* and *B* is *arc elasticity*. See Appendix.

$$Dq = f(P, P_o, I, T, A, TP, PG)$$

where Dq = quantity of a commodity demanded by buyers, be they producers or consumers anywhere a market exists

P = price of the commodity or commodities

P_o = population of people, firms

I = income of consumers or purchasing power of firms

T = net trade

A = alternative consumer commodities or other goods

TP = tastes and preferences of consumer

PG = policies of government

Some of these factors are social and traditional in origin. Several are more centrally important for the study of policy: (1) the size of the population, both domestic and foreign; (2) per capita incomes; (3) net trade; and (4) policies of government, such as subsidies. These determinants apply to all agricultural commodities, among which food is by far the most important.

Factors Affecting Changes in Demand

Population and Income Historically, population growth and increasing real incomes have been the main determinants of increases in demand for food. Among countries that have a consistently large supply of food for export (such as the United States, Canada, Australia, and New Zealand), the aggregate demand for food depends on demand in both domestic and foreign markets. In the United States, because the average consumption of food per capita has increased very little since the 1950s and 1960s, relatively small increases in average consumption per capita may be expected in the future. When most of the population is very well off or affluent, as in the United States and in some other highly developed countries, a further increase in real income will make little difference in food consumption per capita. However, when the real income of poor people increases, their demand for food can expand considerably. Food consumption may also be affected by rising concerns about cholesterol, obesity, and other dietary considerations.

The low rate of population growth in most of the developed countries, indicates that most of the growth in world food demand in the twenty-first century will occur as a result of economic development and population growth of the currently less developed countries. For the immediate future, the average annual increase in demand resulting from growth in the U.S. population and consumer income is projected to be about 1.0 percent or less.

Long-range forecasts of the U.S. population, which help to set the future demand parameters related to food and agricultural policy, are the responsibility of the U.S. Bureau of the Census. Such forecasts are based on specific variables, the most important being changes in the population's fertility or birth rate, net immigration into the United States, and death rates.

Between 1989 and late 1992, based on changes in each of these variables, the Bureau increased its forecast of the U.S. population for the year 2050 by some 83 million people (Figure 4-3). If the slight increase in the rate of growth in the 1992 forecast of the Census Bureau is realized in the future, this will result in a marginally greater increase in domestic demand for agricultural products than was forecasted earlier, but the average annual increase in the U.S. population would still be less than 1.0 percent compounded.

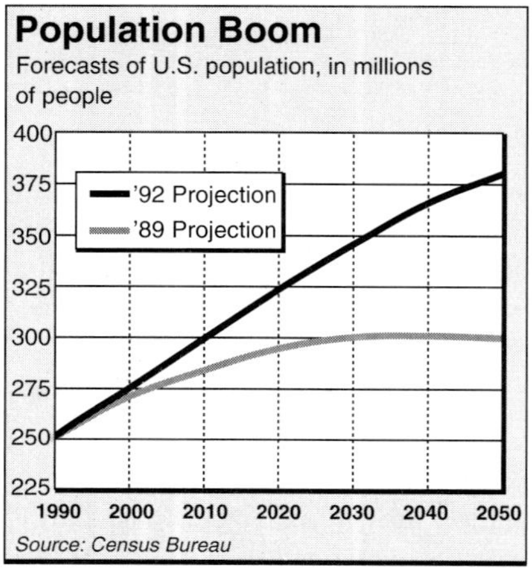

FIGURE 4-3
U.S. Census Bureau population forecasts, 1989 and 1992.

In addition, another important unknown for the future is the nonfood industrial uses for agricultural products, which will be affected by technology, consumer acceptance, and regulatory approval. Technology can, of course, create substitutes for these products, as with fibers and sweeteners, but also expand their markets, such as with ethanol, plastics, and printing ink.

Net Agricultural Trade A surplus in net trade adds demand for domestic production while a deficit does the opposite. Although net trade is usually viewed in nominal currency terms, it should be measured in quantity, or in some real terms, in order to accurately reflect trade-generated demand. Furthermore, even if agricultural exports are substantial, such as the $40 billion level reached in 1991, and net surplus trade at the $17 billion level, no additional demand may be created unless there is an *increase in net trade.* When such increases or decreases result from trade changes, they must still be related to the total value of production in order to assess ultimate effects upon demand. For the past several decades, the increase in trade-generated demand for U.S. agricultural production estimated in these proportional value terms has oscillated wildly, but averaged about 0.5 percent annually. Such increases may be an even more important factor in the future.

Taxes and Subsidies When state governments levy sales taxes, a general policy issue nearly always arises concerning whether sales of food should be taxed. Most states either exempt food sales or tax them at a reduced rate. The arguments for this policy usually rest on the assumption that a tax on food is regressive and unfair to the poor. Certainly, this argument has merit, especially because few state tax systems are progressive and most in fact are regressive.

A still more important issue for policy concerns the desired role of food subsidies to help the poor. Subsidies of this nature include food stamps, which are financed in the United States through the federal Treasury. Private food assistance is also provided by various voluntary programs organized by churches or charitable groups. Helping poor people to have access to more food can be justified in humanitarian terms, and this also has generated much political appeal. Without question, however, shifting purchasing power from the general tax-paying public to the poor who would qualify for food stamps will increase the aggregate demand for food. For similar reasons, a general income tax that redistributes income from people with higher incomes to those with lower incomes will increase the domestic demand for food.

Elasticities of Agricultural Supply and Demand

Computation of Elasticities of Supply and Demand Relevant to Policy For a review of the underlying theory and simplified procedures for computation, see Appendix to this chapter.

Elasticity of Agricultural Supply The supply of total agricultural output tends to be very inelastic in the short run (taken to be two years), or between two harvest seasons. For example, Luther Tweeten and Leroy Quance estimated short-run (two-year) elasticities by using three methods.[1] (1) By estimating the best fit for a supply schedule, their best single estimate was 0.155. (2) By using separate yield and production components for crops and livestock, they estimated short-run elasticities of 0.17 for crops, 0.38 for livestock, and 0.25 for aggregate crops and livestock. (3) By using eight separate categories of inputs, with each weighted according to its relative importance in contributing to total output, the short-run (two-year) elasticity was 0.26 for periods of rising prices and 0.10 for periods of falling prices.

In a later study, Mike Woods, Tweeten, Daryll Ray, and Greg Parvin concluded that supply response (elasticity) is low in the short run for both rising and falling prices.[2] Short-run response of output to price depends on variable inputs such as fertilizer and pesticides. Most agribusiness inputs are readily reversible between planting seasons (as quickly reduced as increased). The costs of some other inputs such as farm machinery and land are more fixed in the two-year short run, which suggests that some supply curves may be more inelastic when prices fall.

The aggregate agricultural supply function is very inelastic in the short run, while the supply of individual commodities tends to be more elastic, especially where commodities can be easily substituted in the production process.[3] A general implication for policy is that the aggregate quantity supplied will respond slowly to either very high or very low agricultural prices. Good weather or the rapid introduction of new producer technologies shifts the supply to the right resulting in relatively low prices, which can last for several years, unless offset by increases in demand.

Price Elasticity of Demand for Food *Price elasticity of demand* measures the relative change in quantity demanded of a product that is associated with a relative change in its price. It can be represented as a measurement for a specific price or an average over a range of prices. In general, the price elasticity of demand for food is very inelastic. For example, the U.S. domestic price elasticity for all foods is listed as −0.237 in Table 4-3. With this demand, a 10 percent increase or decrease in price without any change in supply will result in only a 2.37 percent change in quantity demanded. Or a 2.37 percent change in quantity supplied without any change in demand will result in a 10 percent change in price in the opposite direction.

In cases of general price increases, commodities of low-value may be substituted for those of high value. With general price decreases, the reverse substitution may be made. In other words, when food prices rise, people are not as likely to eat fewer calories as they are to substitute a cheaper food (hamburger) for a more expensive one (steak).

The extent to which one commodity is substituted for another depends on how close the substitutes are. For example, pork and chicken are close substitutes, and people switch rather quickly from one to the other when prices change. This willingness to switch is reflected in relatively more elastic demand for both pork and chicken. Beef is also a close substitute for other meats, and the demand for it is more elastic than for most other foods. Apples are close substitutes for some other fruits. As a result, the demand for apples is more elastic than the individual demands for potatoes, wheat flour, rice, or sugar. Among the important commodities not shown in Table 4-3 is fluid milk; having no close substitutes it is highly inelastic in demand.

When demand is inelastic, any change in the quantity supplied will result in relatively greater changes in price, other things being equal. For commodities that are most

TABLE 4-3
ESTIMATED U.S. INCOME AND PRICE ELASTICITIES OF DEMAND AND PER CAPITA CONSUMPTION OF SELECTED FOOD COMMODITIES

	Estimated elasticities of demand at retail		Annual per capita consumption (weight in pounds at retail)			
	Income	Price	1962–1966	1971–1975	1976–1981	1982–1987
Beef	0.290	−0.644	72.1	85.2	84.2	77.3
Pork	0.133	−0.413	57.8	60.1	60.5	60.3
Chicken	0.178	−0.777	37.2	41.3	47.7	56.7
Fish	0.004	−0.233	10.7	12.2	13.0	13.9
Eggs	0.055	−0.318	40.3	37.6	34.3	32.5
Cheese	0.249	−0.460	14.0	13.6	17.0	21.9
Butter	0.318	−0.652	6.6	4.8	4.4	4.7
Margarine	0.000	−0.847	9.8	11.2	11.4	10.8
Apples, fresh	0.140	−0.720	17.0	15.9	17.2	17.8
Potatoes	0.117	−0.309	82.5	74.2	73.6	74.3
Wheat flour	0.083	−0.300	108.2	108.2	116.8	121.2
Rice, milled	0.055	−0.320	7.2	7.3	8.4	10.7
Sugar, refined	0.032	−0.242	96.8	98.6	88.6	66.5
All foods	0.176	−0.237	1,373.7	1,385.7	1,403.1	n.a.
Nonfoods	1.243	n.a.	n.a.	n.a.	n.a.	n.a.

Sources: P. S. George and G. A. King, *Consumer Demand for Food Commodities in the United States with Projections to 1980,* Giannini Foundation Monograph no. 26, University of California, Berkeley, March, 1971; *Food Consumption, Prices, Expenditures, 1966–87,* Statistical Bulletin 773, Economic Research Service, U.S. Department of Agriculture, Washington, D.C., 1990.

inelastic in demand, price fluctuations generally tend to be the widest unless there are effective storage programs to stabilize supply or compensate for changes in output due to noneconomic factors such as the weather. The very inelastic domestic demands for major crops—such as wheat, cotton, sugar, rice, and tobacco—are a fundamental economic factor that provides incentives for the political efforts of U.S. farmers to support federal programs to control or reduce supply and also to subsidize exports. For example, reduced production of an inelastic product results in increases in both price and total receipts, *ceteris paribus.* A highly inelastic demand for fluid milk is the basic economic factor accounting for dairy farmers' political efforts to get federal and state milk-marketing orders to regulate the price and market flow of fluid milk.

The demand for total farm output, as well as for each food commodity, is more inelastic at the farm level than at the retail level. Price fluctuations at the farm level are in general relatively greater—not absolutely greater—than at the retail level. They are relatively greatest for the major crops that have few close substitutes. This characteristic of variability adds to the political pressure exerted by farmers, agribusiness organizations, and some others to get the government to support the prices of these commodities or to compensate farmers for low prices through compensatory payments.

Income Elasticity of Demand for Food *Income elasticity of demand* measures the relative change in quantity demanded (or real expenditures) of a product associated with a relative change in income of consumers, *ceteris paribus.* Income elasticity of demand in the United States was estimated to be 0.176 for all food (Table 4-3). The estimate for food includes many services, which are more income elastic than the demand for raw food. The relatively higher income elasticities of demand for beef and other animal products indicate that, as income per capita increases, *ceteris paribus,* relatively more animal products will be bought, displacing other foods. Declining incomes have the opposite effect.

Price Elasticity of Demand for Exports The above elasticities, when combined with the projected slow growth for demand in the domestic market, create a setting for policy that magnifies the importance of the price elasticity of demand for exports. Because U.S. agriculture relies heavily on export markets, the price elasticity of export demand is a crucial factor in considering options in farm price and income policy.

Changes in the way international trading is done suggest that the price elasticity of export demand for agricultural commodities is not fixed, but becomes more elastic, or less inelastic, depending on the trade policies of importing countries. For instance, the price elasticity of demand for exports of food and feed from the United States could be as high as -1.91 in the short run of three years and even more elastic, such as -6.42, in the long run of four or more years, provided other countries do not change their trade policies.[4] However, because tariffs and nontariff barriers (NTBs), such as import quotas and other types of governmental controls, have often been imposed when imports grow rapidly, the prices received by exporters may not be fully transmitted to importing countries.[5]

Under these conditions, the *elasticity of price transmission,* which is the importing country's price response to the price received by exporters, may range from near zero, when all trade is controlled by intervention policy, to $+1.0$ when trade is free.[6] For example, due to U.S. import quotas, the export price received for raw sugar in the Caribbean or the Philippines has little if anything to do with the much higher prices that U.S. consumers pay for sugar in all its various forms, and the elasticity of price transmission may be near zero. Similarly, the Australian or the U.S. export price for beef is not transmitted to consumers in Japan.

Some agricultural commodities are traded rather freely around the world, but many are not, which greatly restricts the exports of the major food commodities. In addition, it keeps food prices high in many importing countries, depending on their own import policies. Hence, the elasticity of export demand faced by U.S. agriculture can vary from high elasticity when trade is relatively free, to nearly zero elasticity when export prices are not transmitted to those of the importing country. The trade policies of most major importing countries involve a mixture of protection against imports of some commodities and relatively free trade in other commodities.

For many years, the European Community (EC) under its Common Agricultural Policy (CAP) has held prices for feed grains and wheat well above world levels by means of a variable import levy. The variable levy closed most if not all of the gap between the fixed-target support price within the EC and the lower world price. The levy increased as world market prices fell relative to internal target prices, and it decreased as world prices rose relative to internal EC prices. For several years, the EC maintained free trade for soybean imports, however, and a mixture of free trade combined with a trade agreement on cotton. As the internal protected prices of the leading countries of the EC have been supported substantially above world prices, some of these countries have gradually increased their per capita production of wheat, soybeans, and feed grains. When countries in the EC developed a net exportable surplus, they begin subsidizing their exports, a policy to which most other exporters—including the United States—have strongly protested.

Japan's import demand for wheat, feed grains, soybeans, and cotton has been inelastic, because imports are controlled by a government agency under the general policy goal of stabilizing domestic prices. This suggests that Japan has tended to import the same amounts of these commodities almost without regard to the world price, but could choose among exporters for political reasons or on the basis of small differences in price or quality among the exporting countries.

In summary, the various restrictions that the United States, the EC, Japan, and many other countries place on agricultural imports, combined with the inelasticity of price transmission, reduces the elasticity of demand faced by most exporters.

Since the EC has emerged as a major competitor in world markets, new fears have been aroused in other exporting nations about their own export markets. These fears are widely shared, which accounts for the emphasis on trade negotiations in the General Agreement on Tariffs and Trade (GATT). World supply and demand analysis suggests that the success of any country in increasing its export levels will depend on a freer and more peaceful world trade environment; the continuing economic growth of developing countries; a strong domestic economic policy to support the necessary research and to facilitate the resulting farm resource shifts; a flexible resource policy for agricultural production and marketing; and in the absence of some of these conditions, an aggressive export subsidy program.

SUPPLY–DEMAND BALANCES AND POLICY

Importance of Net Agricultural Trade

The supply and demand characteristics of the U.S. food and agricultural sector have important policy implications for trade. Since the end of the 1940s, U.S. agriculture has increased aggregate output at a rate of about 1.7 percent per year. Fluctuating year-to-year changes have ranged from a positive 16.7 percent to a negative 17.2 percent and, along with this, there has been successive ripple effects on farm prices, family incomes, and rural community welfare. Partially absorbing this production growth has been a relatively constant increase in the U.S. population averaging slightly less than 1.0 percent per year, which is projected to continue in the 1992 forecast of the U.S. Census Bureau. This rate of population growth, coupled with an almost minuscule increase in food demand per capita, which is generated by slowly rising incomes, indicates that there has been a greater increase in domestic supply than in domestic demand averaging about 0.5 percent annually. This gap has been filled in some years by the expanding net agricultural export trade. Yet trade, too, has varied widely, and thus, large supply-demand changes have occurred when the fluctuations in production and trade do not coincide. Every time trade expands faster than these trend averages, as it did in the 1970s, the farm economy benefits. However, when the opposite situation occurs, as it did in the first half of the 1980s, an economic farm crisis looms. Hence, net trade serves as a critical safety valve for the U.S. food and agricultural economy.

Influences of Governmental Programs

The U.S. supply–demand balance, as well as the farm prices and incomes, are influenced directly and indirectly by governmental programs. On the supply side, production technology and quantity have been shaped for a century by the research and education flowing from institutions established by public policies. The high productivity record of this nation's farm sector is attributable in large measure to those policies. Furthermore, since the 1930s, public policies have been used periodically, when supplies seemed excessive, to attempt to bring growth in farm production more in line with growth in demand. As a result, up to 15 percent of the cropland has been held out of marketable production by allotments, quotas, set-asides, and conservation reserves during the late 1960s and early 1970s, the early 1980s, and again in the late 1980s and early 1990s. Without these programs and various credit policies, farm product prices and incomes would have been more variable, farm family incomes generally lower, and shifts in farm structure greater.

On the demand side, a variety of food assistance programs for low income consumers in the United States and foreign food aid programs for needy nations abroad have significantly augmented the commercial market for agricultural products. Program results include marginally higher prices and increased producer returns as well as improved diets for the needy. Prominent among these policies have been the Food Stamp and the National School Lunch Programs. In addition, demand periodically has been strengthened by various public export policies. The most prominent of these policies have been the export subsidy and promotion programs of the 1954 Public Law 480, the Food Security Act of 1985, and the Food, Agriculture, Conservation, and Trade Act of 1990.

In addition, incomes of farm families have been augmented from various governmental programs involving Treasury transfers, such as price supports, deficiency payments, grain storage payments, conservation cost-sharing, debt restructuring, and disaster assistance. The levels of these transfers have varied widely depending on the economic stress in the farm economy, and other factors.

Changing Structure and Economic Size of Agriculture

Economic structure refers to or deals with the organization and ownership of resources. This includes the size and distribution of farm firms and all economic factors relating to organization, ownership, management, and income distribution. Economic structure also includes similar factors for agribusiness firms.

Farm Structure in Terms of Sales and Income In the United States, a *farm* is defined by the U.S. Department of Agriculture and the Bureau of the Census as any operation with actual or potential sales of agricultural commodities of $1,000 or more in a single year. Commodities worth $1,000 or more can be obtained from as little as one to three acres of fruit trees, five acres of corn or soybeans, an acre of some vegetable crops, one good dairy cow, or less than one-half acre of tobacco. Thus farms range from very small enterprises, many of which are of little commercial significance, to very large enterprises with annual sales in the millions of dollars. As a result, within the full range of farm firms, gross cash farm income is highly skewed. For instance, in 1989, some 39,000 farms (1.8 percent of all farms) earned $67 billion of the total $177 billion received by all farms, or 38 percent of total gross cash farm income. At the lower end of the scale, 1,019,000 farms (46.9 percent of all farms) earned just $5 billion gross cash farm income, or 2.8 percent of total gross cash farm income. In between these classes, there were more than 1.0 million so-called family farms (large, medium, and small), totaling 51.3 percent of all farms, and earning 59.2 percent of the gross cash farm income (Table 4-4).

The statistics on farm size can be confusing because many small farms are not fully operated by the owner or the family living there. An unknown number of the rural-residence farms and a significant percentage of the small family farms are counted in the farm income statistics, even though the land and livestock may be managed, farmed, or cared for by a hired person or another farm operator. Many operators of small farms also work part of the time on other farms and off the farm. Most large operators use both land that they own and land that they rent for cash or other types of lease arrangements. Such a farm structure means that the degree of actual concentration in farm (and ranch) operations is greater than is suggested by the data in Tables 4-4 and 4-5.

Farm Structural Changes and Trends A sweeping change has occurred in farm numbers and economic structure. Around 1935, the number of farms, defined somewhat differently than now, reached its peak at about 6.8 million. The number dropped to little more than 2.0 million by 1990 (under the modern definition of a farm). Total farms could drop to about 1.25 million by 2000.[7] Since 1935, there has been a slight increase in the amount of tillable land readily available for growing crops and for livestock grazing, and a fourfold increase in land used per farm.

Each year, however, the farm sector has tended to become more capital intensive, so the real increase in size of the farm firm is much greater than may be inferred from studying changes in land used. The real change implies greater concentration in the future, with a higher proportion of total output concentrated in fewer farms, using more advanced technology, more nonland capital, and more advanced management techniques. Because of rapid technological advances and continuing farm resource adjustments, the trend toward larger commercial farm sizes continues unabated. Unless values are presented in constant dollar terms, inflation of values by itself results in a given size of operation shifting into a larger size classification on strictly a value basis.

Many important gains from technology and productivity have been realized, resulting

TABLE 4-4
DISTRIBUTION OF FARMS BY VALUE OF FARM SALES AND GROSS CASH FARM INCOME, 1989

Farm size and sales class	No. of farms (thousands)	Percent of farms	Gross cash farm income (billions)	Percent of gross cash farm income
Very large: $500,000 and up	39	1.8	$ 67	38.0
Large family farm: $250,000–$499,999	74	3.4	28	15.9
Medium family farm: $40,000–$249,999	514	23.7	63	35.2
Small family farm: $10,000–$39,999	525	24.2	14	8.1
Rural residence: less than $10,000	1,019	46.9	5	2.8
All farms	2,171	100.0	177	100.0

Source: Economic Research Service, USDA, *Economic Indicators of the Farm Sector, National Financial Summary, 1989.* (Washington, D.C., January, 1991).

in substantial benefits for consumers, for the U.S. balance of trade, and for important parts of the farm and agribusiness sectors. For these reasons, the policy emphasis on improving technology and increasing productivity may be fully justified, but the adjustment problems of farm families should not be ignored in public policy making. In view of the projected increases in world population and the prospects for continuing economic development, a policy to encourage improved technology can be visualized as crucial for the future productivity and prosperity of U.S. agriculture.

FARM PRICES AND INCOMES

Since the founding of this nation, public policy has been designed to encourage the growth of agriculture. Especially in the twentieth century, there has been additional policy emphasis on programs to encourage technological advance, while protecting or directly supporting farm prices and incomes. After allowing for inflation, there has been a general decline in prices received by farmers, which may be attributed to generally more rapid increases in farm productivity compared to most other sectors, to pure competition among farm producers, and to highly inelastic and relatively slow-growing demand for aggregate farm output. Since the early 1920s, average consumer food prices have increased less than the general consumer price index. Consumer food prices have increased faster than farm product prices partly due to increased marketing services. Finally, farm input prices have increased more than prices received by farmers. As a result real prices (prices adjusted for inflation) received by farmers declined by an average of 3.0 percent per year in the 1950s, 1.6 percent in the 1960s, and 3.5

TABLE 4-5
TRENDS IN DISTRIBUTION OF FARM SALES BY SIZE GROUPS

Farm size and sales class	1974[a]	1978[a]	1982[a]	1985[b]	1989[c]
			(percent)		
Very large: $500,000 and up	25.2	29.7	32.5	30.2	40.5
Large family farm: $250,000–$499,999	11.5	13.2	15.1	18.6	16.1
Medium family farm: $40,000–$249,999	46.0	43.6	41.5	40.7	33.9
Small family farm: $10,000–$39,999	13.6	10.4	8.2	8.0	7.3
Rural residence: less than $10,000	3.9	3.1	1.7	2.5	2.2
All farms	100.0	100.0	100.0	100.0	100.0

[a] In constant 1982 dollars; excludes abnormal farms.
[b] In 1985 dollars.
[c] In 1989 dollars.

Sources: Census of Agriculture, 1974, 1978, and 1982. (Washington, D.C., United States Department of Commerce, Bureau of the Census). Economic Research Service, USDA, *Economic Indicators of the Farm Sector, National Financial Summary, 1985.* (Washington, D.C., November 1986): 45; *Ibid.* (January 1991): 45, 49.

percent in the 1980s. Prior to the 1990s, significant increases occurred only in the 1970s and the last half of the 1980s.

The total income of farm families from all sources has varied in recent decades. Even though total net farm income for all farmers has generally declined in real terms, the continuous out-migration of the farm population, rapid adoption of new technology, and expansion of farm size has protected an otherwise predictable decline in average operator's income. In real dollar terms, net farm income per farm declined during the 1950s, 1960s, and 1970s but rose in the 1980s. A significant part of this improvement in the latter decade is due to governmental price supports and income assistance. In addition, because of gradually increasing off-farm income, the total farm family income picture is somewhat brighter than the data indicate.

The incomes of farm families from farm production are highly skewed and concentrated toward the largest farms as may be seen in the farm structure depicted in Tables 4-4 and 4-5. Similarly, because of particular choices of policies to stabilize agriculture, government payments are deliberately tied to market sales, and are also skewed, although less skewed than market-generated farm income. The income of farm families from off-farm sources is much more evenly distributed, so the total net income of all farm operator families is not nearly as skewed as income from either farms sales or government payments.

People who live on farms experience great variations in incomes. Some have obtained average or above family incomes, mainly from off-farm sources. Many farm families, however, perhaps as many as one fifth to one fourth of all farm operator families, have well below average incomes and substandard social services, including less than average educational opportunities and relatively modest health facilities, or insufficient incomes to take advantage of health services.

Most landowners and a high percentage of farm operator families are financially solvent. Since rapid technological advance began, however, a significant proportion of the farm population has been in financial difficulty, or on the verge of it. The most adversely affected are those producers who have inadequate capital and management or are caught in unusual weather and macroeconomic shocks. Perhaps this state of affairs is necessary to get the adjustments in scale and labor inputs that advancing technology requires to provide the most efficient production for consumers. We may conclude that a broader social problem exists for the rural society and the nation's farm sector, which can be approached through public policy.

A FUTURE OF INCREASING TECHNOLOGY

The New Age of High Technology [8]

For much of agriculture, a new age of high technology has arrived. High technology involves important advances in both *biotechnology* and *information technology.* Biotechnology includes any technique that uses living organisms to modify inputs or outputs, to improve plants or animals, or to develop microorganisms for specific uses. New information technologies involve significant improvements in communication and information management for organizing and monitoring production processes, marketing, and other activities. Let us examine briefly four general methods of biotechnology followed by an introduction to information technology.

Recombinant (rDNA) Technology Because of the power of rDNA technology to alter plants and animals (including humans), it is one of the most important but controversial achievements of biological science. The basic theory of this technology is that molecules of rDNA formed from two different species can be inserted into a variety of bacteria, yeasts, and animal cells, where they replicate and produce useful proteins and other nutrients such as amino acids and single-cell protein feed supplements.

One application of the pharmaceuticals created by this procedure is to manufacture growth hormones that can be injected into animals to increase their productive efficiency, that is, their rate of gain in weight per unit of feed. For instance, injections of bovine growth hormone (bST) into dairy cows at the rate of 44 milligrams per cow per day have resulted in increases of 10 to 40 percent in milk yield. Of greater importance, genes for new traits can be inserted into the reproductive cells of livestock and poultry. This procedure permanently endows future generations with desired traits such as disease resistance or accelerated growth. This can have far-reaching consequences.

Monoclonal Antibody Production This process—sometimes called *cloning*—can create a large volume of a single antibody, which can be produced by fusing a myeloma cell (a cancerous antibody-producing cell) with a cell that yields another antibody. The procedure creates a hybridoma, which in turn produces (theoretically and in perpetuity) large quantities of identical (i.e., monoclonal) antibodies. Some important uses of monoclonal antibodies include passive immunization of calves against scours; substitutions for vaccines, antitoxins, and antivenoms; sexing of livestock embryos; post-coital contraception and pregnancy testing; detection of food poisoning; the imaging, targeting, and killing of cancer cells; the monitoring of levels of hormones and drugs; and prevention of rejection of organ transplants.

Embryo Transfer in Livestock and Poultry Embryo transfer is being used to rapidly upgrade the quality and productivity of livestock, particularly cattle. In this procedure, embryos are removed, genetically treated, and implanted in surrogates. Among meat animals, this results in improved rates of gain, lower body fat, improved carcass characteristics, increased milk production, and increased resistance to disease among offspring. Some genetically improved animals may also contain genes that increase their powers to reproduce.

Some new technologies will be used to diagnose and treat diseases that formerly reduced the productivity of livestock and poultry by as much as 20 percent in a single year. For instance, a genetically engineered vaccine for colibacillosis (scours), a disease that formerly killed millions of newborn calves and piglets each year, is being used commercially. Genetic engineering of the microorganisms that permit ruminants—cattle, sheep, and goats—to digest forages permits these livestock to digest various feedstuffs such as high-lignin substances, which until recently have been virtually useless to them.

Biotechnology for Plants As with animals, the effects of biotechnology on plant productivity will be phenomenal, but quite likely less controversial. Applications of biotechnology can modify crops so that they will make more nutritious protein, resist insects and disease, grow in harsh environments, and make their own nitrogen fertilizer. Genetic engineering applications will include microbial inocula and genetic modification. Plant propagation techniques, which have been used extensively to produce new species and varieties such as hybrids, are being extended to new areas. The efficiency of photosynthesis, the fundamental basis for plant growth, is being enhanced over several years as a result of research on several factors that naturally limit the efficiency of the process.

New Information Technologies In various stages of development, new information technologies are greatly improving management, including control over financial affairs and increasing productivity of the biotechnologies. On-farm communication and information technologies often include a central computer system; a radio link with operators of tractors, combines, and other implements; and an on-farm weather station. Many processes in plant and animal production can be organized and monitored by new electronic technologies. Some technologies operate automatically using ongoing information to regulate the flow of irrigation water, monitor and control pests, and regulate the environment for livestock. Other applications include devices to identify individual animal performance for feeding, reproduction, and genetic improvement.

Computer technology is producing significant changes in insect management at all levels.

Telecommunication systems provide information that is crucial for operational and financial management and marketing decisions, and also link the farm with other firms and institutions, enabling farmers to have rapid, inexpensive, and reliable access to a wide array of information on crop prospects, markets, and weather. Finally, remote sensing technology is being used increasingly to detect, process, and analyze a wide range of data collected from various sources.

LOOKING AHEAD

In the new age of high technology, there is a great potential for growth of agriculture in the United States and most other countries. At the same time, there is a growing awareness in the United States and perhaps around the world that the recent pattern of agricultural growth is not indefinitely sustainable without more advances in technology and world trade. Although there is no world food and agricultural policy, the world situation is important for every country, and especially important for a large trading nation such as the United States. Hence, a discussion of the world food and agricultural situation is the subject of the next chapter.

ENDNOTES

1 Luther G. Tweeten, and C. Leroy Quance, ''Positivistic Measures of Aggregate Supply Elasticities: Some New Approaches,'' *American Journal of Agricultural Economics,* 51(2), May 1969, pp. 342–352.

2 Mike Woods, Luther Tweeten, Daryll E. Ray, and Greg Parvin, ''Statistical Test of the Hypothesis of Reversible Agricultural Supply,'' *North Central Journal of Agricultural Economics,* 3(1), January 1981, pp. 13–19.

3 For a review of more studies, see Harold G. Halcrow, *Food Policy for America,* McGraw-Hill, New York, 1977, pp. 133–141.

4 Luther G. Tweeten, ''The Demand for United States Farm Output,'' *Food Research Institute Studies,* 7(3), Food Research Institute Studies, Stanford University, Stanford, CA, 1967, pp. 343–369.

5 Paul R. Johnson, ''The Elasticity of Foreign Demand for U.S. Agricultural Products,'' *American Journal of Agricultural Economics,* 59(4), November 1977, pp. 735 and 736; and Luther Tweeten, ''The Elasticity of Foreign Demand for U.S. Agricultural Products: Comments,'' *ibid.,* pp. 737–738.

6 For a discussion of a suggested methodology, see Maury E. Bredahl, William H. Meyers, and Keith J. Collins, ''The Elasticity of Foreign Demand for U.S. Agricultural Products: The Importance of the Price Transmission Elasticity,'' *American Journal of Agricultural Economics,* 61(1), February 1979, pp. 58–63.

7 United States Congress, Office of Technology Assessment, *Technology, Public Policy, and the Changing Structure of American Agriculture,* OTA-F185:96. U.S. Government Printing Office, Washington, DC, March 1986.

8 *Ibid.*

TERMS AND CONCEPTS

Agricultural supply
Factors determining supply
Technology
Productivity
Partial productivity
Demand for agricultural commodities
Factors determining demand
Elasticity of agricultural supply
Price elasticity of demand
Income elasticity of demand
Elasticity of demand for exports
Farm and retail elasticity of demand
Supply–demand balances
Farm prices
Economic structure of the farm economy
Skewness of farm sales classes
Farm family income
Off-farm income
Distribution of farm family income
Biotechnology
Information technology

QUESTIONS AND STUDY SUGGESTIONS

1 Define *supply*. What factors interact to determine agricultural supply?
2 Define *demand*. What are the determinants of demand for agricultural commodities?
3 Define *income elasticity of demand*. How can it be computed? What is the income elasticity of demand for food in the United States? Would it be higher in Bangladesh? In Japan?
4 Define *elasticity of supply* and *elasticity of demand* (in respect to price). What factors determine the elasticity of the supply and elasticity of demand? How can the latter be computed?
5 What are the main implications for food and agricultural policy of inelastic supply and demand?
6 Suppose the production of some common farm commodity, like wheat, increased 10 percent in a year with particularly favorable weather, but demand conditions remained generally unchanged. What effects would you expect in the market price of wheat and the value of the total sales by producers?
7 Why is supply more elastic in the long run than in the short run? Why is the supply of corn more elastic than the supply for total farm output, and what is the importance of this for policy?
8 What is the policy significance of a low elasticity of price transmission? A high elasticity? How may each affect export policy for the United States?
9 What is technology? What is productivity? What has been the record of U.S. farm productivity?
10 What are some examples of high technology in agriculture?
11 Define *economic structure*. Describe the economic structure of the U.S. farm sector.
12 What have been the trends in real farm product prices? Per-farm off-farm income? Per-farm family income?

COMPUTATION OF ELASTICITIES OF SUPPLY AND DEMAND RELEVANT TO POLICY

When movement occurs from point to point on a demand (supply) curve, the price elasticity of demand (supply) is the percentage change in the quantity demanded (supplied) divided by the percentage change in price. If the percentage change in quantity is greater than the percentage change in price, the demand (supply) is elastic. If the percentage change in quantity is less than the percentage change in price, the curve is inelastic. Unitary refers to equal percentage changes in price and quantity demanded (supplied). For demand, the variables are normally changing in opposite directions, but in the same directions with supply. Thus, the correct sign for price elasticities of demand is usually negative and positive for supply elasticities. The data for computing price elasticity of demand or supply are drawn from either (1) time series in which prices and quantities are changing or have changed, or (2) cross-sectional studies of economic data. The resulting numerical measure of datum is called the *coefficient of elasticity.*

Computing the Coefficients

The coefficient of elasticity is a precise measure of the percentage change in quantity relative to the percentage change in price between two points (or at a point) on a demand or supply curve, with all other conditions unchanged (*ceteris paribus*). For example, the coefficient of the price elasticity of demand (E_{dp}) and supply (E_{sp}) may be written as follows:

$$E_{dp} = \frac{\text{\% change in quantity demanded}}{\text{\% change in price}}$$

$$E_{sp} = \frac{\text{\% change in quantity supplied}}{\text{\% change in price}}$$

Coefficients of elasticity can be measured either (1) at a point on the curve (*point elasticity*), or (2) between two points on the curve (*arc elasticity*). In Figure 4-2, the slope of the tangent to the demand curve at point P indicates the elasticity at that point.

Price Elasticity of Demand To illustrate arc demand elasticity, let us assume that, if the world price of grain drops from \$150 to \$50 per metric ton, then the quantity demanded increases from 1,000 million to 1,500 million tons in a year. In Figure 4-2, the straight line AB joins these two points, and the coefficient of elasticity may be computed by using prices and quantitites of these two points. Thus:

$$E_{dp} = \frac{\Delta Q}{\Delta P} \bullet \frac{P_1 + P_2}{Q_1 + Q_2}$$

where ΔQ = change in quantity
ΔP = change in price
P_1 = first price
P_2 = second price
Q_1 = first quantity
Q_2 = second quantity

In the example mentioned above the coefficient of the elasticity of demand is:

$$E_{dp} = \frac{500}{-100} \bullet \frac{150 + 50}{1{,}000 + 1{,}500} = -5.0 \bullet \frac{200}{2{,}500} = -0.4$$

This formula is equivalent to another one, which is sometimes preferred because it emphasizes that changes are relative, where:

$$E_{dp} = \frac{\Delta Q}{(Q_1 + Q_2)/2} \div \frac{\Delta P}{(P_1 + P_2)/2}$$

Then:

$$E_{dp} = \frac{500}{(1{,}000 + 1{,}500)/2} \div \frac{-100}{(50 + 150)/2}$$

$$= \frac{500}{1{,}250} \div \frac{-100}{100}$$

$$= .4 \div -1 = -0.4$$

The normal relationship between change in price and quantity in a demand schedule is negative, but the minus sign of the price elasticity coefficient is often dropped for convenience. Thus, such a price elasticity of demand of .1, without the sign, is more inelastic then .4. However, this convenient practice must not confuse an understanding of income elasticity of demand, which exhibits positive coefficients for most food (Table 4-3).

The monetary value of expenditure at any given point on a demand curve is equal to price times quantity. Hence, on this basis, it is possible to construct a receipts curve showing total receipts from marketing in relation to either price or quantity. When the price elasticity of demand is -1.0, unitary, the receipts remain constant as price and quantity change; the percentage change in price exactly offsets the percentage change in quantity. When elasticity is less than -1.0, or inelastic, such as -0.3, receipts fall as quantity increases, or rise when quantity decreases. The opposite relations exist for elasticities greater than -1.0, elastic, such as -1.5.

Price Elasticity of Supply The coefficient of the elasticity of supply is computed in the same way as that for demand, except that supply curves are generally upward sloping or positive, the *law of supply.* Time is the most important factor affecting the elasticity of supply, which increases as the time available for adjustment increases. Also, a direct relationship exists between the elasticity of supply and the ease of substituting one commodity for another in the production process. The elasticity of supply for a commodity such as corn or soybeans will be much greater than that for total grains or for total farm output. This is important in determining the effectiveness of some farm programs designed to improve farm incomes by controlling supply. For instance, if a wheat supply control program does not also limit the growing of substitute crops, the program may be ineffective as a general price-raising measure. The only gain to farmers might be just the compensatory payments for complying with the program.

Income Elasticity of Demand If the percentage change in the quantity of a commodity purchased is the same as the percentage change in real income, *ceteris paribus,* then the income elasticity of demand for the commodity is 1.0 or unitary. If the percentage change in quantity is half as much as the change in income, then income elasticity is 0.5, which is inelastic. If quantity changes more than income, such as 1.5, it is elastic. Conceptually, income elasticity of demand is the percentage change in quantity purchased divided by the percentage change in real income, computed similar to price elasticities earlier, for either a point or arc.

For example, $$E_{di} = \frac{dQ}{Q} \div \frac{dI}{I}$$

where E_{di} = income elasticity of demand
I = real income
Q = quantity purchased

In a rich country, the income elasticity of demand for food is very low, generally less than 0.2. This means that a general increase in real income will have only a minor effect on the demand for food compared with the demand for nonfood.

CHAPTER 5

WORLD ECONOMIC SITUATION FOR FOOD AND AGRICULTURAL POLICY

The world economic situation affects the food and agricultural policy of every nation. It helps to determine how each nation will react to the policies of other nations, and how effective each nation will be in achieving its own national goals. There is no world food and agricultural policy, only policies of independent nations. However, nations meet together to try to agree on policies that will improve their individual situations, and also be mutually beneficial. At issue are questions concerning the freedom of international trade, including trade in food and other agricultural commodities; the emphasis to be placed on economic development of agriculture; the moral responsibility of each of the major exporting nations (such as the United States, Canada, Australia) to help alleviate hunger and starvation in other countries; and the political responsibility of all nations to adopt policies that will provide for adequate food supplies, thus lessening the need for international aid and relief programs.

IN THIS CHAPTER

By studying Chapter 5, you will learn how to answer the following questions:

1 How does the world economic situation affect food and agricultural policy?
2 How are individual nations affected by changes in the world supply–demand balances? What are the major differences in policy problems between the developed and the developing countries? The net food exporters and net food importers?
3 What have been the trends in world agricultural trade, the conflicts among trading nations, and the implications of the international negotiations for U.S. policy?

POPULATION GROWTH, FOOD PRODUCTION, AND FOOD AVAILABLE PER CAPITA[1]

The changing world food situation highlights the importance of the policy issues. In the 1970s, the world was beset with acute food shortages. In the mid-1980s, it appeared to be awash in food surpluses. Currently some poor countries and regions face acute short-term food shortages. In most developed countries, which already have an abundance of food, supply has been growing more rapidly than demand, but in most developing countries the situation is reversed.

Such shortages are likely to accelerate unless moderated by policy. How can we address the world problems of population growth, food production and distribution? First, let us discuss the prospects for world population growth.

World Population Growth in Perspective

The world's population is increasing by about 1.7 percent per year. This rate of growth is somewhat lower than the recent peak of 1.9 percent, which occurred in the baby-boom years following World War II, but the total number of people added each year is still greater than it was the year before. This number has been projected to peak at about 90 million annually. It is not certain that growth will stabilize at this level. However, even if it does, as many people will be added to the total world population in the next 60 years as lived on the planet in the early 1990s (5.4 billion). Moreover, about 4.5 billion of these will be added to the currently developing countries, where food deficiencies are most apt to occur. The other 0.9 billion will be added to the developed countries.

Population growth will present an unprecedented challenge for world agriculture of far greater magnitude than anything experienced before. Two thousand years ago, the earth's population was only 250 million. It took more than 16 centuries—to about 1650—to add another 250 million, or for the world population to double. Then in just two centuries, by 1850, population doubled again, reaching one billion. The second billion was added in 80 years, by 1930; the third billion in another 30 years, by 1960; the fourth in 15 years, by 1975; and the fifth in just 12 years, by 1987. The sixth billion likely will be added well before the beginning of the twenty-first century. If so, the world population will have doubled in less than 40 years. Although the rate of world population growth has been slowing down, the numbers being added each year suggest that starting from the early 1990s the population can double again well within the next 60 years.

The world is in a unique stage of demographic transition, as illustrated in Figure 5-1. Typically, before industrialization and the advent of modern medicine, both birth and death rates were high, 40 or more per 1,000 people annually. With economic development and improvements in diets and health care, death rates fell first. For a short time, birth rates remained high. Finally, with further development both rates have fallen, until some societies have reached a stage of slow growth, or even no growth. Generally, in the developing countries, the rate of population growth has declined as educational levels have increased, nations have become more industrialized, and more

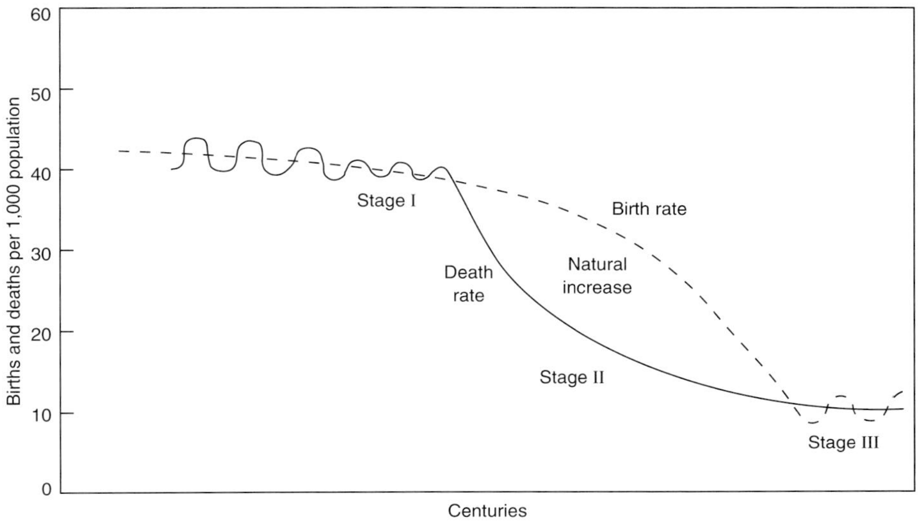

FIGURE 5-1
The three stages of demographic transition. Hypothetical: Stage I is slow growth with high birth and death rates, such as existed before the industrial revolution. Stage II is rapid transitional growth, with death rates declining before the decline in birth rates, such as began with the industrial revolution and continued to recent times in most developed countries. Stage III is slow growth or zero growth, based on low birth and death rates, which is increasingly characteristic of the modern era in developed countries. In general, the developing countries have not completed Stage II, although some may be near the end of it.

women of childbearing age have become gainfully employed. How does this population growth compare with that of food production?

World Food Production in Perspective

While world population has been growing at a steady pace since the 1950s, the rate of increase in world food production (which was well above the rate of population growth in the 1950s and 1960s) has been dropping so that in the 1990s the two rates are nearly equal. As a result, world food production per capita grew steadily during the decade of the 1950s, at about 1.2 percent annually compounded. During the 1960s, this rate declined to 0.7 percent, and since 1970 it has remained at about 0.5 percent.

Without policy-imposed restraints on agricultural production in the United States and stifling restraints on food trade by many countries, world food production could have been much higher. However, with these restraints, the world food situation has improved only modestly over what it was two or three decades ago. Major imbalances in food available per capita between the developed and developing countries generally have not decreased and, in many cases, some of the imbalances have actually widened.

Because the world is still experiencing rapid technological advances in agriculture, which can be accelerated by political and economic reform, significant increases in world food production over the next generation may possibly exceed those of the past. It is also possible that, with world peace and freer international trade, major differences in available food per capita can be narrowed.

Whether or not these changes will occur, however, depends largely on future successes or failures in public policy. How to formulate the appropriate policies and how to get the necessary political support are the two great challenges for policy making. How do the trends in population growth and food production relate to intercountry and regional imbalances in available food?

Population Growth Rates by Regions

The population growth rates vary significantly among regions. Europe has the lowest rate of population growth, at only about 0.3 percent annually. Several European countries are near zero growth, and some with the highest standards of living have seen declining populations, or negative growth. North America in general has a growth rate below 1.0 percent annually, in spite of much higher growth rates in Mexico and Central America.

The regions with the highest annual growth rates are Africa (3 percent), South America (2 percent), and South Asia (1.9 percent). China, the world's most populous nation, has been growing at only 1.2 percent annually. Yet China's population is increasing by 12 million or more each year, which brought China from about 800 million people in the mid 1980s to one billion before the beginning of the twenty-first century. In comparison, as late as 1850, only one billion people inhabited the entire world.

Growth rates such as these will widen the food imbalances among major world regions. These rates will also increase political pressures on rich countries to help the poor countries and will spawn political crises sufficient to tax the imagination. In some developed countries, managing food abundance is a major policy problem, whereas in some developing countries, the main struggle is to avoid famine.

In the sections that follow, the classification we use to distinguish developed from developing countries is based on criteria of the United Nations and its Food and Agriculture Organization (FAO).

The Developed Countries

In the developed countries—which include mainly the United States, Canada, Western Europe, Japan, Australia, New Zealand, and the Republic of South Africa—the available supply of food is more than adequate for nutritious, wholesome diets. Nevertheless, we should not ignore malnourishment due to poverty in some of these countries, and other problems such as obesity and poor food choices by those who are not impoverished. Most of these countries are net exporters of food and other agricultural commodities, and production per capita has been increasing. Diets are generally good and improving. The quantities available for export are increasing and can be increased much more. Although some developed countries, such as Japan and Great Britain, are large net importers of food, they have adequate foreign exchange to pay for food imports, so a general problem of food scarcity does not exist in those countries.

The Developing Countries

The developing countries may be grouped in various ways for identifying food policy problems. The first group might include Russia and the other countries of the former Soviet Union, the countries of Eastern Europe, North Korea, and the People's Republic of China. The common characteristic of these nations up to the early 1990s was central planning and government ownership of most food-producing and marketing resources. Glaring failures in food and agricultural policy may be regarded as a major factor—if not the most influential factor—in bringing about the various political revolutions and economic changes in these countries.

From the early 1950s to the late 1980s, a general increase in per capita food supplies was reported by several of the governments of these countries. However, the average amount and quality of food remained far below that of the developed countries. Very large investments in agricultural technology and the supporting infrastructure will be required to bring food production per capita up to the level of the developed countries. How this can be financed and accomplished is a major problem for policy in these countries and for those countries who may aid them. During the four decades from 1950 to 1990, almost all centrally planned countries became net food importers. Many of them turned to the developed countries for imports to increase their food supplies, but their purchases varied widely from year to year. The Soviet Union was by far the largest food importer among these countries. Nearly every one of them, however, relied to some extent on food imports. After the breakup of the Soviet Union, many still required food aid, at least in the short term.

The developing countries also include South and Central America, nearly all of Asia except Japan, and Africa except the Republic of South Africa. This group includes most of the poorest countries of the world. Even though the average food production growth rate of these countries is higher than that of the developed countries, the average population growth rate is more than double, and diets are generally much poorer. Their per capita food production increased slowly, about 0.9 percent annually in the 1950s and 1960s, 0.7 percent in the 1970s, and then 1.0 percent in the 1980s. Consequently, since the end of the 1970s, per capita food supplies have improved slightly on the average. However, major regions of shortages still persist. Only part of this deficiency has been offset by increasing imports, some of which are heavily subsidized or given as aid. Many of these countries, particularly in South Asia, have made impressive progress in adopting some new technologies. Some have substantially increased production, but on average, rapid increases in population have tended to cancel per capita food gains in many of the less developed countries.

Although changes in international trade and food aid have helped to alleviate some food shortages, these changes have not been sufficient to stabilize the food supply in many developing countries. Even so, changes in international trade are important and hence must be appraised as part of the world economic situation for food and agricultural policy.

CHANGES IN INTERNATIONAL TRADE

Because of changes in economic conditions, there have been massive changes in international trade patterns. Since the early 1950s, the total volume of world trade in agricultural commodities has increased about seven or eight times. The trade balance between the developed and the developing countries changed even more. For instance, in the early 1950s, the developed countries provided only 51 percent of the total food imports of both developed and developing countries. By the early 1980s, they were providing nearly 65 percent of a much larger world volume. By the early 1990s, the developed countries were supplying more than 70 percent of the imports among the developed countries and almost 60 percent of the imports of developing countries. During the same period, the developing countries were falling behind as a source of food imports for the developed countries, and most of the developing countries had lost their former roles of net food exporters.

These changes in the world trade pattern can be linked to several supply and demand factors. Characteristically, in nearly all the developed countries, technological advances in producing and marketing food and other agricultural commodities have continued to outpace the growth of demand in these countries. In comparison, in most developing countries, economic development and population growth have tended to bring greater increases in population than in food supply. The result is expanding world trade and increases in exports of food and other agricultural commodities from the developed to the less developed countries. If this trend continues, the developed countries will have to buy more of some other commodities from the developing countries to provide the resources for developing countries to pay for their imports of food and other agricultural commodities.

Several lessons for policy are imbedded in this trade experience, not the least of which are lessons for a broad range of policy in the United States, which is by far the largest world trader. What are some of these lessons?

How International Trade Affects U.S. Agriculture

Throughout the twentieth century, U.S. agricultural prosperity has been linked to exports, which have been affected in turn by changes in world economic conditions as well as U.S. policy. Specifically, these changes relate to the value of the dollar with respect to other currencies, to policies of other countries, and to changes in U.S. policy. During the two decades just before World War I, U.S. agriculture boomed as a result of increased exports and strong industrial growth in the United States. This export gain was dramatically reversed after the war. Throughout the 1920s, the issue of subsidizing exports dominated U.S. policy debates, but attempts to establish this policy ended in failure. During the 1930s, as a result of U.S. farm price supports and production controls, the historic worldwide depression, and prohibitive trade barriers frantically executed throughout the world, total U.S. agricultural exports declined to near zero.

During World War II and the first three or four post-war years, the strong demand for exports, enhanced by U.S. aid and the massive post-war recovery program, was the major factor accounting for increases in exports of U.S. agricultural commodities and the resulting relatively favorable U.S. farm income. Through the 1950s and 1960s, in spite of export enhancing programs, exports of farm commodities languished at levels generally below $5 billion a year, which was approximately 16 to 17 percent of total U.S. farm production.

In the 1970s, a large, unfavorable balance of trade for the United States resulted in unprecedented devaluation of the dollar, which declined significantly in exchange with other major currencies. In addition, poor crops in several countries contributed to food shortages and increasing demands for food imports by those countries. U.S. exports of agricultural commodities boomed, far exceeding former levels. In 1981, U.S. exports reached an all-time high of more than $43 billion, with a record value of farm exports over imports of more than $26 billion.

Then the growing strength of the U.S. dollar, turbulent world economic conditions, and U.S. farm price supports set above competitive market prices, combined to reduce purchases of U.S. commodities by most importing countries. By 1986, U.S. agricultural commodity exports had dropped to a low of $26 billion, with a value of these exports over imports of less than $5 billion. This combination of events, plus a decade-long expansion of farm debt and high interest rates, brought on the farm financial crisis of the mid-1980s.

Finally, in the last half of the 1980s, the weakening of the dollar again made U.S. commodities more price-competitive in foreign markets. Lower farm price supports helped to unblock grain stocks for export. In addition, greatly increased export subsidization and promotion schemes throughout the late 1980s and early 1990s encouraged a rapid recovery of the market share lost in the early 1980s. In the early 1990s, about one fourth of the total U.S. farm output was exported.

INFERENCES FOR POLICY

We can draw several important inferences for policy from the experiences of the twentieth century. The world market has been expanding at a steady pace commensurate with population growth. The potential for demand and market growth is much greater in the developing countries than it is in either the U.S. domestic market or the markets of other developed countries. An accurate appraisal of the situation in these markets is critical for designing future U.S. production and trade policy.

The U.S. Domestic Market

Forecasts of the U.S. domestic market for food and other agricultural commodities are based on forecasts of population growth and on estimates of changes in per capita food consumption. As we discussed in Chapter 4, the projection of 383 million people by

2050 will be an increase of less than 1.0 percent annually compounded. Among most of the U.S. population, increasing emphasis on diet and other health-related activities, along with an already high per capita consumption of food and high efficiency in food manufacturing, handling, and preparation suggests that the per capita food requirement or the quantity of food demanded per capita will increase very little. However, if real income and purchasing power increase among the poorest one quarter to one third of the U.S. population, the total domestic demand for food could rise significantly. Much depends on economic growth, increasing employment, distributive income policies, and the political support for food-aid programs.

U.S. Exports to Other Developed Countries

This market also provides limited possibilities for growth. Traditionally, Western Europe has been the largest export market for U.S. agricultural commodities, but this situation has changed. After the European Economic Community (EC) was formed, a new Common Agricultural Policy (CAP) was designed to maintain relatively high target prices for some farm commodities. This policy was implemented with a variable import levy, or tariff. It largely filled the gap between the internal target prices and the generally much lower prices outside the EC. This variable import levy increased agricultural incomes within the EC and stimulated agricultural development and growth. It also tended to decrease the agricultural incomes of major exporting countries, which lost some of their traditional market. As a result, since the late 1950s, the EC has evolved from a major net food importing region to become a net food exporter. The major EC countries have been generous in providing import protection and compensatory payments to their farmers, and the EC has continued to decline as a market for the other major agricultural exporters. The prospect is that the EC's total demand for food imports will expand very little if at all because of the region's increasing domestic output, very slow population growth, and the already adequate diets for a large percentage of the total population.

Japan is a more dynamic market, and relaxation of some import restrictions has encouraged more food imports. However, because the Japanese diet has already improved greatly and population growth has been slowed to low levels, further increases in food imports are apt to be small.

The other developed countries—Canada, Australia, New Zealand, and the Republic of South Africa—are all net food exporters. The general result is that the developed countries offer little room for expansion of U.S. agricultural commodity exports. This leaves the developing countries as the major market for expansion of U.S. food exports. Among these countries the successes or failures of economic developmental policy will be critical.

The Developing Countries' Future Food Import Demands

Based on current trends, about five sixths of the total world's population growth will occur in the developing countries over the next few decades. If the per capita food supply in these countries is not to shrink below its already low levels, then (1) birth rates must fall more rapidly than they have, or (2) food production must be increased by greater amounts than it has, and/or (3) more food will have to be imported into these countries.

1 Can Population Growth Rates Be Lowered? Among the very poorest people, population growth may be constrained according to a theory advanced two centuries ago by Thomas Robert Malthus[2] that population tends to increase until it presses against the physical means of subsistence. This condition may be happening in some countries in Africa and to some extent among the poorer people in India, Bangladesh, and some other countries in Asia and Latin America. Not all classes of people in these countries are at risk, however. For the great majority of the world's population, the rate of growth is more a function of economic development, the availability of health care and medical facilities, people's desire for children, family planning or the lack of it, and government economic policy. It is not certain that most governments in developing countries will adopt policies that will result in growth rates much below current levels, but it is possible that some will do so.

Experience and economic analysis reveal that, in country after country in all stages of economic development, birth rates have fallen as education improved, economic development occurred, and levels of living increased. The economic reason for this is that opportunity costs of raising a large family increase as off-farm employment opportunities grow, especially as professional and other employment opportunities improve for women in the labor force. Thus parents benefit from limiting the size of the family. In poor countries, where children generally work to help support the family, having more children rather than fewer is often the goal of parents, so families resist government programs to limit family size. The keys to lower birth rates are education, economic development, increasing employment, and rising levels of living. Do we have significant evidence that this is happening?

In many countries, population is approaching a new stage of slower growth. The processes at work can be seen most dramatically in the fast-growing cities of developing countries—Mexico City, Cairo, Calcutta, Bombay, Lagos, Jakarta, São Paulo, Rio de Janeiro, Caracas, and others. The explosive growth taking place in these cities is of very recent origin, with the flood of migration into the cities consisting largely of poor people trying to participate in the growing market economy, gain access to government assistance, or seek their livelihood within the urban economy.

The second response, often seen in the second generation if not in the first, is to have fewer children, as old customs gradually—or sometimes quickly—disappear. Poor people living close to cities absorb new traditions; ambition for a better life becomes the driving force for them to spread their resources over few dependents. There is little charity from governments or other sources in these economies, and what is available seldom meets basic needs. So people are driven to observe how others are saved from want or starvation through struggle and seeking work in the growing market economy. In such an economy, people often adapt very quickly despite the hardship, and lower birth rates are the result.

The dominanting prospect is that the population growth rate of the developing countries will reach its peak and decline as lower birth rates overcome the effects of increasing longevity. If the world population approaches a significantly lower growth level, economic development can create a level of affluence that will bring more increases in the per capita demand for food. These increases in demand must be matched by increases in food production if greater hardships among poor people are to be avoided.

2 How Is Food Production Affected by Economic Development? The most rapid gains in food production have occurred in those countries experiencing the most rapid economic development, provided they have the basic conditions for exploiting comparative economic advantages. Several countries in the early stages of rapid economic development, such as Taiwan and South Korea, have demonstrated this trend, just as the developed countries did earlier. Food production nearly always increases under economic development, while food demand per capita and total food consumption increase more. Increasing real income through economic development encourages people to shift away from diets relying largely on grains and other basic food groups to diets that use more animal products. This shift increases the grain equivalent as income per person increases. Although smaller and smaller amounts of grain are consumed directly as human food, much more is consumed in the form of animal products and other processed food products (Figure 5-2).

Among countries experiencing economic development, however, some may not become greater food importers or offer larger export opportunities for the United States and other food exporting nations. When economic or political conditions are not favorable for trade, some countries develop their own resources to feed themselves. Countries such as India may even become agricultural exporters. Sometimes this involves a significant social cost to their own citizens.

3 Why Does the Market for Food Imports Usually Grow Under Economic Development? Under economic development with rising levels of employment, several trends occur: (1) a decline in birth rates, but a continual rise in total population; (2) increasing production of food commodities in which a comparative advantage exists; (3) increases in food consumption in total and per capita, and (4) improvements in food quality.

Much of the increased food demand can be met in some countries by developing their own food-producing resources. All countries gain, however, by producing those commodities in which they have comparative advantages. As a result, economic development generally brings significant increases in demand for food imports, and increases in international trade if food supplies are readily available in world markets.

With the aid of supportive policies this is what has happened in many countries over the last four decades. Trade in food and other agricultural commodities, however, has been slowed or interrupted by trade-restricting policies that have been adopted in efforts to increase food self-sufficiency, protect farm producers, increase national security, or save on limited foreign exchange. Efforts to combat such policies through freer trade have had a high priority in trade negotiations among nations, especially multilateral negotiations carried out under the General Agreement on Tariffs and Trade (GATT).

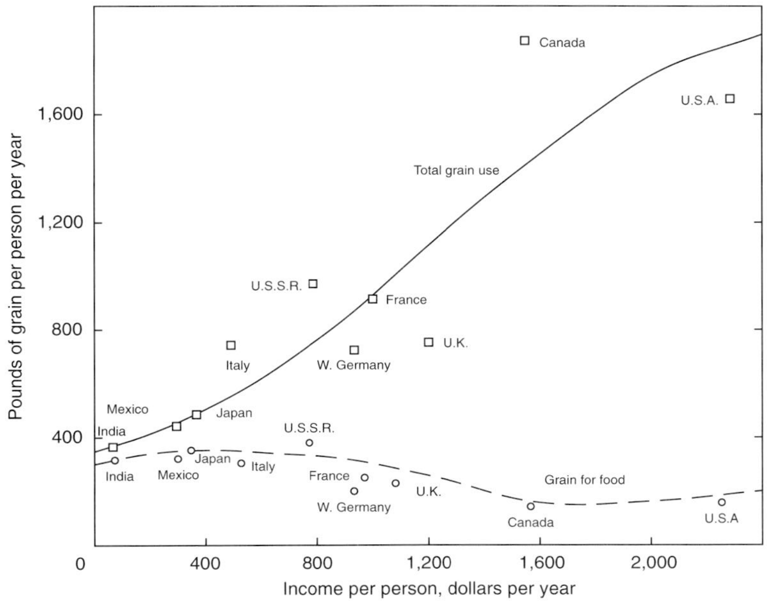

FIGURE 5-2
General relationship between income per person and pounds of grain per year in selected countries. (*Source:* Orville L. Freeman, "Malthus, Marx, and the North American Bread Basket," *Foreign Affairs,* July 1967, p. 582. Reprinted by permission of *Foreign Affairs,* April 1993. Copyright 1967 by the Council on Foreign Relations, Inc.)

Agricultural prosperity in the major food exporting countries (the United States, Canada, Australia, New Zealand) is linked most directly to expanding international trade. Meeting the most urgent food requirements of the food-deficient countries requires both economic development in these countries and more trade. Almost universally, increasing trade is necessary for optimum development and growth. What are the major policy issues in modern trade policy?

Background to Future Trade Policy

Traditionally, since the time of Adam Smith and publication of *The Wealth of Nations* (1776), economists have recognized that trade based on comparative advantage can enhance the productivity and wealth of all nations. Most industrialized nations, however, have imposed tariffs and nontariff barriers (NTBs) either to raise revenue or to protect certain industries from foreign competition. The benefits to individual industries and labor groups usually seem clear-cut and immediate, while the hidden costs of higher prices to consumers are difficult to detect and more diffused. This results in a constant political struggle between advocates of free trade, and producer groups desiring protection.

The recent trade experience of the United States mirrors much of the world's

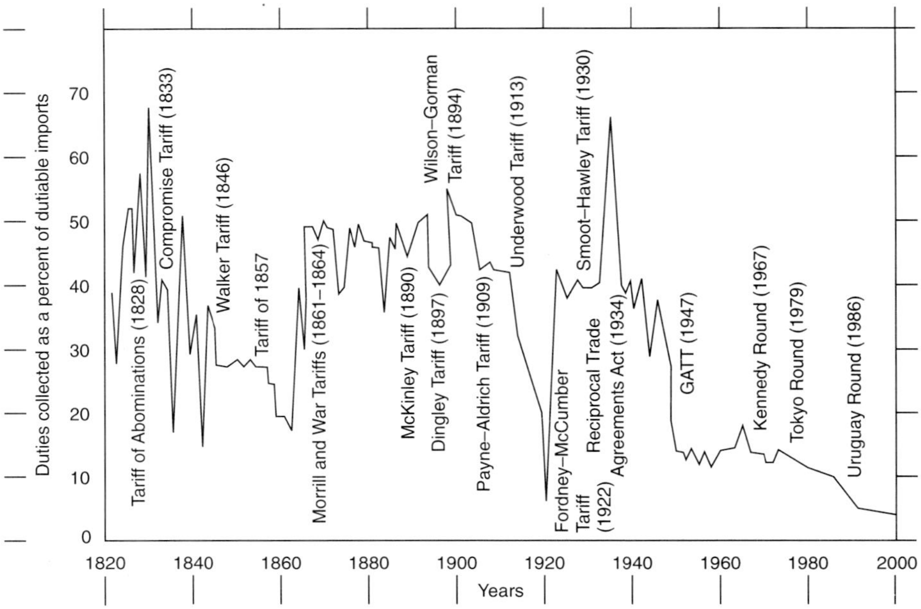

FIGURE 5-3
Tariff rates of the United States, 1820 to 1991. Average tariff-rate figures understate the importance of tariffs because they do not reflect the fact that some goods are excluded from U.S. markets because of tariffs. Also, some imports are limited by quotas, such as those on sugar and dairy products. (U.S. Department of Commerce data).

experience with trade. From the early 1820s to the 1920s, the United States was a relatively high-tariff nation (Figure 5-3). The Smoot-Hawley Tariff Act of 1930 imposed some of the highest rates ever adopted. This Act, which was supported by major industries seeking protection, was widely opposed by economists. It greatly restricted all trade.

The Reciprocal Trade Agreements Act of 1934, specifically aimed at tariff reductions, was adopted in the hope of reviving trade. It contained two main features: (1) negotiating authority for the president to cut tariffs as much as 50 percent, provided other nations reciprocated by lowering their tariffs on U.S. exports by equivalent amounts, and (2) the most-favored-nation clause, which could extend these provisions to all nations that made similar tariff reductions.

New Trade Emphasis under the GATT

In 1947, the approach to tariff reductions was broadened when 23 nations, including the United States, signed the General Agreement on Tariffs and Trade (GATT). The GATT was founded on three cardinal principles: (1) equal, nondiscriminatory treatment for all nations; (2) reduction of tariffs by multilateral negotiations; and (3) elimination of import quotas. Since 1947, the GATT has grown to serve as a forum for nearly 100 nations to negotiate according to these principles, and also to serve as an umbrella for the formation of regional trade pacts among a small number of nations seeking freer trade among themselves, or with the rest of the world. Eight forums called "rounds," which have lasted from less than two years to four years or more, have been held. The reductions in tariffs and NTBs are generally regarded as impressive because they have reduced impediments to trade, but agriculture stands out as the most resistant of all major industries to such trade liberalization.

The first objective under the eighth round of GATT negotiations, which began in Uruguay in 1986, was to eliminate all trade barriers and domestic subsidies to agriculture. Substantial agreements were achieved on other objectives, but in the early 1990s, the Uruguay Round was slowed largely because nations failed to agree on the issue of domestic subsidies to agriculture. This issue came to a head over the inability of the major EC nations to agree to the amount of reduction in domestic subsidies proposed and agreed to by many other nations in the GATT. At that time, the total cost of the EC's CAP for European farmers was substantially above the aggregate amount planned in the federal budget for U.S. farmers. Hence, as the EC countries increased their food-exporting capacities, the large domestic subsidies to EC farmers from the governments of individual EC countries, plus some export subsidies, became the focal point for disagreement between the EC and other food-exporting countries. These disagreements stalled a more general agreement under the GATT to lower and eventually eliminate all agricultural trade barriers. In place of a general free-trade agreement, the United States, Canada and Mexico negotiated for the North American Free Trade Agreement (NAFTA), and some other nations also formed regional free-trade agreements.

Producer subsidies of various kinds to farmers enable a nation to subsidize and stabilize the income of its farm sector, but this also encourages farmers to produce more unless they are restrained by programs such as land retirement, marketing quotas, or other policy measures. If domestic demands are met at the chosen intervention price level, then the additional output must be exported through subsidies or other means, it must be absorbed by other programs such as donations or a food subsidy to low-income consumers, or destroyed.

Producer subsidy is a primary issue that has caused discontent among exporting nations involved in trade negotiations for freer trade, and this discontent applies to both the less developed and developing countries. The effects of U.S. policies, however, have been softened by lower price supports, compensatory payments coupled with massive land retirement programs, more limited use of restrictive marketing agreements, and new trade agreements.

The main policy problem in achieving freer trade in food and other agricultural commodities, which is essential for agricultural prosperity for most trading nations, hinges on the domestic policies adopted for each nation's farm sector. Because of its political ramifications, this problem will not be easy to settle, so it will remain important. The problem must be solved, however, if agricultural development and growth are to match the potentials for population growth. Much will depend on maintaining world peace, which can create confidence in a reliable food supply, and reduce global political uncertainty. If U.S. agriculture is to profit as much as is possible from its strategic position in the world food economy, the U.S. government must be successful in its negotiations for freer trade.

LOOKING AHEAD

Our appraisal of the national and world economic situation for food and agricultural policy provides a perspective for developing a clearer vision of the economic foundation for this policy. The next chapter is an overview of the political foundations for U.S. farm price and income policy. It includes a chronology of the major laws that have been adopted to deal with situations discussed in Chapters 4 and 5, the reasons for such laws, and a brief preview of how they may be analyzed.

ENDNOTES

1 Most of the data in this section are from the *Economic Report of the President of the United States* (recent years), various statistical reports of the United States Department of Agriculture (USDA), the United Nations' *Monthly Bulletin of Statistics* (selected issues), the *United Nations Statistical Yearbook, United Nations Conference on Trade and Development* (UNCTAD) *Handbook of International Trade and Development Statistics,* and reports of the U.S. Bureau of the Census.

2 Thomas Robert Malthus (1798), *On Population,* Gertrude Himmelfarb (ed.), The Modern Library of the World's Best Books, New York, 1960.

TERMS AND CONCEPTS

World food situation
World population growth
World food production per capita
Stages of demographic transition
Population growth by regions
Food availability: developed and developing countries
Food trade balance between developed and developing countries
Changing patterns of world agricultural trade
Effects of international trade on agriculture
Prospects for growth of world food demand
Factors affecting population growth
Effects of economic development on trade alternatives in trade negotiations

QUESTIONS AND STUDY SUGGESTIONS

1 How do the prospects of world population growth compare with those faced 30 or 40 years ago? What are some of the implications of this for food and agricultural policy?

2 How does the growth of world food production compare with the growth of world population? What has happened to the quantity of food available per capita?

3 How does food availability per capita differ among major world regions? Between developed and developing countries?

4 What have been some of the important effects of international trade on U.S. agriculture in the twentieth century? Have patterns of agricultural trade changed? What are the implications for policy?

5 What is the general effect of economic development and rising levels of living on food consumption? How is food production affected by economic development? Why has there been a growing market, or increasing demand, for food imports among nations experiencing economic development?

6 Even though all nations can benefit from international trade, why do most of them impose tariffs or restrict trade through nontariff barriers to trade (NTBs)? What are some of the implications of this for food and agricultural policy?

7 Why do some agricultural producer groups resist free-trade policies that might be negotiated under the GATT? What has been the negotiating experience? Why have producer subsidies become a policy issue under the GATT?

8 Now that the EC is no longer a major food deficit area, why is the EC CAP policy an important policy issue under the GATT? How has the U.S. government handled this issue? Why does the U.S. policy also come under criticism? What are the U.S. policy alternatives?

CHAPTER 6

FOUNDATIONS OF PRICE AND INCOME POLICY

In the 1920s, after some 150 years of policy that encouraged land settlement and helped develop agriculture's input markets, the dominant political interest shifted to agriculture's output or product markets. The primary impetus behind this shift was the farm price and income crisis that followed World War I. Throughout the rest of the twentieth century, the federal government and much of the nation have struggled with issues relating to the output markets. The issues have revolved around low farm incomes, farm price and income instability, perceived weak foreign demand for farm commodities, and the need for food aid to low-income consumers in domestic and foreign markets. These issues have been commonly referred to as price and income policy, agricultural price and income policy, or even sometimes just farm price policy. However, we emphasize that such policy, by whatever name, focuses on direct intervention in such diverse areas as agricultural output, price, income, trade, and food-aid policy.

The purpose of this chapter is to help you understand the economic, social, and political foundations of this type of policy, which we will call price and income policy.

IN THIS CHAPTER

By studying Chapter 6, you will learn how to answer the following questions:

1 How did price and income policy begin and how has it evolved?
2 What are the major economic problems underlying this policy?
3 What major economic, social, and political values have influenced its development?
4 What is an appropriate alternatives framework for analyzing this policy?

EVOLUTION OF PRICE AND INCOME POLICY

Agricultural price and income policy grew out of acute farm income and financial crises, which led to widespread political beliefs that the marketing system was not adequately rewarding farm people for their agricultural commodities. It could be interpreted that society had deliberately chosen to develop an efficient and productive agriculture through various developmental policies, and then gradually decided to develop policies to provide what came to be regarded as fair returns for farmers, stability for farm resource allocations, and an equitable distribution of income.

The U.S. policy experience contrasts sharply with the experience of many other major countries that intervened in prices and incomes early while they were still grappling with how to develop and support an efficient agricultural economy and food distribution system. The emergence of farm price and income policy in the United States paralleled other important, interrelated changes, among them being the industrialization of the economy, the increasing commercialization of farming, the displacement of animal power on farms by machine power and new technology, and the failure of the nation's agricultural export markets.

How Price and Income Policy Began

Throughout the nineteenth and early twentieth centuries, many people voiced concern about agriculture's economic problems, and sometimes large numbers joined in political protests. Not much was done to develop national price and income policy, however, until the farm crisis of the early 1920s brought matters quickly and urgently to a head. At the end of World War I, the destruction from the conflict, coupled with the surrender burdens imposed by the Allies on the Central Powers of Europe, bankrupted these previously importing nations, and thus closed major export markets for U.S. farm commodities. As a result, in 1920, prices for the major export commodities dropped sharply. Both grain and livestock producers were among those adversely affected. One of the first public policy responses was enactment of the Packers and Stockyard Act in 1921 to provide for government inspection and regulation of livestock commission firms and packers in public stockyards.

Next, as a result of farmers' efforts to gain more control over marketing, the Capper–Volstead Act of 1922, sometimes called the magna carta of cooperative marketing, authorized farmers' cooperatives to issue capital stock and engage in interstate trade so long as they did not violate antitrust laws by combining or conspiring in restraint of trade. The Grain Futures Act of 1922, which was later amended and expanded in a number of Acts, provided the foundation for government regulation of agricultural commodity exchanges. This Act prohibited manipulation of markets, or attempts to manipulate prices of commodities traded on futures markets. Commission merchants and floor brokers, as well as boards of trade, were subject to regulation under the jurisdiction of the secretary of agriculture.

As a result of the depressed farm prices of the 1920s, political support for a new farm price and income policy developed, and grew in prominence until it dominated much of the national policy agenda. Early in the 1920s, some leaders of agribusinesses and farm organizations identified the depressed export market as the source of their woes. Following a large meeting of prominent political leaders called by the secretary of agriculture, a plan was proposed to subsidize exports of major crops to be financed by a federal sales tax on the farm commodities involved. On five separate occasions, a bill to do this was introduced in Congress. These bills, named the McNary–Haugen bills after their sponsors, Senator McNary of Oregon and Representative Haugen of Iowa, were passed by Congress twice. President Calvin Coolidge vetoed each bill, however, as unfair, unworkable, and an improper intervention by government into the nation's markets. Finally, adoption of the Agricultural Marketing Act of 1929, which is explained later in this chapter, signaled the beginning of the federal government's involvement in farm price and income policy. The government's role has continued to evolve in terms of policy content, comprehensiveness of issues, and groups of participants.

Evolution of Policy Content

The evolution of price and income policy content has been incremental, continual, persistent, slow, and somewhat unpredictable through a succession of more than 20 major public laws and many more special Acts and appropriations. The policies encompass both continuity and change. A complete reversal of policy direction has not occurred, although changes have been made in each new law. The changes have been precipitated by disruptions in economic conditions, shifts in political priorities often arising out of perceived problems, or dissatisfaction with the results of current policies.

The major laws have been:

- Agricultural Marketing Act of 1929
- Agricultural Adjustment Act of 1933 (the first AAA)
- Soil Conservation and Domestic Allotment Act of 1936
- Agricultural Marketing Agreement Act of 1937
- Agricultural Adjustment Act of 1938
- Steagall Amendment of 1942
- Agricultural Acts of 1948, 1949, 1954, and 1958
- Agricultural Trade Development and Assistance Act of 1954 (more popularly known as Public Law 480, or P.L. 480)
- Soil Bank Act of 1956
- Emergency Feed Grain Act of 1961
- Agricultural Act of 1961
- Food and Agricultural Acts of 1962 and 1965
- Food Stamp Act of 1964
- Agricultural Act of 1970
- Agriculture and Consumer Protection Act of 1973
- Food and Agriculture Act of 1977
- Agriculture and Food Act of 1981
- Food Security Act of 1985
- Food, Agriculture, Conservation, and Trade Act of 1990

We will explain several of these laws when we examine selected alternative policies in the following chapters. Some of them launched a new policy or program, such as setting procedures for nonrecourse loans, which served as minimum prices for farmers; food aid for low-income consumers, as well as food aid to other countries; target prices, which set standards for direct payments to participating farmers; grain reserves; and soil conservation programs. Other more restrictive provisions have been discontinued, such as marketing quotas, parity price supports, purchase requirement for food stamps, and the soil bank program. The incentives to participate in most programs are limited by constraints of the federal budget for agriculture.

Increasing Comprehensiveness of Issues

Through many years of evolving policy, issues and provisions have become increasingly comprehensive. For example, the Agricultural Marketing Act of 1929 dealt primarily, although not exclusively, with wheat and cotton, while later installments of the major provisions have included other commodities. The Food, Agriculture, Conservation, and Trade Act of 1990, the broadest of all price and income policies, included 25 separate titles covering most of the major crops, domestic and foreign food aid, grain reserves, a massive soil conservation reserve, major farm credit provisions, agricultural education and research, marketing promotion, and many special issues sometimes reaching across the entire food and agricultural system.

This trend toward comprehensiveness appears to be related to three factors. First, the increasing importance of world markets has buffeted the farm sector with far-reaching and interrelated problems. Second, policies that were once treated in distinct and sometimes fragmented actions, such as food stamps or conservation, are now combined into omnibus laws. Third, there has been a growing concern about environmental problems and food safety, and the public now seems more responsive to policy issues raised about food and agriculture. For instance, during some earlier policy-making cycles there were serious discussions about separating domestic issues of food aid and its distribution from other issues of agricultural commodity programs. In recent years, this possibility is rarely discussed.

Expanding Groups of Participants

As price and income policies have evolved in content and become more comprehensive, an expanding array of individuals and interest groups has become involved. Several interest groups began to represent concerns about the environment, food safety, and water quality, and many people expressed their concerns individually and through new and growing organizations.

According to popular belief, early farm policy was negotiated in the "exclusive club" of the secretary of agriculture, chairs of congressional agriculture committees, and presidents of two or three of the leading farm organizations. This portrayal is only partly true. The Agricultural Marketing Act of 1929 drew political fire from both major parties. The Agricultural Adjustment Act of 1933 gained much of its political support as an emergency measure to save agriculture from financial disaster, and its adoption was preceded by a strenuous, professionally directed political campaign. However, early decision making about a cotton program or wheat export policy was indeed much less of an organized process than were recent decisions, some of which have been a drawn-out process lasting for years of alternative policy proposing, extensive committee hearings, and much compromising on policy issues. Farmer members of the well-funded, influential agricultural commodity organizations have not been inclined to leave policy decisions to the discretion of the more general farm organizations nor to consumer groups or city professionals. Many other groups have also been active, such as consumer representatives, futures traders, religious relief agency personnel, farm input suppliers, retirees, environmentalists, and taxpayer organizations. The adage that food is too important for food and agricultural policy to be left to farmers can also be applied to many consumer organizations and several other interest groups. A participatory society leaves its policy in the hands of all citizens who choose to become involved.

PROBLEMS UNDERLYING PRICE AND INCOME POLICY

Public policies are responses to perceived problems. It is therefore important to understand the problems that underlie price and income policy. Some problems seem to be chronic, while others seem to arise overnight. The origin of a unique policy provision or an entire policy is often rooted in a particularly troubling situation. For instance, U.S. President Richard Nixon once embargoed soybean exports to stop sharp increases in domestic prices, and President Jimmy Carter embargoed grain exports to the Soviet Union to protest that nation's invasion of Afghanistan. Both of these embargoes were immensely unpopular with farmers and exporters, who perceived that loss of exports was the inevitable result. Hence export-embargo exclusion provisions were adopted in the 1977 and 1981 Acts. In another instance, following the bumper crop years of 1981 and 1982, an unusually large volume of grain was taken into the grain reserve, which was financed through the Commodity Credit Corporation (CCC). This was done primarily to bolster sagging market prices for grain. Sometime later, this action was recognized as excessive and too costly. Hence a provision was included in the 1985 Act that put a maximum limit on the quantity of commodities that could be deposited in CCC grain reserve. These examples of policy responses to specific problems do not explain the continuing government intervention with price and income policy, however.

A study of policy history reveals persistently troublesome conditions and periodically resurging public concerns about farm prices and incomes. Whether the public's perceptions are entirely accurate or not, the perceived problems can be grouped into six general areas: (1) the variability and uncertainty of agricultural commodity prices and production; (2) rural community welfare related to farm income; (3) the periodic divergence of farm prices and incomes from trends in other industries; (4) problems of farm resource adjustment, poverty, and food aid; (5) unique values of the farming sector, which sometimes revolve around the concept of agricultural fundamentalism; and (6) the favorable proportional political representation of agriculture. These six problem areas are discussed in the following sections.

Variability and Uncertainty of Agricultural Commodity Prices and Production

One of the often-stated goals of agricultural price and income policy is to reduce the variability and uncertainty of prices received by farmers. Prices of agricultural commodities vary more widely than prices in the economy as a whole. Given a competitive market system and the inelasticity of supply and demand for agricultural products, variability in agricultural production or demand is translated quickly into greater variability of prices than would occur with a different market structure and elasticity.

Historically, the average annual price variability of all agricultural commodities has been much greater than the variability of all other industrial commodities (Figure 6-1; this historic series has been discontinued, but current similar data appear in Figure 6-3). In addition, prices of some individual agricultural commodities buffeted directly by weather conditions fluctuate even more than the industry generally. Agricultural prices also vary more than industrial prices from month to month, week to week, day to day, and hour to hour. Finally, as shown in Figure 6-1, prices of agricultural commodities are more sensitive to major changes in macroeconomic conditions, such as the depression during the 1930s, the inflation following World War II, and the inflation of the 1970s.

The greater variability in prices of agricultural commodities relates to the inelasticity of both supply and demand and to the many factors that cause a change in either supply or demand. Also, because farm production is related to the biology of plants and animals, output may change with changing weather or outbreaks of diseases and pests. Small changes in either demand or supply will bring relatively large changes in price, especially at the farm or wholesale level.

FIGURE 6-1
Historical comparisons of wholesale prices of agricultural and industrial products. *Source:* USDA, Economic Research Service. Series discontinued in 1977; see Figure 6-3 for trends since 1970.

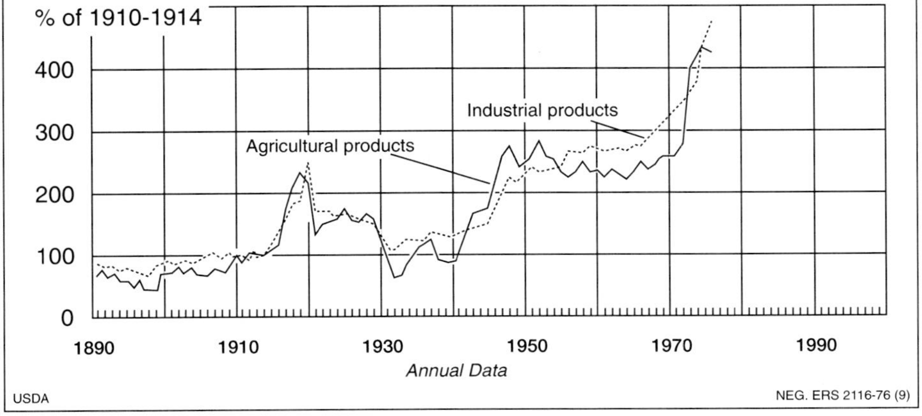

Part of the variability in supply has been moderated by new and improved agricultural inputs and through technological advances in production and marketing. For example, disease-resistant grain varieties help to stabilize grain production, large-scale machinery enables farmers to rapidly counteract adverse production conditions, and futures markets smooth some price trends.

In summary, the instability problem is partly attributable to the economic structure of farm production, which results in inelastic supply, and partly to inelastic demand. Most agricultural commodity markets reflect the model of pure competition, which means that individual producers can neither affect the price nor individually protect themselves entirely through production curtailment or individual marketing strategies. Furthermore, many agricultural products are relatively perishable, which usually prevents inventory balancing to even out market supply. As a result, the highly inelastic demand results in large shifts in prices to accommodate even small shifts in supply. The uncertainty and the variability of commodity prices increase the difficulty of making consistently sound and profitable decisions in investing, managing, and obtaining satisfactory returns on farm production. Hence reducing variability and uncertainty is an important motivation underlying price and income policy.

Rural Community Welfare Related to Farm Income

The prosperity of many rural communities is directly related to the level and the variability of farm prices and incomes. High prices for agricultural commodities mean high farm incomes and high profits for most rural businesses that sell commodities and services to farmers. Also, high farm prices and incomes generally bring more output, which helps firms that sell farm inputs and firms that market farm commodities.

The average annual variation of personal income of the farm and the nonfarm populations is shown in Figure 6-2. During recent decades, the variability in total farm family income has been moderated by relative increases in farm families' off-farm income, which is generally more stable than farm-derived income. However, as farms grow larger and the farm population declines, the relative importance of off-farm income may not increase much, if any, and it may actually decline. Uncertain income affects the farm family's level of living and efficient use of its resources. Therefore, the problem of farm income variation continues to highlight the importance of farm income stability as a goal. Business firms and nonfarm people in rural communities support this goal as well.

This troublesome periodic low level of farm income may well be intrinsically related to the land factor as being unique primarily to the agricultural sector. It is largely nondepreciable in economic terms and it has exhibited chronic low annual returns to its owners from its productivity.

Periodic Price and Income Divergencies

Prices of agricultural commodities and incomes of farm families are not only variable, but they also periodically diverge substantially from other relevant trends. Often, prices received by farmers diverge from prices they pay for production inputs and family

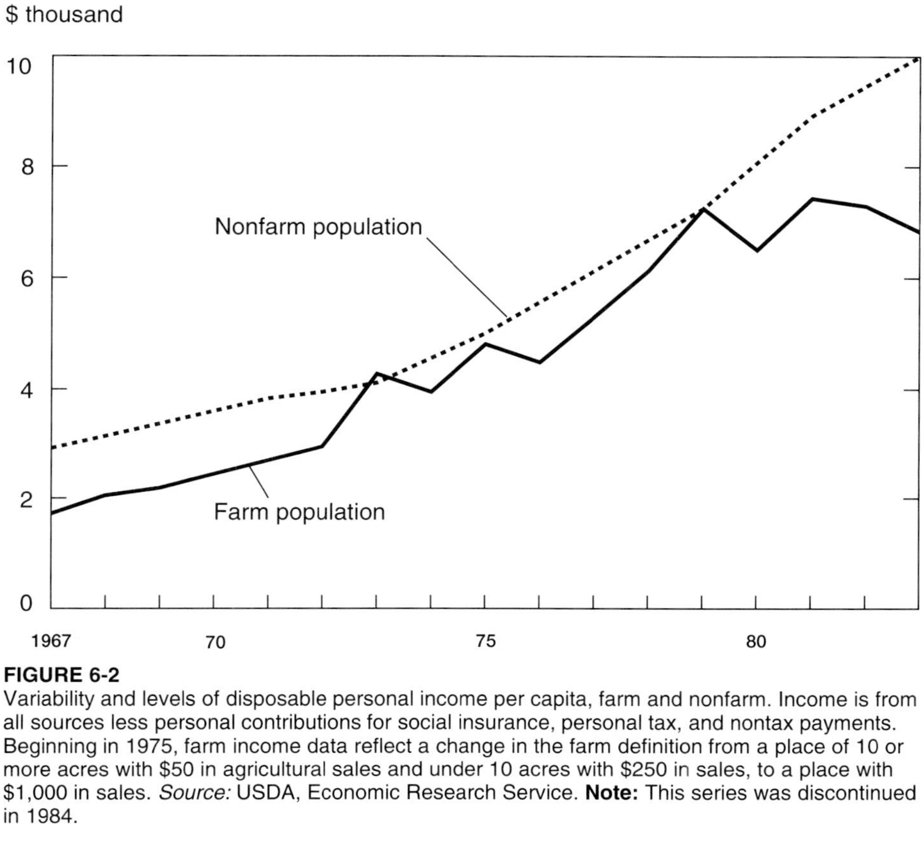

FIGURE 6-2
Variability and levels of disposable personal income per capita, farm and nonfarm. Income is from all sources less personal contributions for social insurance, personal tax, and nontax payments. Beginning in 1975, farm income data reflect a change in the farm definition from a place of 10 or more acres with $50 in agricultural sales and under 10 acres with $250 in sales, to a place with $1,000 in sales. *Source:* USDA, Economic Research Service. **Note:** This series was discontinued in 1984.

living, and their incomes diverge from those of nonfarmers. These divergences are not endless, because economic conditions change, counteracting market forces set in, or public policies alter the course of economic events. Occasionally, however, wide divergencies have persisted for several years, becoming a problem for price and income policy. For instance, throughout much of the 1920s, while most of the nation experienced widespread abundance during the so-called "Coolidge Prosperity," the farm sector in general was in deep depression. This set the stage for the first major efforts to involve the federal government in price and income policy. In the mid-1970s, declines in the value of the U.S. dollar, the boom in exports of agricultural commodities, and the expansion of farm credit created prosperity for the farm sector that was generally unmatched by the nonfarm population. Then, in the early 1980s, the situation reversed itself, and the "farm crisis" came to dominate policy.

These troublesome gaps in farm-related prices and incomes occur often enough and with sufficient severity to be one of the most important problems underlying agricultural price and income policy. As Figure 6-3 shows, for some 12 years from 1979 to 1991 prices received by farmers diverged sharply from prices paid. This discrepancy resulted

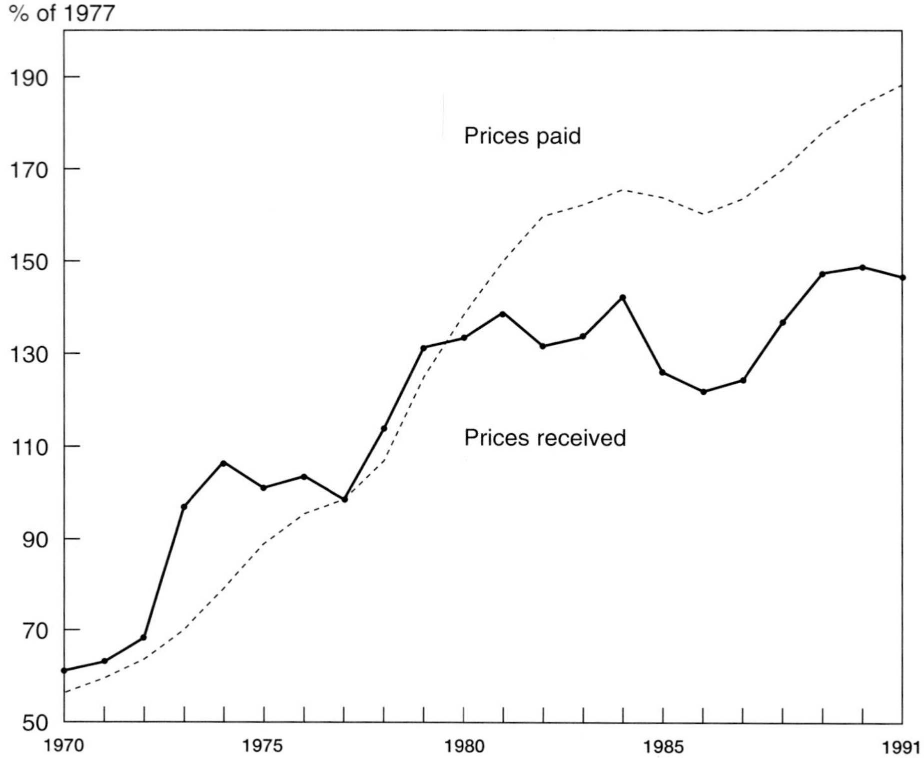

FIGURE 6-3
Prices received and paid by farmers. Prices include commodities, services, interest, taxes, and wage rates. *Source:* USDA, Economic Research Service.

in adverse returns to farm family resources and contributed substantially to the widely recognized ‘‘farm crisis’’ of the mid-1980s. The downturn in the farm economy was due largely to the pincer effect of lost agricultural export markets and rising interest rates, both of which were in part the result of monetary policies and currency appreciation. Ironically, the reverse sequence of depreciated currency and easy credit, which set off the short-lived agricultural export boom of the 1970s and the unprecedented prosperity for the farm sector, did not continue into the next decade. Hence periods of farm price and income divergence have figured prominently in forming price and income policy. How to react—or not to react—becomes an important test for price and income policy making.

Problems of Resource Adjustment, Poverty, and Food Aid

Most of the resources used by the farm sector are not easily liquidated or rearranged in times of economic stress, and many farm families are too poor to make some of the large-scale adjustments that economic analysis might indicate would be profitable.

Hence a major problem for policy is how best to facilitate profitable resource adjustments and ameliorate the conditions of farm and other rural families living in poverty. For a society concerned with the welfare of all families and individuals, and for a government organized to promote the general welfare, the question is whether to attempt some improvements or not; and if so, how much public support should be attempted through price and income policy and how much by other means.

Farm price and income policy has been promoted as necessary to provide stable prices and equitable incomes for farm families, to assure increasing productivity and growth, and to assist the poor, children, and the elderly in obtaining an adequate and nutritious diet. Furthermore, it has been recognized that, apart from the periodic effects of fluctuating market conditions, a fundamental contributor to the price and income problems confronting farmers is their continual burden of resource adjustments. All resource managers in a market system must respond to economic forces that accompany changes in technology and consumer demand. But producers of primary commodities such as grains, which have highly inelastic demands, must often adjust more rapidly to achieve comparable resource returns. In some industries, these adjustments involve labor layoffs. Price and income policy has been supported as a way to help in such adjustments.

In times of rapid technological change, characteristic of U.S. agriculture since the 1920s, efficient use of resources requires several adjustments simultaneously. When new technology is adopted, much of the excess farm labor must find other employment. Additional capital in new forms must be absorbed. Generally the scale of operations must be enlarged for the farm to remain economically viable. Because attractive alternative opportunities are often not readily available to the members of farm families, they are not perfectly mobile. Thus reallocating human capital and other resources involves some costs that intensify the income problems of the farm sector. In addition, technological advance that leads to lower real prices for farm commodities exacerbates the farm income problem. A high priority for resource adjustment was recognized in the first major farm price and income legislation, the Agricultural Adjustment Act of 1933. Succeeding price and income legislation has generally continued this emphasis.

The extent to which economic forces affect poverty in the farming sector is depicted in Figure 6-4. On average, the poverty rates were higher for farm residents than for residents of metro and nonmetro areas during the early 1970s, and then again during the first half of the 1980s. The real arguments in farm price and income policy are not about whether farm poverty exists. Clearly it does. Nor are these arguments about whether resource adjustments are easy or difficult to make. They are easy when capital, economic knowledge, and management ability are efficiently combined, and difficult when these resources are unavailable.

Food availability and assistance have increasingly become focal policy issues. Undernutrition and hunger in the United States result primarily from a lack of adequate resources, although other factors are also important. Food aid programs can ease the problems inherent in the adjustment process as human resources exit farming. These programs free up income that can then be used to purchase shelter, health care, and other nonfood items. The food and income supplementation effects of the food-aid programs are also important to poor, rural, nonfarm families and to the urban poor.

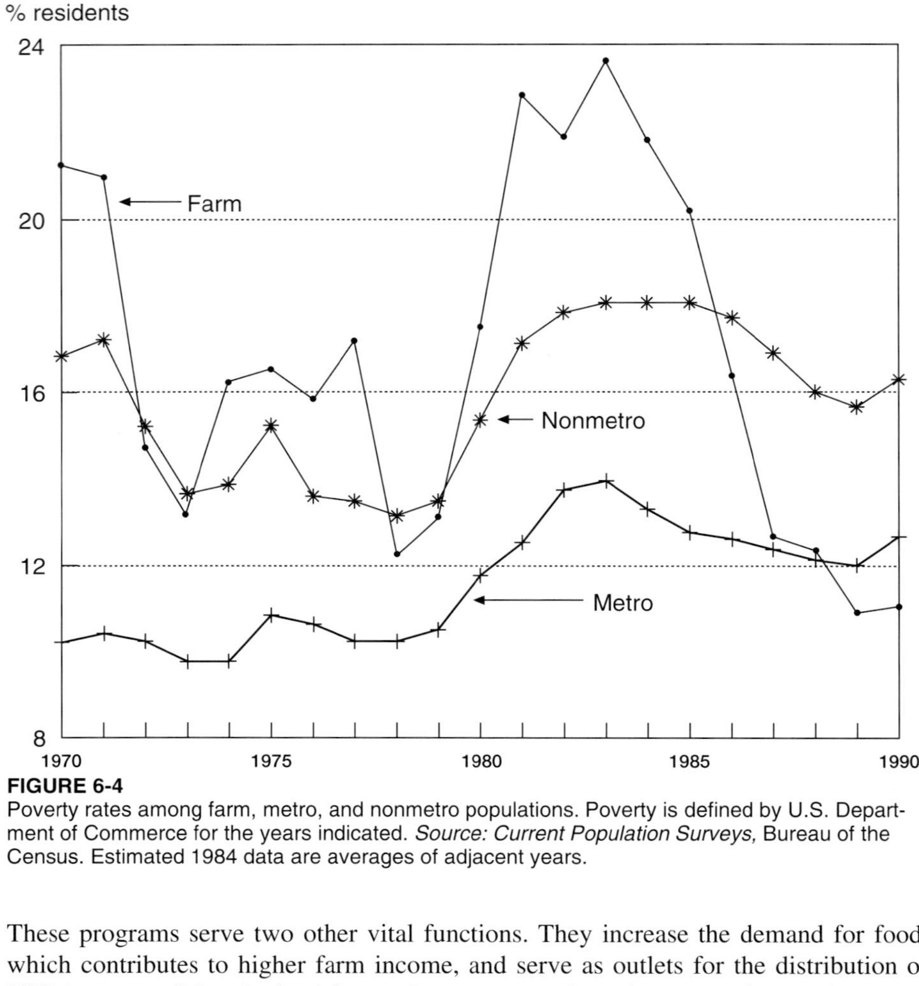

FIGURE 6-4
Poverty rates among farm, metro, and nonmetro populations. Poverty is defined by U.S. Department of Commerce for the years indicated. *Source: Current Population Surveys,* Bureau of the Census. Estimated 1984 data are averages of adjacent years.

These programs serve two other vital functions. They increase the demand for food, which contributes to higher farm income, and serve as outlets for the distribution of USDA commodities obtained from price support and surplus removal operations.

The recurring question for policy analysis is to what extent farm price and income policy can contribute to the agreed on goals and at what cost compared with other available policies. To understand the issues, it may help to examine some of the beliefs and unique values that pertain in policy making, and the political structure that underlies policy in this area.

Unique Values of the Farming Sector

Understanding price and income policy requires knowledge of the social and political values as well as the economic values that underlie farming and its supporting economic sectors. These values intertwine to contribute to price and income policy decisions. Three important values are noteworthy.

First, many individuals identify agriculture with food, a prime ingredient in any nation's level of living. Even though innumerable goods and services make up this level of living, food is the primary consumer item that is largely biological in its production and utility. Food is associated with personal and family security. It is crucial for national independence and survival and is the first line of military defense. Food and agriculture strike a strong positive popular note that translates into support for price and income policy.

Second, the family farm is popularly identified as (1) vital to the nation's food supply, (2) helpful to our democratic political ideals, and (3) central to our honored creeds of private enterprise and freedom. The family farm has received unique policy support in this nation as an ideal system for food production and family living, an ideal that was fostered by the early policies for public land distribution and education. Although the truth may be a mixture of economics, sociology, and nostalgia, the ideal is still a prominent force in price and income policy.

Third, agriculture has been perceived throughout history as the "fundamental" and only truly natural source of value and wealth to the nation. The agricultural product has been thought to be more fundamental in the total production process, farm people more fundamental to democratic ideals, and farm income more fundamental to the creation of the nation's income. Although these beliefs may only partly reflect reality and their political force may be fading, they still influence price and income policy decisions.

Favorable Proportional Political Representation

Price and income policy is supported in the Congress by a built-in favorable representation for farming and other related agricultural interests. Irrespective of the economic and social characteristics of agriculture, which may give impetus to direct public intervention in its markets, the organization and functioning of the political system assure a strong, direct political voice for those concerned with agriculture's welfare.

All public policy with legislative origins must be acceptable to both houses of Congress. The Senate in particular provides rural people, agriculture in general, and the farm sector in particular with a strong political presence because each state, regardless of population, is represented by two senators. Senators from sparsely populated or primarily rural states are likely to be more sensitive to the concerns of farmers, rural communities, agricultural businesses, programs serving people in rural settings, and resources of particular importance to agriculture, such as land and water. To illustrate, two senators from the state of California represent a population twice as large as the combined populations of 10 rural states west of the Mississippi River that are represented by 20 senators.

For similar but less extreme reasons, the lag in reapportionment of the House of Representatives helps assure that people in rural or less populated regions of each state will have a strong voice. Since 1962, when the historic Supreme Court decision of *Baker v. Carr* led to the "one-person, one-vote" principle, there has been mandated reapportionment of the 435 congressional districts of the House after each decennial census. Representation in the House is proportional to population in and among all states for only a few years immediately following these reapportionments. During the interim years, as the inevitable population shifts continue from farm to nonfarm and from rural to urban, there is the potential for an increasingly favorable representation for farm and nonfarm rural people. Furthermore, since House reapportionments are determined jointly by the state governors and legislatures, which are also a product of the same persistent lag in proportionality, there nearly always is a favorable tilt toward farm and rural interests. Thus price and income policy is influenced by legislative bodies that are usually more sensitive to economic concerns of farm and rural people than to people in any other sector of the economy.

LAUNCHING PRICE AND INCOME POLICY

From its beginning in the 1920s to the present day, the development of price and income policy has been a process of trial and error in which economic policy analysis is combined with the political pressures coming from the more vitally interested parties. Although important benefits have accrued to some parts of the agricultural economy, seldom have the results been satisfactory to all concerned. Yet the public commitment to some type of agricultural price and income policy has persisted. It began with a short-lived policy approach.

Price Support Without Supply Control or Stimulants to Demand

The Agricultural Marketing Act of 1929 initiated price and income policy in the United States. The Act established the Federal Farm Board with limited intervention, but was destined to fail almost from the start because the board had no means to control production or stimulate demand. Simply, it was given $500 million on ''loan'' from the Treasury, a relatively large sum for that period, to raise and stabilize market prices for grains and cotton by financing agricultural cooperatives through a new system of banks to buy and store these commodities until prices reached the desired levels.

This first price and income policy can be characterized best as *price supports without production controls.* The economic implications of intervening in pricing without basically changing supply or demand are as follows. A higher policy-driven price restricts the quantity demanded and simultaneously increases the quantity produced. Surplus commodities are inevitable. They must be either redirected outside the primary market into another one, such as a foreign market, or stored indefinitely until market conditions improve; also, they can rot, be given to the poor, or be destroyed. Thus, a price support without effective production controls or stimulants to demand is an incomplete and generally unworkable policy alternative unless supply or demand conditions are altered by other means. The Federal Farm Board used its funds to purchase and store commodities in order to raise prices. Soon all its funds were committed. Then, in the face of its insolvency, policy makers declined to appropriate more money. The acquired surpluses were dumped back into declining markets, depressing prices even more.

In the late 1920s, it may have been naive although politically expedient to perceive the economic problems of agriculture as transitory, rather than chronic and more fundamentally rooted in the nature of the industry, its markets, and technological revolution. In one sense, the short-lived attempt at price intervention with the 1929 Act failed. In another sense, however, the public had to face the need to carefully examine the challenge if it were to choose further intervening and developing policy more appropriate to the problem. All agricultural sectors, especially the farm sector and the agribusiness input industries, were in a deep depression. New policies and programs were urgently needed to deal with the rapidly worsening economic conditions. What policies and programs have been adopted is the subject of several future chapters.

Measures of Economic Well-Being: Cost-of-Production, Parity, and Others

Among the many difficult questions of price and income policy is: *How do we measure economic well-being?* That is, if policy is to alter the prices and income distributions determined by markets, what is the desired level of price or income to be sought? Or by what standard is the goal and achievement of public intervention to be measured? Earlier, during the decades of developmental policies, similar questions had to be answered; for example: How much land should be given to homesteaders? What segment of the rural population should be offered free public education? How much credit should be extended to individual settlers or farm families and on what terms? In price and income policy, as with these developmental policies, no single measure of economic well-being, or welfare, can serve as an universally acceptable standard.

Cost-of-Production Versus Parity In the 1920s and early 1930s, there were continuing concerns about the low prices for agricultural commodities, and many arguments about what should or should not be done about these prices. Some suggested that farmers were entitled to receive their "cost-of-production." However, economists pointed out that such costs could be determined only as averages, that the procedures for computing costs were fraught with internal contradictions, and that such imputation of costs into prices conflicted with market determination of value. Yet, this policy approach persists and is encouraged by such pricing and by using a "cost-of-production" standard in the economy generally.

Gradually, in this early period of policy history, interest shifted to the concept of "fair prices" and to indexing price-support levels to past prices or standards. Finally, the language in the Agricultural Adjustment Act of 1933 expressed the goal of raising "agriculture to a parity with other industry," which had emerged as a rallying cry during the difficult 1920s. Hence the goal of parity prices for agriculture, or some percentage of parity, became the early cornerstone of farm price and income policy.

Parity This was the first measure used in developing price and income policy, and it survived for many years because of its assumed fairness and convenience as a target or standard for price-support programs. Parity was used by agricultural leaders as a price standard, a certain percentage of parity becoming a goal for policy. Borrowed from a commonly used root word, *par (equal),* and with an emotional value put on it, parity expressed the level of prices that farmers desired in order to obtain the same purchasing power that they had during the more prosperous period preceding World War I.

The most nostalgic years of the Golden Age of American Agriculture were the four years from August 1910 to July 1914, and these years became the initial base for measuring parity. Hence attaining parity prices for all agricultural commodities became a major policy goal. During the years that parity was used in policy, however, the fixed and gradually more distant base became increasingly less defensible, so the base was changed by various adjustments. On January 1, 1950, one of the more significant changes used a 10-year moving average of prices received to adjust parity up or down for a specific commodity, in comparison with parity for all farm commodities combined. This adjustment, then called ''new parity,'' tilted the measure in the direction signaled by more recent market forces and provided for automatic yearly adjustments. An improvement but still backward-looking, new parity was not always consistent with current market prospects, and the percentage of parity to be supported was always in contention.

In addition, the concept of parity ratio gradually emerged to provide some of the same technical and emotional meaning for the composite of prices of all agriculture commodities. Hence, beginning with the Agricultural Adjustment Act of 1933, parity was codified by statute with a technical meaning, which as modified continues to be computed and published in USDA's *Agricultural Prices.* For more than half a century, ''parity prices'' was a policy goal for many farm leaders and most farm organizations, and price-support levels usually were specified as some percentage of parity, such as 75 or 90 percent.

In recent usage, the technical meaning of parity is *the price per unit that gives a commodity the same purchasing power it had in 1910 to 1914, adjusted for relative price changes in the last 10 years between that commodity and all farm commodities.* Based on this definition, the current *parity price* for a unit of a commodity is equal to the average price received by farmers for that commodity over the last 10 years, divided by the index of average prices received for all commodities over the same 10 years, and multiplied by the index of prices paid in the current month. Complete data for this calculation may be found in the USDA publication *Agricultural Prices,* each January issue.

The current parity price per unit of a commodity equals

$$\frac{PR_{sc10}}{IPR_{ac10}} \times IPP_{ac,cu}$$

Where PR = price received by farmers
PP = prices paid by farmers
I = index (1910–1914 = 100)
sc = single agricultural commodity
ac = all agricultural commodities
cu = current month
10 = last 10-year period

The *parity ratio* is an expression of the relationship between prices received by farmers for all commodities and the prices paid by farmers for all farm inputs and consumer supplies, with 1910 to 1914 being the base.

The current parity ratio for all agricultural commodities equals

$$\frac{IPR_{ac,cu}}{IPP_{ac,cu}} \times 100$$

Parity has both advantages and disadvantages as a measure of well-being and as a basis for policy. On the plus side, parity: (1) indirectly reflects changes in costs of production, in part an indicator of input-commodity prices; (2) relates pricing of commodities produced to changing efficiencies of production; and (3) embraces wide participatory recognition as a policy term. On the minus side, there is: (1) the difficulty of obtaining the extensive, detailed data necessary for reliable computations; (2) the increasing lapse of time since 1910 to 1914, the period used partly in computing the base, and hence an incomplete reflection of productivity changes; and (3) the confusion between the emotional meaning and the technical content of the concept. As a policy guide, parity could fail to reflect current market conditions and prospects. Although parity was used for almost four decades as a major policy goal, it faced increasing criticism from professional and lay leadership, and it gradually lost political support as a basis for price supports.

Cost-of-Production In the meantime, cost-of-production had emerged again as an alternative measure of economic well-being. In the Consumer Protection Act of 1973 and the Food and Agriculture Act of 1977, pricing provisions of the policies were related to selected measures of production costs, with price supports increasing as the selected indexes of costs increased. These ideas were carried over into the Agriculture and Food Act of 1981, with provisions that price supports would increase at fixed increments, influenced by the higher rates of inflation that had been occurring. Then in the early 1980s, when inflation was substantially reduced and recession occurred, increases in price-support levels were delayed. Hence, in the Food Security Act of 1985, price supports were lowered to facilitate exports, reverse the build-up in government stocks, and reduce the Treasury costs related to price supports. However, target prices were reduced slowly, and the terms of trade for agricultural commodities worsened, leading to rapidly escalating federal budget costs for direct payments to farmers.

Ad Hoc Measures and Budget Concerns As farmers adopt technological innovations to reduce the real costs of production, the price of individual inputs is not a reliable measure of costs. Increasing productivity is an important factor in reducing costs. This was recognized to a limited degree in both the 1981 and 1985 Acts, when new measures began to be substituted for either parity or cost-of-production. These measures have involved various ad hoc compromises based on economic analysis of market prospects. Although price supports and target prices are by and large politically determined, federal budget costs have become a critical factor in determining the levels;

therefore, the role of economic analysis has increased. In other words, there is a significant budgetary constraint in setting the terms for price and income policy, and the economic costs and benefits of any policy or program must be scrutinized in evaluating policy alternatives. For example, the Congressional Conference Committee responsible for the final compromises in the 1990 Act was required by a Congressional agreement to reduce its projected cost by some $13 billion.

Inferences for Policy

Price and income policy will continue to loom large in controversies over food and agricultural policy. In a dynamic economy, the economic problems of agricultural growth and adjustment will not disappear. Yet it is almost certain that the farm sector will continue to increase productivity and shrink as a portion of the total national economy. This will pose the difficult public question as to whether the strong productivity prospects imply reduction of further supportive policy or whether the certain shrinkage in relative economic size requires continued public assistance. The function of policy analysis is to identify some of the alternatives and to show how some of the costs and benefits of alternative policies can be determined.

Accordingly, we need what may be called an alternatives framework for policy study. What should be in this framework, and how is it to be developed?

Identifying an Alternatives Framework

Policy education is rooted in understanding and analyzing a policy, rather than in promoting or advocating it. To understand price and income policy, we need to be aware of the nature of various types of policies and to recognize the problems for which each type may be a response. To apply economic analysis to the various actions that have been taken or could be taken in the future, we need to identify a framework in which policy alternatives may fall. The first step is to become familiar with price and income policy options or with alternatives that could be adopted. Next, it is critical to develop an ability to analyze and compare the probable economic consequences of each alternative, with particular attention to the most relevant economic sectors of the economy. Finally, we should examine the use of each alternative in U.S. policy experience.

This understanding of price and income policy alternatives, along with the ability to analyze and compare consequences, is the subject of the next four chapters as follows: Compulsory Production and Marketing Controls, Voluntary Production Control and Compensatory Payments, International Trade Policy, and Food-Aid Policy: Domestic and Foreign. Each chapter describes one of the four general price and income policy approaches, which are charted with customary economic supply and demand curves, analyzed in terms of their consequences, and further amplified by reference to past or present policies.

The four policy alternatives represent innumerable possible packages of provisions related to price, production, trade, food aid, and income. The criteria for selecting the alternatives include the prevalence of the policy in the United States and in other representative political and economic market systems, the coverage of the broad spectrum of different types of policy interventions in either supply or demand, and the opportunity to consider the differential consequences to various economic sectors of the economy. Each alternative is presented in "pure economic form," that is, by specific alterations in the usual assumptions of the functioning of perfectly competitive markets at the firm and industry levels. This approach allows a clearer understanding of the economic meaning of each policy.

As is customary in economic analysis, the presentation abstracts from actual public policy experience, where combinations of several different techniques, provisos, and approaches constitute the policy compromise. For each alternative, the formal analysis is followed by an examination of present and past-enacted policies. Thus an understanding of future policy alternatives is developed by combining "pure economic analysis" with a study of actual policy experiences.

LOOKING AHEAD

The policy alternatives in the next four chapters, Chapters 7 to 10, are discussed generally in the order in which similar policies appeared chronologically in the United States. In Chapters 7 and 8, the formal analysis is developed largely as if the United States were a closed economy (that is, closed to international trade). This assumption is dropped in Chapter 9 to deal with the effects of foreign trade on the United States and other trading countries, and to develop policy alternatives relevant to international trade policy. It is appropriate to start with compulsory production and marketing control.

Chapter 7 Compulsory Production and Marketing Controls This chapter (1) outlines the basic assumptions and features of compulsory production control, (2) discusses some of the expected economic consequences of the policy, and (3) briefly reviews some of the U.S. experiences with such policy.

Chapter 8 Voluntary Programs and Compensatory Policy This chapter considers another policy approach in which production is voluntarily restrained in response to inducements, with farmer returns enhanced by a combination of high prices and direct monetary incentives.

Chapter 9 International Trade Policy This chapter (1) sets forth the basic principles of comparative advantage in international trade, which is the general basis for all voluntary trade transactions; (2) shows how to identify the mutual benefits from trade; and (3) considers policy alternatives for dealing with trade.

Chapter 10 Food-Aid Policy: Domestic and Foreign This chapter discusses (1) the economic and political rationales for a food-aid policy; (2) the important economic effects of food aid on producers, consumers, the federal budget and taxpayers, and foreign trade; and (3) the general experience with food aid in the United States.

Each of these chapters follows a similar format for the policy alternatives:

- description of the policy
- pure economics of the policy
- expected economic consequences that the pure policy alternative may have on farm producers, agribusinesses, consumers, foreign trade, and the U.S. Treasury (taxpayers)
- Examples of similar past or existing policies

TERMS AND CONCEPTS

Output market policy
Price and income policy
Increasing comprehensiveness of policy
Expanding participant groups
Variability of agricultural production, prices, and incomes
Divergence of prices and incomes
Nonmetro, (rural) poverty
Agricultural fundamentalism
Favorable proportional political representation
Policy alternatives framework
Measure of economic well-being
Parity price
Parity ratio
Cost-of-production

QUESTIONS AND STUDY SUGGESTIONS

1 Why should public policies focus on developmental problems of the agricultural sector before those in the price, income, and output areas?
2 What general, persistent problems of the food and agricultural sector have tended to bring forth price and income policies?
3 When did the first public price and income policy emerge in the United States? What was it? How well did it work?
4 Why do developed, high income countries generally have extensive price and income policies to protect their agricultural producers?
5 Why have price and income policies become more comprehensive in content?
6 Why is favorable, direct representation of agricultural interests built into most participatory political systems?
7 What is the relevance of a standard or measure of economic well-being in price and income policy?
8 What is the definition of *current parity price?* What are its advantages and disadvantages as a policy concept?
9 Why is an approach emphasizing policy alternatives important in education and research in the policy area?
10 Do you believe that there is a fundamental difference in interventions of participatory governments in the market to ensure universal education or an interstate highway system compared to a minimum price or supply for farm commodities? Why?

PART THREE

ALTERNATIVES FOR PRICE AND INCOME POLICY

CHAPTER 7

COMPULSORY PRODUCTION AND MARKETING CONTROLS

Compulsory production or marketing control is one among many alternative approaches to price and income policy and is the first of four examined. This policy is designed primarily to increase and stabilize a market price above a commodity's purely competitive equilibrium price. Thus, it might be chosen to respond to problems of instability of farm product prices and incomes, their low levels, and periodic divergence of those prices and production costs. The policy achieves this goal by using mandatory producer quotas to restrict total production or amounts marketed. Before being implemented in the United States, approval must be obtained from a two-thirds majority of the eligible producers who vote in a referendum, after which penalties can be assessed for violations of the quotas. Some form of price supports, possibly with a nonrecourse loan, may also be used in conjunction with this policy to stabilize prices during the marketing year.

Since 1933, compulsory production or marketing control has been used for varying periods in the United States for commodities such as tobacco, wheat, cotton, rice, peanuts, sugar beets, various fruits and nuts (for instance, lemons and walnuts), and a few vegetables. Even though this policy approach has been replaced by other alternatives for most field crops, the policy idea continues to be a central issue in policy-making controversies. Thus the purpose of this chapter is to examine the characteristics, rationales, provisions, economic relationships, consequences, and U.S. experiences with this policy alternative.

IN THIS CHAPTER

By studying Chapter 7, you will learn how to answer the following questions:

1 What are the underlying assumptions and characteristics of compulsory production or market control?

2 What are the general economic consequences of such policy?
3 What are some of the experiences with this policy in the United States?
4 How is this approach applied to fruits, vegetables, and dairy products?

BASIC ASSUMPTIONS AND CHARACTERISTICS

Compulsory production or marketing control is an alternative used primarily to support farm prices and incomes when other programs are considered to be too expensive for the U.S. Treasury or inefficient in achieving the desired policy results. Generally, it is associated with policy makers' views that prices of agricultural commodities are too low or producers' incomes are unsatisfactory because the supply is too high, too seasonal, or unstable. Furthermore, supporting the price of certain commodities is seen as justified through supply management instead of supplementing demand, through the marketplace instead of the Treasury, and through compulsory instead of voluntary means.

Procedures for Policy Implementation

Program development and implementation are carried out in several steps:

1 The desired price, income, or level of commodity supply is determined, usually by using some measure of economic well-being such as cost-of-production, parity, or trends.

2 The estimated aggregate quantity of production to achieve the desired economic result in the market is allocated among the individual producers qualified for the program, usually by using historical bases of production or sales.

3 Before proceeding with this kind of compulsory policy, approval—often a two-thirds majority—of producers or their marketing organizations is sought by a referendum vote.

4 If the policy is so approved, allotments to all producers are made, and enforcement of the production or marketing quota is usually implemented through compliance checks and economic measures such as loss of allotment, a prohibitive tax on sales in excess of the quota, or fines.

5 Subsequent decisions about production and marketing made by farm and agribusiness firms are left to market forces.

6 Imports at prices below the established minimum level are prevented by import quotas or tariffs.

7 A reassessment is made at the end of each production and marketing cycle to adjust future quotas based on previous experiences and projected market demands.

8 Special approval by an elected farmer committee is usually authorized for quota additions, changes, or transfers.

Assumptions and Necessary Conditions

Several conditions or assumptions are necessary for the effective design and implementation of this policy approach: (1) a satisfactory means for determining quotas; (2)

a public commitment to implement the necessary interventions in respect to producers' decisions, trade, and market pricing; and (3) an inelastic demand for the commodity. Policy makers and administrators must make many decisions about which commodities to include, the level at which to set the total quota, how to allocate the individual quotas, which import restrictions to apply, and whether or not to use export subsidies to maintain a targeted level of exports. Thus many variations of this policy alternative can be envisioned.

Economic Rationales

The underlying economic rationales for compulsory production or marketing control policy need to be examined. This policy has been used regularly for some commodities since 1933 (tobacco being the best example). It has also been used intermittently for some other important crops, and it continues to be proposed for broader applications. This compulsory supply control approach first received attention after the 1929 Federal Farm Board failed in supporting prices without such control. When the approach has been used, import restrictions and export subsidies have usually followed. It has been the policy approach traditionally favored by some farmer organizations (The National Farmers Union and the National Farmers Organization, NFO, for example). Also, the policy has been supported by some political leaders and administrations (the Kennedy Administration in the early 1960s, for example) as a major part of their approach to solving farm income problems. In each case, adopting the policy would have shifted the main expense of farm income support away from the federal budget and toward U.S. consumers.

The basic motivating force for producers is the inexorable quest of participants in the market system to gain market power, maintain or increase their market share, or establish a monopolistic advantage. A successful program can help farm producers or agribusiness firms in oligopolistic situations to ration quantities supplied to the market, and hence protect pure profit against its erosion from competing producers or marketing firms. Nonfarm owners of factors of production throughout the economy commonly seek these economic goals by using size, market share, advertising, licensing, franchising, unionizing, accrediting, and administered pricing. The main objective is to effectively control supply by private means in which the ultimate consumer bears the cost in the market system.

Even though markets with some of these features can be found in many sectors of the economy in which competition is imperfect, the economic structure of farming and most of the agricultural marketing system does not permit this kind of market control unless it is supported by government policy. Only for specialized, localized commodities is this possible. As a result, some farmers tend to look to this kind of public policy as a means for making competition "fairer" or "more effective" for them. Thus, to enhance their prices and incomes through the market rather than the Treasury, supporters of this approach look to public implementation through compulsory control as the only realistic way to attain such a goal.

The policy assumes a highly inelastic demand for most agricultural commodities, such as those discussed in Chapter 4. With such commodities, a decrease in quantity precipitates such a large increase in price that the aggregate return to producers actually rises, other things remaining equal. That is, low production in any year is likely to bring a higher aggregate revenue to farmers than high production does. Even though this result may appear highly unlikely to individual producers who are viewing the prospects of a bumper crop, this is the economic reality facing most farmers who market a particular agricultural commodity. Hence the characteristic inelastic demand is the crucial economic rationale for the policy alternative of using publicly mandated production or marketing control to raise or stabilize farm prices and incomes.

Political Rationales

There are several political considerations. Price supporting through compulsory production or market control is commonly associated with chronic instability or excess supply. Sometimes the policy has been started when the prices of a particular commodity persistently diverge from the prices or costs of commodities used in production, thus confronting producers with relatively low or declining factor returns. These circumstances have provided the political rationale for the public to sanction this kind of market intervention, especially when producers have a favorable record of increasing productivity, technological advances, and continual structural adjustments with out-migration of farm labor and increased capitalization.

Another rationale for this policy is the unique variability of farm production, as discussed in Chapter 6, and the public's concern for stable food supplies. Under the normal forces of competition, farmers often produce more than is profitable, and consumers have abundant supplies at low prices. Sometimes when farmers cut production or adverse weather reduces supplies, consumers face shortages and high prices in certain years or seasons of the marketing year. This policy alternative has usually been coupled with a parallel approach involving a storage or commodity reserve program to stabilize supplies and reduce market fluctuations.

Finally, a political rationale that has been dominant at times is related to minimizing government costs of supporting farm income. If demand is inelastic and supply can be controlled by economic measures such as a tax or legal quotas, then the real costs of a program can be shifted largely, if not entirely, to consumers. This has been the case for tobacco, because it has been easy to control both production and marketing. The domestic demand is highly inelastic, and taxes on tobacco have more than covered the costs of subsidizing exports. It must be readily noted, however, that tobacco taxes cover only a small fraction of the nation's welfare costs resulting from the loss in productivity, disease, and premature death arising from the use of tobacco and its products. The policy is also widely applied to some fruits, such as lemons and nectarines for example, where marketing associations operate under federal marketing orders to control both the quantity and quality of the commodity marketed.

Policy Perspective

The characteristics of agricultural production, market supply, and demand, and the unique economic problems of the agricultural sector have provided both economic and political rationales for continuing considerations or adoptions of policies to support farm prices and incomes through compulsory production or marketing control. Each variant of the control program has implications for the effectiveness of control, administrative feasibility, cost of implementation, efficiency of resource use, farm structural effect, and acceptability. Also entering into policy making are issues about who will administer the program and how much discretion administrators will have. For each commodity, some control measures are more appropriate than others.

Compulsory production or marketing control has been applied most effectively where it has been possible to monitor both the quantity and quality of the commodity coming to market. Thus the program has been attractive for commodities that can be differentiated according to quality, such as fruits and vegetables; some commodities that can be stored, such as tobacco; and some commodities that are highly seasonable in production and perishable, such as fresh fruit and fluid milk. Policy makers have been responsive to the political power of large numbers of producers, for example, tobacco farmers who face a small number of oligopsonistic buyers, or fluid milk producers who, before milk marketing controls, periodically and seasonally faced chronic problems of excess supply and low prices.

What are the expected economic consequences of such policy?

ECONOMIC CONSEQUENCES OF THE COMPULSORY PRODUCTION AND MARKETING CONTROLS

The expected economic consequences involve producers, consumers, federal budgets or taxpayers, and international trade. Also, the effects will vary, depending on the commodities, the degree of control, and the elasticities of supply and demand. Some of these effects can be illustrated with a basic economic model.

Basic Economic Model

In Figure 7-1, the economic relationships implicit in a policy alternative of compulsory production or market control in a closed economy are presented at both the industry and the farm firm levels. In the short run, the farm firm will experience a windfall gain. In the long run, however, windfall gains may diminish or disappear as producers bid up the prices of certain resources such as land and as costs rise until profit margins are equalized among competing commodity or firm enterprises. That is, all costs including land become more variable in the long run. In other words, land prices or costs move up to equilibrium levels, and other costs are readjusted until a new long-run equilibrium is established at the output level of Q_{pc}, where $MC_{pc} = AC_{pc} = P_{pc}$ (same as MR_{pc} and AR_{pc} would be at the firm level).

The basic economic model in Figure 7-1 may be applied to a wide range of compulsory production or market control policies. Let us emphasize, however, that this graph represents only a model of a ''pure'' policy compared with purely or perfectly competitive conditions. Just as individual producers and the industry may be operating at other than optimal levels, thereby achieving different economic results, this policy also might be modified by provisions differing from such a pure form, thereby yielding different values.

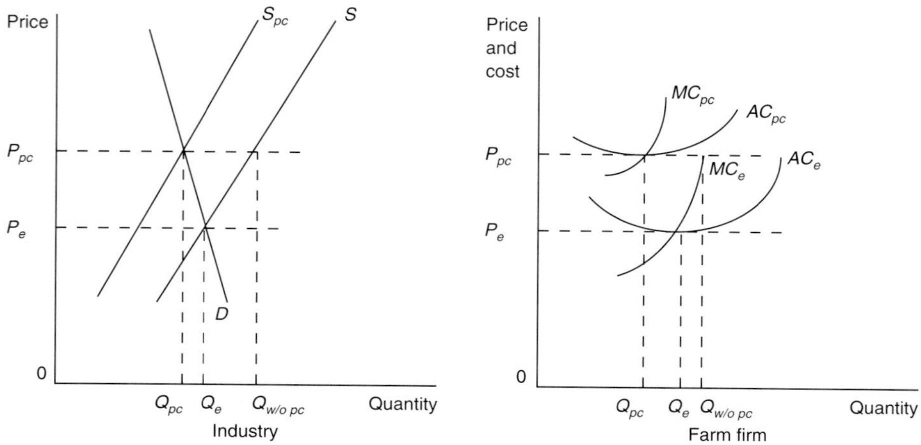

FIGURE 7-1
Price supporting policy through compulsory control (closed economy). For the industry, S and D are the initial supply and demand, P_e and Q_e are the equilibrium price and the quantity without production control.

For the industry with production control, if the desired market price is P_{pc}, the industry supply is decreased to S_{pc} and quantity is Q_{pc}. The effective production control program must withhold $Q_{w/o\ pc} - Q_{pc}$ of the industry's production potential.

For the farm firm, without production control, equilibrium is at P_e and Q_e. With production control, equilibrium is P_{pc} and Q_{pc}. In the short run, if AC does not change, the firm would have profits equal to $P_{pc} - AC$ at the output of Q_{pc}. But over the long run, where all costs are variable, the AC would shift to AC_{pc} because of the higher variable costs and rising imputed fixed costs per unit of output. For products with inelastic demands, such as most farm commodities, the total returns at Q_{pc} exceed those at Q_e, but any net profits are gradually reduced with long run competitive forces.

Control Techniques and Effects on Producers

Farmers are the major economic beneficiaries of a policy of compulsory production and marketing controls in a closed economy setting; yet most do not like the associated controls, regimentation, and restrictions on their decision making. The gains occur from the higher total receipts and lower total variable costs. Producer gains are more prominent in the short run and begin to disappear over a longer period with rising real fixed costs, or economic rent, accruing to present owners through capitalization of the higher returns. Increased receipts result from higher prices precipitated by the reduced aggregate production of a highly inelastic product. Decreased total costs flow from the mandated reduced production.

Trade implications cannot be avoided, and as the higher factor returns are gradually capitalized into farmers' fixed resources such as land, the initially attractive net returns are gradually eroded or transferred into appreciating assets. With this policy, however, returns to all farm factors of production in a dynamic economy are likely to be as great as or greater than returns at equilibrium in a free market, even though the relative gains by each factor will be uneven. Finally, even as the higher incomes reward farmers who are already established, the associated higher cost of fixed factors deters entry by farmers with limited initial equity. Just as every potential entrepreneur aspiring to hold a franchise of a major, brand-name industrial corporation cannot gain entry, also potential farmers confront a new economic barrier under this policy.

The control techniques warrant special attention. Producer quotas can be allocated on a strictly proportional basis, or alternatively they can be tilted in favor of some groups of farmers selected according to size, location, political influence, and other factors. Quotas can be related to some historically determined base, which may be either fixed or moving. For example, a wheat quota might be equal to a 10 percent reduction from the average harvested acres on an individual farm for the three years preceding the policy. The program might exempt cuts for bases of fewer than 15 acres, or 150 acres, for example, with the proviso that the administering agency could shift acreage from farms discontinuing wheat production to new producers. Further, the farm production or marketing quota can be applied to various identifiable measures such as acres planted, acres harvested, quantity produced, quantity sold, quality of product, and even quantity of inputs such as fertilizer. Finally, quotas can be identified with a farm, a landowner, or a farm operator and can be either nontransferable or transferable and negotiable. For most of these control measures, supply becomes perfectly inelastic until a high threshold price is reached.

Another differential effect that this policy has on producers occurs when another farm sector uses the commodity that is being controlled, as when livestock or poultry producers buy feed grains. Initially, the grain farmers' gains from higher prices translates into higher variable costs and lower net incomes in the livestock sector. If the higher prices for feed grains are due to the reduced supply rather than an increase in demand, the higher costs will tend to curtail livestock feeding, meat production, or quantities marketed. This will tend to raise meat prices. Consumer demand for total meat is inelastic, and therefore the gross return to livestock producers will generally increase. But because the demand for total meat is not as inelastic as the demand for total feed grains, the increases in meat prices generally will not be offset in strict proportion to the increases in grain prices. However, because of differences in elasticities, livestock producers will likely gain in income, although probably not in proportion to the grain sector.

A final important implication that this policy has for producers involves the place of the controlled commodity in international trade. To the extent that any commodity sector is substantially dependent on the export market or is likely to become so, and is not the exclusive supplier, a policy of compulsory production or market control will thwart the potential for long-term market expansion unless other policy measures such as export subsidies are used to counteract the effects of the higher domestic prices. As discussed in Chapter 9, the market gains from such policies flow directly from the fact that, for most agricultural commodities, export demands are less inelastic than domestic demands. Sometimes, if the U.S. exports are a small part of world exports, the effects on foreign markets may be neutral, or nearly so. But if U.S. exports are a large part of total world trade, as is the usual case for wheat, soybeans, and corn, for example, then the U.S. export and producer subsidy policy is of vital concern to other nations involved in trading the same or competing commodities, and U.S. policy becomes a serious issue for international negotiations. Clearly a policy of compulsory production or market control is most promising for a commodity that has almost self-sufficient levels in the U.S. market but that is unlikely to gain substantially from export expansion.

Agribusinesses selling to producers and some rural communities generally find their economic welfare closely tied to that of farmers. Thus, with a caveat about the implications for export markets, this policy will tend to benefit the agribusiness sectors most closely identified with farming or dependent on the buying power of farmers' income gains.

Effects on Consumers

Just as producers are the major beneficiaries of a policy of compulsory production or market control, consumers of agricultural commodities are usually major losers. This blanket conclusion is softened, however, by the stabilizing effects of this policy on farmers who invest in new technology and on the perceptions of consumers who want a stable food supply. In general, this policy stabilizes supply, but restricts the total quantity of certain agricultural commodities available to consumers and raises the average price levels. Consumers are the sole source of any higher income accruing to producers through markets.

Another way to understand this adverse consequence is to recall that in equilibrium theory, the primary consequence is maximum efficiency of production and minimum prices for consumers. Under this policy of compulsory controls, the firm operates at a level short of optimum, with the normal economies of scale not fully realized (Figure 7-1). The average cost per unit is increased because resources are not used at their optimum level, and some resources are rewarded at higher levels than under free-market conditions.

There are also important differential consequences to consumers. The burden of higher prices and smaller quantities will be greater on low-income consumers than on those with higher incomes, because poor people must spend a higher percentage of their total income on food (marginal propensity to consume). Thus the general consequence of this policy for consumers is an economic loss, although the loss for some groups, such as low-income consumers, could be counteracted by food subsidies and other policies. Further, we must be cautious about the indeterminant effects that any market-stabilizing effect of this policy can have on increasing farm investments for improved productivity, and hence possibly increase the longer term consumer gain.

Effects on Budgets and Taxpayers

Unlike most other farm price and income intervention policies, a policy of compulsory production or market control has more neutral effects on the federal budget and the eventual taxpayer than do policies of voluntary production control, such as land retirement. The major transfer of income is from consumers to producers. The only likely direct budgetary burden is for administration, which may not be trivial for a policy that is difficult to enforce. Yet these costs are minor compared with the costs of most other public income-transfer policies. Moreover, even these administrative costs can be shifted to producers by assessing them on a per unit basis, as was mandated for the tobacco program. As a user tax, however, the tax is more likely to be passed on to tobacco consumers because the demand for that product is more inelastic than supply.

Effects on International Trade

In the above analysis, within the context of a closed economy, the trade consequences of a policy of compulsory production or market control are clearly revealed. On the import side, the policy assumes that agricultural commodities will not be permitted to enter the domestic market at prices below the support price and thereby threaten the effectiveness of the policy. Hence foreign competitors are denied access to this market, just as consumers also are denied access to the imports at lower prices.

On the export side, both the higher prices and the associated reduction in quantities produced and marketed militate against potential gains from selling abroad. The adverse economic effects of this policy on U.S. producers can be moderated by other export-enhancing policies, and the effects on trade may be inconsequential because of the domestic nature of the commodity or other offsetting trading policies abroad. Nevertheless, the generally negative consequences of this policy for trade may have wide repercussions, complicating international negotiations especially.

Political considerations beyond economics may strongly influence trade policy. For example, it may not be in the national interest to rely heavily on agricultural imports. U.S. negotiators interested in expanding exports have been hampered by this feeling in other countries. Apologists for the U.S. sugar program, which drastically restricts imports into this country, often use this argument in support of the sugar policy. Also, the level of prices that export markets offer may be below the minimum deemed to be in the public interest for the use of its nonrenewable resources or for returns to its human capital resources. Finally, concerns about foreign exchange savings or earnings and about foreign relations may overshadow any identifiable economic consequences of this price and income policy.

U.S. EXPERIENCES WITH THE POLICY ALTERNATIVE

Now that we have discussed the definition, economic relationships, rationales, and economic consequences of a ''pure'' policy of price supporting through compulsory controls in a closed economy, let's consider how such a policy has been applied in the United States. Over the years since 1933, a program of compulsory controls was the dominant approach of price and income policy for some commodities until the early 1960s. Since then, it has been largely replaced by other policy approaches for all except the few commodities mentioned earlier. Although the policy has been reconsidered several times for major crops such as wheat, corn, or cotton and has been strongly promoted at times for other commodities (turkeys, for example), its expansion has been rejected by political decision makers, or in a few instances has been voted down in a farmer referendum.

Initial Experiences

The 1929 Federal Farm Board's initial attempt at direct market intervention, which involved price supporting without any production control, did not stabilize commodity prices. However, it did pave the way for new and stronger proposals for production adjustment to be enacted into law. When a new administration was launched in March 1933, the nation was in deep economic depression. The plight of agriculture had deteriorated into a national crisis. In record time, an entirely new price and income policy featuring a different approach—production curtailment—was developed and adopted as the Agricultural Adjustment Act (AAA) of 1933. During the emergency for that production year, which was well under way, some crops already planted and growing were plowed up (every third row of cotton, for example) and some livestock were purchased and destroyed (little pigs, for example). Both of these actions were voluntary, having been induced by direct payments to producers. The public was so averse to these actions that they were soon halted and never repeated, but marketing quotas were authorized in 1934 for both cotton and tobacco, provided that producers voted for them in referendum.

Originally, under the AAA, inducements to help curtail production were to be financed by a new processing tax on commodities included in the program. In January 1936, however, the U.S. Supreme Court declared the Act unconstitutional in that it was a program for regulating and controlling production, which, the Court ruled, was a right residing in the governments of the individual states rather than in the U.S. government. The processing tax was declared null and void, the Court concluding that it was inseparable from the scheme to control production. Soon, a hastily drawn substitute program, the Soil Conservation and Domestic Allotment Act of 1936, approached production control through conservation. The program offered payments to farmers for planting specified soil-conserving crops such as legumes and grasses, as well as for limiting the so-called soil-depleting crops to each farm's total assigned base. In effect, this Act, which launched the Agricultural Conservation Program, has survived as a conservation measure, as will be discussed in Chapter 12.

When voluntary, half-hearted efforts to control production proved unable to stem the worsening economic tide for farming, control of marketing was authorized in the Agricultural Marketing Agreement Act of 1937, and a broader version of the earlier AAA policy was adopted as the AAA of 1938. This reaffirmation of a public choice, which was later tested and found to be constitutional, set the course of a definitive policy of price supporting through a combination of voluntary and compulsory production and marketing controls. Subsequent revisions and amendments in 1941, 1948, and 1949 implemented price supporting and stabilizing by use of nonrecourse loans and compulsory production control by means of acreage allotments and marketing quotas.

Nonrecourse Loans

Begun in 1934, nonrecourse loans have continued through the years as a way to link price supports with production and marketing controls. These loans are offered by the federal government's Commodity Credit Corporation (CCC) to eligible producers as credit, usually for a marketing year. The loan level is the established price-support level (discussed in Chapter 6 as one of the measures of economic well-being). The loan is to be repaid if the farmer sells the crop, or is to be redeemed by the forfeiture of the stored commodity used as collateral. Thus the producer can pay debts incurred for production and marketing costs, while holding the crop in anticipation of higher prices throughout the marketing year and being guaranteed at least the minimal price per unit established in the program. When the supply of a commodity was judged to be burdensome, a national acreage allotment specifying each farmer's share was declared. The only penalties for noncompliance were the denial of nonrecourse loan benefits and acreage allotment payments, if any. This modest effort at production control was seldom effective and led to a more effectual technique, the marketing quota.

National Marketing Quotas

Under the AAA of 1938, the secretary of agriculture could declare a national marketing quota in the event of continued market-depressing supply. Each producer would be informed of the farm's individual quota, usually the amount of a commodity that could be produced on the reduced acreage allotment. Then the secretary would announce a referendum that would be open to all registered producers. This referendum would offer two options: (1) voting for the program, with prospective benefits of specified price supports or payments, but with penalties for exceeding a quota; or (2) voting against the quota, with the expectation of free rein to produce, but also with lower price supports and payments, or perhaps none at all. If two thirds of those voting were in favor, the quota was enforced for all producers. If two thirds did not vote in favor, price policy efforts would be discontinued for that year, or the level of support would be lowered according to what the law required. Thus this policy with its degree of compulsion was never implemented without the additional approval in referendum by producers who were eligible to vote.

Although these provisions for compulsory controls through marketing quotas were retained in the general agricultural acts from 1938 through the early 1960s, they were used intermittently as the supply conditions warranted, and then only for a few basic crops such as wheat and cotton. Only tobacco has continued this policy every year to the present, while the program for peanuts has been continued in a modified form. The referenda for these two commodities have received the strongest farmer approval. Price supports for commodities covered by marketing quota generally ranged from 65 to 90 percent of parity until recent Acts, which generally have replaced parity with specified price levels and more administrative discretion granted to the secretary of agriculture.

Quotas have continued to be of interest for some commodities, but since the early 1960s, price and income supporting through voluntary land retirement has replaced compulsory controls for most crops. As agricultural export commodities have continued to play a larger role in the nation's international trade, political pressure has mounted for a policy that will not only help to stabilize farm income, but will also be compatible with the national benefits to be achieved through trade. Compensatory payments to farmers in lieu of price do provide this support of farm incomes without directly hindering export trade.

Administrative and Policy-Making Experience

Farm price-support policy, such as compulsory production and marketing controls, has been administered by various USDA agencies with direct ties to the state and county and under various names, from the initial AAA committees to the recent Agricultural Stabilization and Conservation Service (ASCS) committees. Staffed by professional employees, these agencies have been advised in their functioning by farmer-elected committees.

Throughout the experience with this policy, publicly held stocks of storable commodities have built up periodically through forfeitures of the nonrecourse loans. This situation was emerging in the late 1930s, but at that time the stocks provided a valuable reserve for the war effort. Then, when they reappeared with a gradual build-up throughout the 1950s and in the absence of any national grains reserve, people became concerned about wasteful surpluses, unplanned budget costs, and ineffective policies. The stockpile problem provided repeated evidence that the policy makers' desire to support farm prices and incomes was not backed up by their resolve to enforce compulsory production or marketing control when opposed by the fiercely held tradition of independence among farmers. This was abundantly clear for corn and wheat. Quotas were not even authorized for corn because of the complications associated with such a policy, especially because many corn producers also feed livestock.

In the 1950s, wheat commodity programs emphasized raising and stabilizing wheat prices and incomes above world market levels by means of storage, export subsidy, and production control. They used acreage allotments, marketing quotas, nonrecourse loans, usually in terms of parity, and production payments as implementers. Among the three regular major world wheat exporters—the United States, Canada, and Australia—the United States took the lead in restricting wheat acreage, subsidizing wheat exports, and accumulating stocks as a means of price support. For several years, the minimum national allotment was set at 55 million acres, and the seeded acreage seldom fell below this level. Farmers were allowed to grow 15 acreas of wheat without having an allotment history, and they were permitted to share in price-support and acreage allotment payment programs. About 100 million bushels a year were grown on these 15-acre allotments, most of which were in the Corn Belt, but it was still only about 8 percent of total production. By the end of the 1950s, however, carryover stocks of wheat and corn were near or at an all-time high, and the stage was set for a crucial political test of compulsory production or marketing-control policy.

In spite of strenuous political efforts led by the new administration at the beginning of the 1960s, compulsory production and marketing controls were not extended to wheat and corn. The Emergency Feed Grain Program of 1961 authorized new, voluntary land retirement and payments to reduce feed grain production. Then, late in 1961, the administration forwarded a draft of a bill to Congress to authorize the secretary of agriculture to set quotas or allotment levels for individual commodities. The quotas would go into effect within 60 days, provided Congress did not vote against the program within that period. Subsequently the Food and Agriculture Act of 1962 provided for a referendum by wheat growers on whether to adopt a compulsory program for wheat.

The program proposed by the administration would have restricted wheat production and sales on the domestic market enough to attain a market price equal to 65 to 90 percent of parity, which then would have been some 50 to 100 percent above the free-market equilibrium price. Production or sales in excess of the quota could have been used for feed or sold in the export market at a free-market price. The alternative to this was continuation of the existing acreage allotment program with a price support for wheat at 50 percent of parity, then about $1.25 per bushel. One other unique provision in the legislation, very critical as it turned out, was to enfranchise all wheat producers to vote in the referendum, meaning that the economic interests of the very small wheat growers could be quite different from the interests of commercial producers under quotas.

Although the administration's proposal was strongly supported by several farm organizations (for example, the Farmers Union and the National Farmers Organization, plus some commodity organizations), the American Farm Bureau Federation rejected it on grounds that it would restrict or destroy farmers' traditional freedom to farm and would eventually bring most, if not all, of agriculture under government control. In the political campaign that followed, one of the most intense in the history of U.S. food and agricultural policy, more than one million votes were cast in the referendum by the expanded eligible body of wheat growers. The proposal for the compulsory production control lost, getting only 48 percent of the total vote, compared with the two-thirds majority required. Although the proposal carried in some of the primary wheat-growing areas, it was doomed by a negative vote in other areas where there were many 15-acre allotments or where wheat was a minor crop in a crop rotation.[1]

This critical period in farm price-support policy was followed by the Wheat–Cotton Act of 1964, which established a voluntary two-price certificate program for wheat and became the basis for the policy for the next several years. Certificates were issued to farmers who complied with the program to cover the amount of wheat in their farm allotment that would be sold in the domestic market. Farmers then cashed the certificates and could sell their wheat or store it under the CCC loan program. Complying farmers were also required to divert land from wheat to soil-conserving uses, for which they received land diversion payments.

Under the 1964 Act, all cotton farmers complying with their regular allotments were to have their cotton crops supported at 30 cents per pound, which was above the free-market price, and farmers who reduced acreage to a smaller allotment level were to receive support at 33.5 cents per pound. To keep domestic handlers and textile mills competitive with foreign manufacturers, the secretary of agriculture was authorized to make subsidy payments to handlers or millers sufficient to reduce the effective domestic price of cotton to the export price level.

Experience under the 1964 Act served as a foundation for the Food and Agriculture Act of 1965, which turned the nation toward price and income supporting through voluntary land retirement, replacing compulsory production control as the predominant price and income policy approach. In view of the preference of producers who resist compulsion and a society that is reluctant to use it, the evolution from compulsory to voluntary programs probably represents an improvement in the general welfare. Compulsory controls, either publicly mandated or sanctioned through marketing orders, still prevail only for tobacco, to some extent for peanuts, and for certain fruits and nuts.

MARKETING AGREEMENTS AND ORDERS FOR FRUITS AND VEGETABLES

A major part of the U.S. policy experience involves marketing agreements and orders, which have been used extensively to regulate the marketing of fruits and vegetables as a supplement to competitive markets rather than as a substitute for them. An *agreement* is a voluntary statement of understanding between the secretary of agriculture, the handlers, and, indirectly, the producers of a commodity that covers certain marketing rules and regulations. An *order* is a binding ruling by the government that covers all those producers and initial buyers operating in a specified market.

To establish an order, a proposal is generally developed at the request of industry representatives, discussed in various industry groups, and reviewed in the U.S. Department of Agriculture. Public hearings are held to hear evidence and arguments for or against the proposal. If the results appear to be favorable and consistent with the purposes of the Agricultural Marketing Agreement Act and have substantial industry support, a proposed order may be submitted for an approval vote by producers.

Each order establishes an administrative organization and specifies the controls to be used and other regulations to be followed, including the mechanics of operation. For each order, a control board appointed by the secretary of agriculture from nominees chosen by the industry serves as an intermediary between the secretary and the industry, proposes regulations, and recommends them to the secretary for issuance as an order. For some programs, a second committee, composed largely of handlers, serves in an advisory capacity and furnishes information on marketing conditions. The control board selects a program manager and staff to collect information and report alleged violations for possible enforcement action, collects assessments, and disburses monies to pay expenses of the program.

Provisions of Orders for Fruits, Nuts, and Vegetables

Marketing orders have been in effect for many different fruits, nuts, and horticultural specialties. They provide for any one or all of the following:

1 Specify grades, size, quality, or maturity of the commodity that handlers may ship to market.

2 Allot the amount that each handler may purchase or handle on behalf of any and all producers.

3 Establish the quantity of the commodity that may be shipped to market during any specified period.

4 Establish methods for determining the extent of any surplus, for controlling and disposing of the surplus, and for equalizing the burden of surplus elimination among producers and handlers.

5 Establish a reserve pool of the product.

6 Provide regulations and procedures for inspecting the commodity.

7 Fix the size, capacity, weight, dimensions, or pack of the container used in handling a commodity.

In addition, a major component of many orders provides for funding of research and promotion, primarily the latter, by means of a check-off, which is equivalent to a small sales tax on the commodity covered by the order. The major policy goal of this component is to expand the market for the commodity or to increase its demand. Research may be directed toward finding new markets or improving the quality of the commodity.

Some orders give considerable emphasis to "orderly marketing" to avoid waste by shipping uniform products. Standardizing grades, containers, and so forth is an acceptable protection for consumers, provided that standardization does not become overly restrictive and simply an excuse for market control, as has occurred in a few orders. Some orders authorize actions to prohibit certain unfair trade practices in interstate commerce or to coordinate federal orders with those of individual states. Although controls over marketing vary among the states, those with enabling legislation generally do not attempt to control the volume of supplies going to market, but rather limit their activity to regulating quality, size, or packaging; controlling advertising and sales promotion; and supporting research and investigation. The federal legislation, however, is very comprehensive in terms of the commodities to which it applies and the marketing rules and regulations it covers. Fruits and vegetables for processing are only two of many exclusions requested by industry groups. Enforcing sanitary and other standards for consumer protection is not, of course, dependent on an agreement or order.

When marketing agreements and orders were first authorized in the Agricultural Adjustment Act of 1933, many producers and marketing firms looked upon them as temporary devices to boost farm prices and incomes in the Great Depression. Later, they were regarded more as aids in dealing with marketing problems through good as well as bad times. The Agricultural Marketing Agreement Act of 1937 put agreements and orders on a more permanent footing and served as the parent legislation for succeeding amendments. In the early years, control over volume was stressed, but in later years less than one third of the orders used volume control or provided for diversions to secondary markets. Some agreements have done little more than keep commodities of inferior quality off the market. Some California and Arizona citrus agreements simply allocate shipments by weeks throughout the shipping season, thus avoiding waste and spoilage and excessive bulges in market flows. However, these agreements and orders, as with most policies, continue to be highly controversial and likely to be vigorously debated in the future.

Evaluation of Orders for Fruits, Nuts, and Vegetables

Most orders involve some transfer of economic gain from consumers to producers, because the volume of a particular commodity is reduced below what would be sold in freely competitive markets. An order deliberately grants monopoly power to a private sector of the economy, but with strict limitations, including government and industry approval for each action that is taken. Generally, incomes of producers are improved to some degree in the short run, depending primarily on the elasticity of demand for the commodity. Orders are usually most applicable and popular among growers under the following circumstances: the quantity or quality of a perishable crop varies from season to season; production is highly localized; demand is relatively inelastic; production is not dominated by a few large firms that may combine as oligopolists or vertically integrate through the marketing chain; a strong cooperative has much influence on the marketing of the commodity, and growers want the protection of an order. Where surplus diversion is to be undertaken, a marketing order is sometimes helpful for attaining stability in prices.

A general study of two decades of experience with orders for fruits and vegetables was published in 1956,[2] and since then literally hundreds of studies of individual programs have been made. Orders have brought increased returns to growers primarily when demand is inelastic or when no close substitutes exist for the controlled commodity. Most orders are limited when there is more than one producing area, when a fruit or vegetable has one or more close substitutes, and when improvements in processing have made the processed food a closer substitute for the fresh. Accurate, quantitative evaluation is difficult because most orders do not stay in continuous operation for more than a few years. More important, few analyses provide a dependable basis for judging what prices and marketing conditions would be if no program were undertaken. Market conditions, demand, production, and timing vary greatly from year to year and even from season to season. Hence any attempt to relate prices and other factors to conditions that might have prevailed in the absence of an order is beset with great difficulties. A frequent contention has been that order programs engender greater certainty and tend to induce larger plantings than would occur otherwise, but the proof of this belief has not been developed with any consistency. To the extent that it is true, of course, consumers would gain. Producers might not be hurt; in fact, they also could gain over time through wider acceptance of their commodity and a larger market.

In addition, the future role of marketing agreements and orders must be related to questions of producer organization and market structure. An evaluation of whether an order is good or bad may depend on the effect it has on the structure of the market. Often an order favors marketing cooperatives. At issue is whether it helps smaller, independent firms to survive and whether it helps the family farm in competition with large-scale producers, some of whom are highly integrated in producing and marketing and for whom an order may be of little or no benefit.[3]

DAIRY MARKETING POLICY

The dairy industry has been highlighted throughout the history of U.S. price and income policy. Not only have dairy products received more public intervention than any other commodity, but the industry has continually been the center of controversy. It promises to remain so because of several unique characteristics of the commodity, its distribution, and consumption.

Milk is universally recognized as inherently nutritious, vital to human health, highly perishable in its fluid form, and also very inelastic in demand. Milk is produced throughout the nation, although with widely differing efficiencies and, until recent decades, has been difficult to mechanize and transport beyond local milksheds. It has never entered much into commercial exports, but has been characterized by strong farmer organizational support. Furthermore, dairy production involves an unusually high investment in fixed assets in plant and equipment that have little salvage value. As a result of these characteristics, dairy problems have remained on the forefront of the policy agenda.

Important policy issues relating to dairy technology have emerged as consumers have become more sensitive to health and food safety issues such as irradiation technology for milk preservation, fat intake and the amount of butterfat in milk, and production-stimulating hormones such as bovine somatotropin (bST). These developments have not escaped the attention of dairy producers and processors because they have dramatic effects on efficiency, market prices, and economic structure. Fundamentally, producers want protection from erratic, unprofitable markets; consumers want a secure, stable, safe, low cost supply of dairy products; and the marketing industry prefers minimal regulations.

Dairy policy involved one of the earliest and most comprehensive systems of federal and state marketing orders, which are primarily voluntary, yet result in mandated pricing and market segmentation. The industry's manufactured products have been covered by continuous price-supporting policy, but generally have not had effective production and marketing control. The dairy industry has intermittently experienced compensatory payments, as during World War II, and voluntary production control, as during a short-lived, whole-herd buy-out program in 1986 to 1987. Finally, for decades milk, cheese, and dried milk have been at the center of an ambitious food-aid effort in the school and commodities distribution programs. Overshadowing all these issues are the high costs to the U.S. Treasury and difficult administration of surplus storage and distribution programs, which have continued to plague national dairy policy.

The case for dairy policy is unique, however, because there are two distinct types of products, fluid milk and manufactured dairy products, which have quite different economic characteristics. Demand for fluid milk is very inelastic. The product is seasonal and perishable and comes to market in a rush. Producers are purely competitive, supply of milk is inelastic, and storage costs are high. Consequently, in competitive markets, without government regulation of the quantity supplied, prices would be very seasonal and unstable. In addition, fluid milk markets differ from markets for the manufactured dairy products in being dominated by large city markets with transportation costs largely determining the fluid milk marketing areas. In contrast, individual markets for the manufactured dairy products are literally nationwide, and the demands for butter, cheese, yogurt, ice cream, and others are more elastic, because each product has at least one close substitute among the other dairy products, or other substitute products. For example, butter has several close dairy or nondairy substitutes. Yogurt is a close substitute for cottage cheese. Frozen yogurt is a close substitute for ice cream, and so on.

Goals of Dairy Programs

Before the federal government set up dairy programs in 1933, large cooperatives and some state governments had tried to perform stabilization functions, but they met with only partial success. So, when the Agricultural Adjustment Act was passed in 1933, cooperatives and dairy producers asked for a program of federal marketing orders to regulate the flow of fluid milk to market. The policy goals that have developed are intended to assure a reliable supply of fluid milk at all times and places and to stabilize both farm and retail prices so as to provide stable incomes to farmers and maintain a smoothly functioning market. Complete stabilization of prices is not the objective, however, because this would create seasonal and locational shortages and surpluses of fluid milk.

Milk Marketing Orders[4]

In the United States, each major milk market, which draws supplies from an adjacent area called a *milkshed,* is covered by either a federal or a state milk-marketing order. The principal objective and function of an order is to convert the entire milk supply of the market area into a pool within which milk is allocated to alternative uses, yet the price to producers is a uniform, pool-wide average adjusted according to distance from market. About two thirds of all milk is delivered to processing plants under federal orders; most of the balance is under state orders, which are coordinated with federal programs.

These orders are therefore a way to bring distribution by all milk handlers in a market under uniform control, place all handlers under uniform competitive positions in regard to the minimum price for milk, provide for uniform participation by producers, and overcome much of the variability of fluid milk markets.

Terms of Orders The terms of each order are developed through public hearings, and the role of the government is to hear arguments and proposals, to evaluate and resolve differences, and to enforce orders as promulgated under the authority of the secretary of agriculture in the case of a federal order, or under authority of the government of a state in the case of a state order. In regard to prices, order programs establish two conditions. One is the minimum price that regulated handlers in each market must pay for milk, and the other is the price that producers are entitled to receive for their milk. The first is based on a classified pricing system that regulates the price of milk according to the end use or ''class.'' The second is based on an average price or ''blend,'' in which all producers in each specifically designed area or zone who meet the sanitary requirements for fluid grade milk receive the same price, regardless of the end use of that milk.

Classes of Milk The federal-order classified pricing system establishes different prices, depending on how the handlers use the milk. Three classes of milk are identified in most federal-order markets, although a few have only two classes. Class I usually includes all fluid milk products, which are comprised of whole milk, skim milk, low-fat milk, milk drinks, buttermilk, filled milk, milk shakes, and ice-milk mixes. Class II usually includes cream, sour cream, eggnog, yogurt, cottage cheese, frozen desserts, custards, puddings, pancake mixes, and some feeding formulas. Class III typically consists of cheese other than cottage cheese, butter, powdered milk, evaporated or condensed milk, and milk products disposed of in animal feed.

Grades of Milk Both producers and handlers are subject to government inspection, and milk is graded according to the sanitary standards that are maintained. Grade A or ''fluid-grade'' milk is that which meets all of the sanitation requirements of the area for milk to be used for fluid consumption. Federal-order regulations apply only to Grade A milk, which is the only grade permitted to be sold for Class I use. Grade B or ''manufactured-grade'' milk, which does not meet the sanitation requirements for Grade A milk, is used solely for manufactured products and never for fluid products.

Class I Prices The minimum price for Class I milk is established in most markets according to a formula based on what is called the ''Minnesota–Wisconsin price,'' or the ''MW price,'' plus a ''fluid or Class I differential.'' This naturally grew out of the dominant role that these states had played historically in milk manufacturing. The MW price for a given month is the amount paid for Grade B milk by a number of manufacturing plants in Minnesota and Wisconsin in the second month preceding the given month. About 100 of these plants are surveyed each month by the U.S. Department of Agriculture, and an audit is made of the prices they reported paying for milk of manufacturing grade. The MW price is selected as a good indication of the value of milk used for Class III milk. The Class II price in most order markets is the Class III price pus 10 cents per hundredweight.

The minimum price for Class I milk is established in each order as the MW price for the second preceding month plus a ''fluid differential'' specified as a fixed amount in dollars and cents. This differential varies from market to market, generally increasing with the distance of the market center from the Minnesota–Wisconsin market area, or milkshed. In addition, in markets that cover a wide area, concentric zones radiating out from the market center are established. Starting at the center, the processing plants are authorized to reduce their Class I price by a few cents per hundredweight from one zone to the next. Established to offset the costs of transporting fluid milk, such zoning is required for reasons of market efficiency. As a result, the percentage of milk going into Class I use decreases as the concentric zones radiate out from the market center. Zoning also makes it necessary to have a specific formula for determining the blend price to be paid to producers in each zone.

Determining the Blend Price The blend price to be paid to a producer or to a producers' cooperative is a weighted average of all the milk sold in the order marketing area, adjusted by zones as discussed above. All of the milk covered by the order is said to be ''pooled,'' and the market administrator calculates the amount and value of pooled milk going into each class, from which is derived the weighted average value per hundredweight. Each handler is required to pay each producer or each cooperative the average or blend price adjusted for location or zone.

In addition, all regulated handlers in a given order pay money into or draw money out of ''the pool,'' or ''the producers' settlement fund,'' depending on the distribution or sale of their milk, according to use. Most markets operate with a marketwide pool, although a few have been authorized to establish pools of individual handlers in which case the blend price to be paid to producers is calculated for each handler separately. Producers shipping to different handlers receive different blend prices if these handlers have different utilization percentages, regardless of location. Each handler pays the producers the full amount of the obligation only.

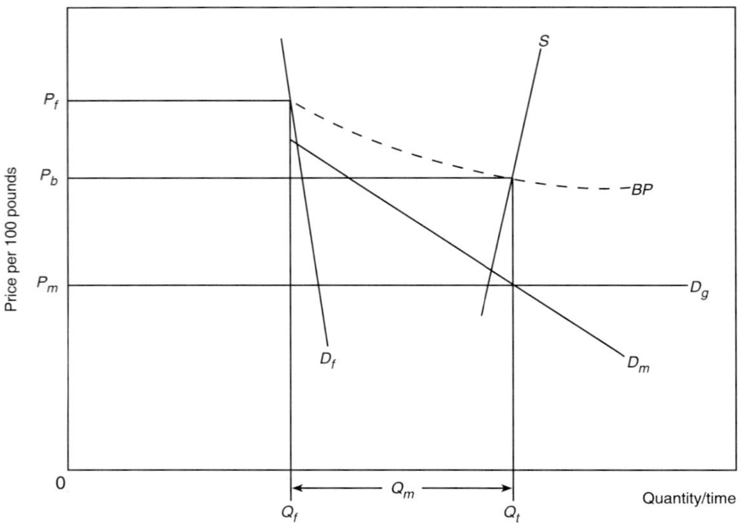

Where: S = total milk supply
D_f = demand for fluid grade milk
D_m = demand for manufacturing milk, as measured from O_f
BP = blend price, with P_b on the vertical axis being the equilibrium price
Note: It must be emphasized that BP is not a demand curve.
P_f = Class I price in federal order markets
P_m = combined Class II and Class III prices, as well as the assumed federal price-support level
$D_m + D_g$ = combined demand for manufacturing milk plus government removal programs, as measured from the vertical line O_f
Q_f, Q_m, Q_t = quantities of fluid, manufacturing, and total milk, respectively.

FIGURE 7-2
A model of classified milk pricing. Theoretical.

A Model of Classified Milk Pricing In Figure 7-2, a federal or state milk-order market is in equilibrium when the blend price is P_b, the Class I price is P_f, and the government's price-support level for the milk that is used for manufactured products is P_m. As long as the supply fluctuates around the average supply curve, or supply and demand move in the same direction and amount over time, prices can be maintained at these levels. But if technological advance increases supply and there is not an equivalent increase in demand, either the government will accumulate surpluses, or price supports must be lowered.

In Figure 7-2, if the price for fluid milk is raised to P_f or above, other things remaining equal, then the widening of the Class I differential will create an unnecessary surplus of fluid milk. An unnecessary surplus is defined as the amount of milk left after deducting from total milk receipts (1) the Class I sales, (2) a reserve of 20 percent, and (3) the seasonal surplus. Typically, in federal-order markets, only a little more than half of the producer deliveries of milk are used in Class I. If the differential is wider than necessary, this will increase the amount of Grade A milk that must be diverted to manufactured products.

Under such conditions, if the level of price support is above P_m (Figure 7-2), then this will be reflected in increases in government stocks. Eventually, these stocks may spoil, be sold in domestic or export markets at a loss, or be given away in aid programs.

Price Supporting in the Dairy Industry

Dairy products, other than fluid milk, have had a price support program since the 1930s. The secretary of agriculture is mandated to use the CCC to buy sufficient quantities of dairy products to keep milk prices at support levels and to dispose of them outside regular market channels. This mandate may involve exporting dairy products at world market prices under authority of P.L. 480, diverting dairy products to school lunch and other food-aid programs at prices below price-support levels, distributing dairy products to U.S. military forces or veterans' hospitals, and using dairy products for emergency relief.

The price supports serve as a lever to raise and stabilize prices for all dairy products. Two important costs are involved in the process. One is the government cost of storage and disposal of products at less than support levels, which in some years has been as much as $4 billion. The other is the added cost to consumers of prices that are higher than necessary to produce the quantity of milk that satisfies the consumers' demands. This added consumer cost—if any—varies constantly, largely because of fluctuations in supply. However, the net consumer cost over several years is uncertain because the greater price and income certainty created by the program tends to be a powerful factor in helping to create the rapid technological advance that is characteristic of the dairy industry. If the program has encouraged a high rate of technological advance in the dairy industry, the benefits to consumers are more real than apparent. The distribution of surplus products to food-aid programs is also a benefit for some consumers and should therefore not be overlooked.

Political Forces and Pressures in Dairy Policy

Political action committees (PACs) associated with the dairy industry are among the most politically powerful of all PACs operating at the national level and they typically give strong support to the dairy price-support program. Also, they generally are very large contributors to the election campaigns of candidates for Congress. Prominent supporters of these PACs are: Associated Milk Producers, Inc., headquartered in San Antonio, Texas; Mid-America Dairymen of Springfield, Missouri; and Dairymen, Inc., of Louisville, Kentucky. Increased political activity of these PACs has brought protests from consumer groups such as the Consumer Federation of America, National Consumers Congress (which merged in 1977 with the National Consumers League), Congress Watch, and the general citizen organization, Common Cause. In addition, labor union activity in dairy marketing has sometimes been accused of raising costs through higher negotiated delivery charges and ''make-work'' restrictive work rules. Finally, the federal government has been under continuing pressure to use tariffs and quotas to limit imports of dairy products.

Implications for Future Dairy Policy

Classified milk pricing and marketing in the United States, as in some other countries, continues to be controversial and troublesome for policy makers because of concerns about restrictions on producers, effects on consumer prices, and possible waste. The dairy policy survives, however, because free-market pricing of fluid milk is highly unstable due to very inelastic demands for milk, high storage costs, and very inelastic but highly variable supply from season to season within a year. Although political compromises have been difficult, an optimum economic solution for classified pricing may be some modification of the government's price-support levels toward equilibrium market levels, with restrictions on the width of the Class I price differential. Experience has shown that failure to limit this differential and the price supports on the manufactured products results in excessively large surpluses, distortions of incentives to produce, restriction of consumption of both fluid milk and manufactured products, and large federal costs.

Dairy policy is continuing to adapt to the increasing concentration in dairy production and to the small number of very large processing and marketing cooperatives. Pricing of products is not likely to revert to a free-market model for fear of handing over oligopsony–oligopoly power to the large marketing firms for the nation's vital milk supply. This does not appear to be a viable alternative for either producers or consumers. Hence the problem will be to find a policy compatible with the level of concentration that exists.

LOOKING AHEAD

The price and income policy alternative of compulsory production or marketing control focuses on restricting supply. In general, it facilitates the transfer of income from consumers to producers and relies on mandated performance of producers and interventions in trade. Although this policy alternative was important for the early decades of U.S. price and income policy and is still used for some commodities, it has been largely replaced by combinations of various voluntary approaches, more representative of the preference patterns of producers and the general citizenry alike.

The viable options for future price and income policy depend on many economic and political considerations, as well as on analysis of other policy options. Chapter 8 continues this exploration of alternatives involving an analysis of voluntary production control and compensatory policies.

ENDNOTES

1 For further discussion, see Don F. Hadwiger and Ross B. Talbot, *Pressures and Protests: The Kennedy Farm Program and the Wheat Referendum of 1963,* Chandler Publishing Co., San Francisco, 1965.

2 Murray R. Benedict and Oscar C. Stine, *The Agricultural Commodity Programs: Two Decades of Experience,* The Twentieth Century Fund, New York, 1956, pp. 380–415.

3 For further discussion, see Harold F. Breimyer, *Individual Freedom and the Economic Organization of Agriculture,* University of Illinois Press, Urbana, 1965, pp. 166–169 and 293–295.

4 This section draws in part on Ronald E. Deiter, "U.S. Dairy Cooperative Service Behavior: An Overview of the North Central Region and a Case Study of the Chicago Market," unpublished Ph.D. dissertation, University of Illinois, October, 1979.

TERMS AND CONCEPTS

Compulsory production or marketing control
Acreage allotment
Marketing quota
Producer referendum
Nonrecourse loan
AAA of 1933
Soil Conservation and Domestic Allotment Act of 1936
AAA of 1938
ASCS committee
Marketing agreements
Marketing orders
Classes of milk
Political action committees (PACs)

QUESTIONS AND STUDY SUGGESTIONS

1 Define the price and income policy alternative of compulsory production or market control.
2 What are the assumptions or necessary conditions for a compulsory approach?
3 Characterize the economic rationales for the compulsory policy. Is there also a distinct political rationale?
4 How is the compulsory production control approach related to the price elasticity of demand of a commodity?
5 What are the necessary steps usually required to implement a marketing quota?
6 What are economic consequences of a compulsory policy approach in a closed economy for producers? For consumers? For budgets and taxpayers?
7 What kinds of problems are created in the realm of international trade by the compulsory approach?
8 For what kinds of commodities is the compulsory production control best suited? Least suited?
9 Why are the marketing agreements and orders more commonly found in the fruits, vegetables, and nuts industries?
10 What has been the U.S. dairy policy? How is *milkshed* defined in dairy marketing? How is a milk-marketing order established? Who sets the price-support level for butter and dried milk?
11 Can milk price supports serve as a lever to raise and stabilize prices for all dairy products? If not, why not? If so, how is this done?
12 What are the economic consequences of the dairy price-support program?
13 How can the federal cost of dairy price-support policy be reduced?

CHAPTER **8**

VOLUNTARY PRODUCTION CONTROL AND COMPENSATORY PAYMENTS

So far we have examined the policy alternative of price supports through production and marketing controls. Now we explore a voluntary approach to production control and income enhancement. This approach has been the foundation of U.S. price and income policy for the past three decades and is emerging in the European Community and some other developed nations. Even though the two intervention techniques of voluntary production control and compensatory payments are commonly combined in the overall U.S. policy, we will consider them separately as well as in combination to highlight the distinctiveness of some of their economic relationships.

IN THIS CHAPTER

By studying Chapter 8, you will learn how to answer the following questions:

1. How can the program techniques of voluntary production control and compensatory payments be combined into a consistent policy approach? Why have these two programs become more prevalent in the policy of developed countries?
2. What are the primary characteristics and economic relationships of a voluntary production control and compensatory policy alternative?
3. What are the general economic consequences to be expected from such a policy?
4. What have been the experiences of the United States and other selected countries with the voluntary production control and compensatory payment policy approach?

BASIC ASSUMPTIONS AND CHARACTERISTICS

Defining the Policy Alternative and Program Techniques

Voluntary production control and compensatory payments can be defined as a policy designed to increase the total returns of producers, through production containment by participants who respond voluntarily to an announced total return, partly from the market but supplemented by a governmental compensating payment. In this policy, compensatory payments are used as an inducement to get voluntary compliance with a program to control production. This policy might be chosen to respond to problems of instable or low product prices, low producer incomes, aversion to compulsory controls, needed resource adjustments, and low-income consumers.

Both the compulsory and voluntary approaches restrict production, but the latter rests on voluntary instead of mandated action, relies more on market allocative forces, and supplements market returns with Treasury payments instead of relying on market price for enhancing producer incomes. Payments or other financial inducements for participating in voluntary production control can take many different forms.

Compensatory payments alone can be defined as a policy technique to increase returns to producers, without raising prices to consumers, by use of a preproduction announced total targeted return, usually per unit of commodity, when producers comply with specified program provisions. The return is assured by means of appropriate advance or final payments at the end of a marketing period. Unlike other production control programs, this one relies on the market to allocate resources and distribute the products once the target return is set. The program looks to the Treasury rather than to consumers for income enhancement. It has been variously referred to as a program of income, production, producer, subsidy, direct, deficiency, and price-support payments. In reality, it is also a consumer subsidy, because it permits a higher level of total production, with associated lower consumer prices, while still assuring a targeted level of total net farm return.

Assumptions, Necessary Conditions, and Procedures

Voluntary Production Control The basic assumptions for voluntary production control are threefold: the demand for the commodity is inelastic, the existing market price or return is unsatisfactory for policy makers, and the voluntary policy approach is preferred over the compulsory approach.

At least two conditions are necessary for the satisfactory implementation of a voluntary land retirement policy:

1 It must be technically and administratively feasible to confirm accurately the compliance of participants, for example by field checks or aerial photographs, in order to grant the appropriate payments or loan incentives. The technique is less feasible for livestock products. It would be possible and perhaps more efficient to implement voluntary production control by restricting or banning certain pesticides, by taxing other inputs such as fertilizer, or by rationing fuel for tractors. But these measures are not as politically acceptable nor as administratively feasible as restricting acreage.

2 There must be enough public commitment to fund the substantial and generally rising costs of the incentives required to achieve the goals of a voluntary production control policy. Because one of the purposes is to raise the market price above the free-market price level, a move that tends to encourage more output, the government may have to increase repeatedly its outlay to maintain necessary compliance.

When these conditions are met, procedures to implement a voluntary production control program are undertaken in several stages:

1 The public policy process determines that a particular market price and associated total farm income level are more desirable than the free-market situation.

2 For normal growing conditions, an estimate is made for the number of acres needed to produce the total commodity that will clear the market at the desired market price. From this estimate is derived a uniform target percentage of cropland to be withdrawn from the national average acreage for the program commodity under a free market. Then the payments or other inducements are determined that are required to assure both the participation and desired reduction (acreage set-aside, such as 10 or 20 percent) in acreage planted.

3 Each farmer with a registered record of production is informed of the acreage base assigned to the farm and the required set-aside acreage.

4 Farmers notify the administering agency, recently the Agricultural Stabilization and Conservation Service (ASCS), of their intention to participate voluntarily in that year's program provisions, and compliance of the farmers is checked after they plant the program crop.

5 Production and marketing decisions are then made at the private firm and industry levels, and payments or other incentives are made.

6 In the case of major export commodities, if prices are raised above world levels, export subsidies must be applied to maintain export volume, and imports must be restricted to protect the higher domestic prices of program commodities.

7 Subsequently, the administering agency evaluates the market prices and farmer income levels and adjusts the program provisions for the following year.

The bases established for individual farms usually are the average acreage planted or harvested for a specified time period, and they may be fixed or a moving average. Each farm will also have a farm yield base, which is an average of the varying land productivities of the farm, but which is also consistent with the yield averages of the township and county. Participants are most likely to be those farmers who have particular resources, goals, and management characteristics that make compliance beneficial economically. Farmers who lack those characteristics usually do not participate.

The voluntary production control approach has several economic rationales. Because most farm commodities have an inelastic demand, at least in the domestic market, a reduction in output for this market implies increased total returns to producers. Furthermore, the returns from deficiency or compensatory payments can augment these improved market returns. Whether producers choose to participate in the program or not, they are able to allocate their committed fixed and variable inputs to maximize net returns under the higher price threshold. The higher market price, the increased total producer returns, and the most efficient resource allocation and management (given the policy objectives) provide the economic rationale for this policy.

There are also political rationales for a policy of voluntary production control. Voluntary land retirement permits restricted production with higher prices and incomes to all producers without directly compelling any farmer to make specified production or marketing decisions. Although the incentive package may be attractive enough to make participation the more profitable choice for a large majority of farmers, participation is still voluntary, just as is the choice of a farmer responding to any free-market price signal.

Furthermore, this flexibility of choice can appeal to motives other than economic ones, such as being part of a joint effort to solve a common problem, or supporting or rejecting governmental programs for ideological reasons. Finally, the voluntary approach is a way of expressing the public's changing preferences to conserve some land resources when demand conditions are less favorable, but later to expand production in response to stronger domestic or foreign markets.

Various adaptations of voluntary land retirement are possible. When combined with temporary price supports through nonrecourse loans, this policy approach can smooth out the market prices both within and among marketing years. The entire incentive package to induce sufficient voluntary participation can be varied to include price supports, land diversion payments, deficiency payments, flexible set-aside options, disaster payments, and access to longer term conservation reserve programs and grain reserves. In addition, the approach can permit haying or grazing set-aside land and offer cost-sharing payments for planting cover crops on the retired land. Finally, limits can be placed on payments per operator or per farm to tilt benefits toward some targeted producer group or to control Treasury costs.

Problem of Slippage with Voluntary Approach ''Slippage'' is a typical problem that confronts policy makers and administrators of a voluntary land retirement program. Because participation is voluntary rather than compulsory, some farmers choose not to comply with the program. Those who do participate generally opt to take their poorest land out of production. Furthermore, farmers with yield bases perceived to be high are more likely to enter into the land retirement program than those who perceive their bases to be low. Finally, prospective higher market prices will encourage both participants and nonparticipants to use more yield-increasing inputs and practices on the land left in production. Because of slippage, a more ambitious and costly set of inducements may be required to achieve the original production control target. In operation, this condition often calls for a further lowering of the national base in successive years as the initial reduction fails to meet the program objectives.

Compensatory Payments A policy technique of compensatory payments rests on several basic assumptions: (1) the market prices or returns are unsatisfactory for the policy-making public; (2) consumers' desire for more product at lower price is as important as producers' desire for higher income; (3) the Treasury is a more desirable source of farm assistance than is a higher consumer price; (4) access to export markets is important, and (5) minimal interference in market forces that allocate resources and distribute incomes is a preferred policy goal. Because this approach combines enhanced farm income with low food costs, and because its total cost is therefore not fully apparent in the market, it has unique appeal to both agricultural producers and consumers.

The compensatory payments policy approach can be adapted to various policy objectives such as stabilizing farm income, increasing or decreasing production, shifting resources to other preferred uses, keeping commodity prices low to enhance exports, and reducing consumer food costs. Thus it was used frequently in the United States during World War II to encourage production of high-protein food commodities, has been a major feature of the national wool program, and, in recent years, has become increasingly a part of the omnibus price and income Acts. A compensatory payments approach can be found in other countries. For instance, it was the mainstay of the United Kingdom's post-war commodity policies before the UK entered into the EC's Common Agriculture Policy (CAP), and it has emerged as a major program in EC's reform of its CAP.

Three conditions are necessary for effective implementation of a compensatory payments program:

1 It must be administratively feasible to identify the quantity of a commodity for which compensatory payments are to be made and to assure adherence to any prerequisite compliance provisions.

2 A sufficient public commitment must exist to provide the considerable magnitude of funds commonly associated with such a policy. The commitment may be related to the relative affluence of the taxpaying public and the size of the farm population.

3 The demand for the program commodity cannot be so inelastic or its supply so elastic that production responses or Treasury liabilities increase to the point at which the approach becomes impracticable.

When these conditions are met, procedures to implement a compensatory payments program are undertaken in the following stages:

1 The public policy process deems as unsatisfactory the market-determined total economic returns to farm producers, the level of farm production, the pattern of farm resource management and use, or the quantity and price of domestically produced agricultural goods available to consumers.

2 Before the beginning of a production year or period, a desired total return is established for farm producers, usually on the basis of a unit of a commodity.

3 During the production and marketing year or cycle, the market supply and demand forces are free to set prices, allocate resources in production by producers, and distribute commodities to consumers.

4 Either export or import policies are adjusted to accommodate any undesirable trade disruptions or retaliations associated with the compensatory payments approach. Alternatively, the supply could be modified through farm production controls.

5 At the end of a specified marketing period, the national average market price is determined, and any deficiency in these returns per unit of commodity below the target level will be compensated by payments made directly to producers on the basis of the quantity verified by market receipts.

There are economic, political, and social rationales for the compensatory payments approach. The approach is proposed and sometimes adopted precisely because it can be adapted to different commodities, economic conditions, and objectives. Given the program incentives, it leaves in place the final determination of prices to supply and demand forces. The policy also permits producer discretion in allocating resources and consumer discretion in purchasing agricultural commodities. Distinct from some other policy alternatives, this approach can be applied to perishable commodities such as fruits, vegetables, and livestock products, as well as to the traditional program commodities, most of which are storable. Finally, its reliance on market distribution can prevent problems of surplus stocks, which tend to occur with most commodities under any price-support or production-control program.

A compensatory payments policy alone usually involves increased quantities of food and decreased price, a situation that provides attractive political and social rationales. For example, some people in the food and agricultural industry, who are concerned about undernourishment and malnourishment among lower income consumers, realize the United States has enough resources to supply more food. Even among farm groups, some prefer to unleash the productive capacity of the food and agricultural industry so that it may supply more food and other commodities for consumers, both domestic and foreign.

ECONOMIC MODELS OF THE TWO POLICY APPROACHES

Voluntary Production Control

In Figure 8-1, the economic relationships implicit in a "pure" policy of voluntary land retirement are presented at both the agricultural industry and the farm firm levels. At the industry level, P_e is the equilibrium price and Q_e is the equilibrium quantity produced under free-market assumptions. With voluntary land retirement and P_{lr} as the targeted market price, Q_{lr} becomes the production that will achieve this price. Although a land retirement program might be formulated to reduce supply to S_{lr}, actually because of slippage the supply curve moves back only to S_{lrs}. As a result, market equilibrium is established with an aggregate production of Q_{lrs} and a price of P_{lrs}, which are short of the program goals. Thus, in initial programming or in subsequent years, enough additional acreage must be retired to move the supply to S_{lr}, quantity produced to Q_{lr}, and market price to P_{lr}.

At the farm firm level, the voluntary nature of this policy produces a dichotomy of responses. Initially, in Figure 8-1, before the program is adopted, it is assumed that the typical firm is in equilibrium at P_e and Q_e, with the intersection of MC and AC being the same as P_e. A participating firm that reduces production through land retirement will have an output equal to about Q_{lr}, the intersection of MC_{lr} and AC_{lr}. The exact production level for each participant depends on how much above the market price each participant expects the total return per unit to be. A nonparticipating firm that responds to the higher price will add enough other inputs to increase its output to Q_{nlr}, which helps account for the slippage observed at the industry level.

At the new equilibrium price, some participating farmers will use more inputs per acre on the acreage that is left in crops, thus producing at a higher average cost.

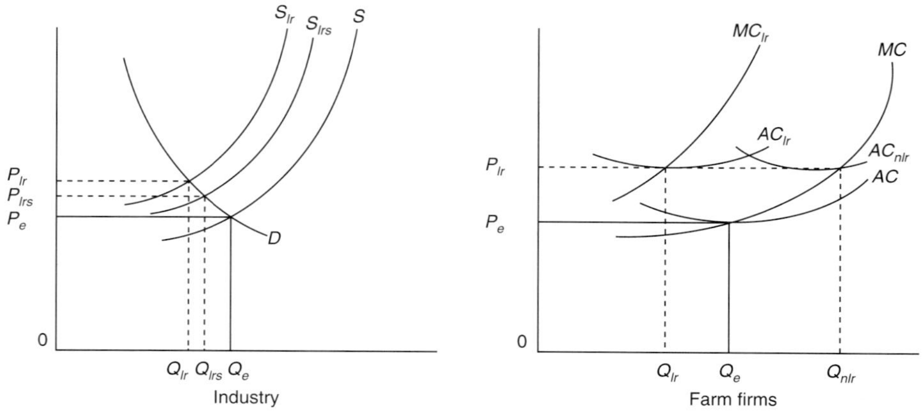

FIGURE 8-1
Voluntary production control through land retirement. For both industry and farm firms, P_e is the equilibrium price and Q_e the quantity. With effective voluntary production control, Q_{lr} is the reduced production and P_{lr} is the desired market price. Among farm firms, however, nonparticipant producers operate at Q_{nlr}. Hence, because of supply "slippage" peculiar to this approach, only successive levels of S_{lrs} may be achieved, short of the policy goal, before sufficient voluntary reduction is induced.

Without land retirement:
- P_e = equilibrium price
- Q_e = equilibrium quantity produced

With land retirement:
- P_{lr} = price with no slippage
- Q_{lr} = quantity produced with land retirement (no slippage)
- P_{lrs} = price with land retirement and slippage
- Q_{lrs} = quantity produced with land retirement and slippage
- Q_{nlr} = quantity produced by a firm that does not join the program

Nonparticipating producers will not only plant all of their land that is available for crops, but will increase some other inputs and produce also at a higher cost and on a scale exceeding the free-market output. Both producer groups are maximizing their economic returns, given the policy provisions, but are producing at different levels of both fixed and variable inputs.

Compensatory Payments Without Production Control

In Figure 8-2, the economic relationships implicit in a "pure" policy alternative of compensatory payment without production control are presented at both the industry and the farm firm levels. At the industry level, P_e and Q_e represent the equilibrium price and the quantity produced under free-market assumptions. With compensatory payments, the average target return per unit for all producers is set by the policy process at TR_{cp} before the production period. Note, however, that this is an average return to all producers including their payments, not a price. Nor is it the same return for particular producers, each of whom will obtain a unique market price and hence a unique total return, depending on quality of product, marketing strategy, and other factors. The response of all producers to this policy signal is increased production to

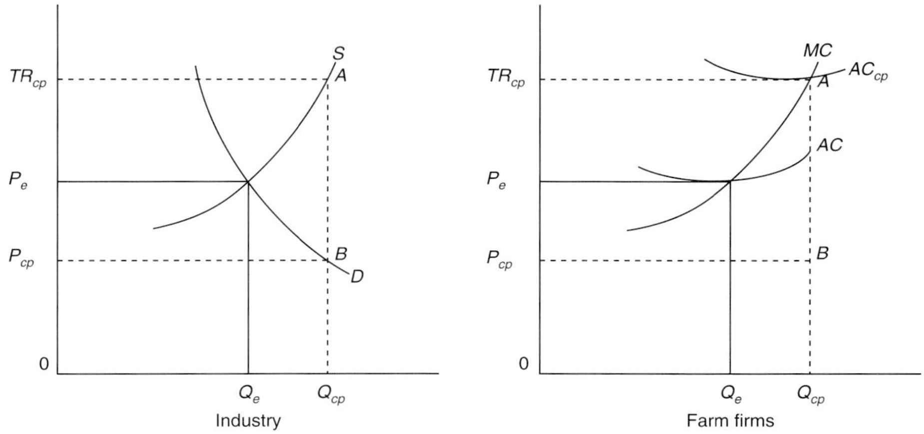

FIGURE 8-2
Income support through compensatory payments. For both the industry and farm firm, P_e is the equilibrium price and Q_e the quantity. With an effective compensatory payments program, given a targeted average total return of TR_{cp}, increased production at Q_{cp} will result in a market price drop to P_{cp}. Farm firms will add inputs to a higher cost threshold AC_{cp}. Total compensatory payments will be $TR_{cp}ABP_{cp}$, or at the rate of $A - B$ per unit.

Under pure competition with no policy:
P_e = equilibrium price
Q_e = equilibrium quantity produced
AC = average cost
MC = marginal cost

With compensatory payments policy:
TR_{cp} = target average return per unit
Q_{cp} = quantity produced
P_{cp} = average market price
$A - B$ = average compensatory payment per unit
$TR_{cp}ABP_{cp}$ = total compensatory payments

Q_{cp}, again somewhat indeterminate because of the individual responses of producers to an average return signal rather than to a price signal. With the increased production, the market price declines to P_{cp}, where consumers now obtain more of the commodity at lower prices.

When the marketing period is complete, program administrators identify this average market price, calculate the difference between that price P_{cp} and the target return TR_{cp}. All producers (or the proportional percentage) are then paid $TR_{cp} - P_{cp}$ for each unit of commodity marketed as determined by records of market receipts, or a total payment of $TR_{cp}\ ABP_{cp}$, clearly yielding higher gross income than before. Farmers expand production along the marginal cost curve MC until expected returns are maximized on the higher cost level AC_{cp} at A. Again, the production response Q_{cp} is indeterminate, emanating from the signal of a natural average total return, not a price, and each producer's price P_{cp} is unique as well. As a result, the national average payment $TR_{cp} - P_{cp}$ will achieve the objective of TR_{cp}, but that uniform payment will not ensure TR_{cp} to every producer, some exceeding the target and some falling short of it. This policy technique preserves the discipline of the market so that each producer can strive, through production and marketing strategies, for maximum economic returns. In general, the consequences of a compensatory payments policy approach in a short-run planning horizon are higher producer incomes, larger quantity at lower prices for consumers, high Treasury or taxpayer costs, and effects on trade that could require some export controls to prevent excessive disruption of world markets.

Important implications flow from these economic relationships of a compensatory payments policy. The price elasticities of supply and demand for the relevant commodities significantly affect how the policy functions. For example, the more elastic the supply, the greater is the production response, and the more inelastic the demand for a commodity, the lower the price must drop for the market to clear. In fact, with highly elastic supply and very inelastic demand, almost the entire TR_{cp} would be governmental payments. The opposite is true with inelastic supply and elastic demand. Then, the compensatory payments would be minimal. Thus the commodity to be produced will help determine the feasibility of this policy and the need for other adaptations to operate a program efficiently.

It also follows from the economic relationships that this policy approach is most compatible with an objective of higher national self-sufficiency, as exemplified in the U.S. wool policy. The approach can also be used to entice producers to alter their performance by, for example, following management practices to reduce erosion or groundwater contamination. It can even provide incentives to set aside acreage and hence decrease production. It also follows that the policy can become very costly to the government and will attract and retain more resources in farming or ranching than would occur otherwise.

ECONOMIC CONSEQUENCES OF VOLUNTARY PRODUCTION CONTROL AND COMPENSATORY POLICY

Effects on Producers and Agribusiness Firms

Farm owners and tenants are the primary beneficiaries of a voluntary production control and compensatory payments policy. Because of inelastic demands, and given the program incentives of a higher aggregate gross income for producers, at as high or higher total net returns, higher capital values from capitalization of fixed inputs, such as land, will gradually erode the increased net returns. Furthermore, the compensatory incentives for participation may augment net farm income. Because the policy is voluntary, producers can choose to participate or not, thus providing opportunities for resource allocations and marketing strategies. If they participate, their only restriction is to set aside a specified portion of their cropland, and to follow required practices on both the land set aside and their cropping acreage. They are, however, now dependent on the public as well as the market for their returns. If the policy is applied selectively to certain commodities, such as corn but not to soybeans for example, patterns of resource use will be altered.

It should be emphasized that this combined policy embodies counteractive economic forces and possible consequences. The policy restricts production thereby raising market prices, but the compensatory payments, which augment market returns to producers, permit lower consumer prices than would occur otherwise with the level of total returns arising only from the market under compulsory controls. Furthermore, as with the policy presented in the previous chapter, the enhanced producer gains, most prominent in the short run, are gradually diminished by capitalized returns to the fixed factors of production, like land, over a longer period. The ultimate effect of these counter forces depends on the magnitude of the compensatory payments, the extent of the production control, and the elasticities of supply and demand.

Because many agribusiness input industries, and many rural communities thrive when the farming sector thrives, they too generally benefit from this policy. However, certain input suppliers, such as those who provide seed, fertilizers, and chemicals, depend directly on volume of acreage and may therefore be adversely affected if the set-aside percentage is substantial for a particular crop. Furthermore, marketing firms such as grain elevators and exporters, whose returns are closely related to the volume of such a product that is handled, also may be adversely affected by the restricted production.

Effects on Consumers

If an effective voluntary production control policy is implemented, consumers will pay higher prices for food as a result of the restricted quantity of agricultural commodities. The increases in consumers' expenditures and the government's incentive benefits for participating farmers are the two sources of economic gains for producers. However, consumers may obtain some longer term benefits from the more stable food supplies and more stable price levels resulting from this program. Any adverse effects on consumers can be moderated or eliminated by maintaining target prices at levels significantly higher than price-support loan rates, or market prices, and then substantially augmenting producers' returns with compensatory or deficiency payments.

Effects on Budgets and Taxpayers

Consumers and federal taxpayers are the source of the additional income flowing to producers. Taxpayers bear the burden for various compensation payments to farmers, CCC loan programs, export subsidies, and the costs of administration. In view of the inelastic demand and supply of most farm crops, the budget costs tend to be larger than costs under compulsory production control, but smaller than for a compensatory payments program without production control. Although the budget outlays to induce farmers to participate in a voluntary production control program are substantial, a number of tradeoffs are available. For instance, if the government chooses to severely restrict production, thereby increasing prices and incomes for participating farmers, then direct payments to farmers may be kept relatively low; however, any export trade will have to be subsidized heavily if markets are to clear. Alternatively, if target prices are kept high but production control is minimal, the main expense for the government will be deficiency payments to farmers. Adaptations that lower the Treasury burden by limiting payments to large farms may be counterproductive by reducing farmer participation in the program and hence weakening production control.

Effects on International Trade

The voluntary production control and compensatory payments policy adversely affects exports and would invite excessive imports without some import barriers. Unless export subsidies are added, exports will be discouraged by the reduced exportable supply and by higher domestic prices. Import restrictions would be necessary to prevent importers from taking advantage of the strengthened U.S. market. Note, however, the achieved purpose of the compensatory payment feature of this policy is to supplement farmers' market returns by payments so as to avoid even higher market prices. Although the adverse effect of this policy on trade is generally less than with compulsory production control, the loss of exports engendered by it is serious compared with compensatory payments without controls.

U.S. EXPERIENCE WITH THE POLICY ALTERNATIVE

Compensatory Payments Alone

A ''pure'' compensatory policy has periodically been discussed in policy deliberations, and it has also been adopted for specific commodities and times. It was the heart of a proposal known as a forward-price program, introduced in 1945 to help stabilize farm incomes against fluctuations of the business cycle.[1] With this plan, equilibrium-type prices under full employment conditions would be estimated and then set prior to production. If the economy operated at less than full employment, compensatory (income) payments would be made to cover the gap between that forward price and the expected lower market price. In 1949, a broader proposal using payments as part of a compensatory program and a stabilization program, known as the Brannan Plan, was advanced by Secretary of Agriculture Charles Brannan.[2] Essentially all products would be covered but with a limit on volume. Both the forward-price plan and the Brannan Plan were vigorously debated, but neither was adopted.

A direct application of the compensatory policy alone first appeared in the United States during World War II. Production needed to be increased for the war effort, but food prices were restrained under regulations of the U.S. Office of Price Administration (OPA). To induce rapid growth in production of high-protein foods, given the technology of the period, a compensatory payments policy was implemented for milk and pork producers. An announced payment per hundredweight was paid on all production in addition to the market price, which was under strict controls. This policy permitted the OPA ceiling prices to be held while still encouraging substantially more food production.

Soon after World War II, another major experiment with a compensatory payments policy was launched by the UK, which applied especially to England and Scotland. During the war, the British Isles had been threatened continually with trade interruption by enemy submarines. Traditionally, the UK had relied on cheap food sources from domestic production and foreign countries—the United States especially—which complemented the UK's need, as a major industrial exporter, to hold down the costs of its labor.

Commencing in 1947, a policy of compensatory payments was adopted and followed until the UK entered the European Community, which functioned under an almost diametrically opposite policy of price supports without production control. In the earlier years, at the annual reviews of the UK's agricultural economy and policy, the ministry of agriculture and UK farmer organizations examined relevant data and negotiated a target return for major crops and livestock. Production and marketing decisions were left to the market, with the necessary deficiency payments being made from the UK

Treasury to all producers on the basis of their production. The results were relatively favorable farmer incomes, a relatively low-cost food supply coming from domestic production and from continued imports, and increased agricultural production. This increase within the UK, along with production-encouraging policies of other European countries, helped the EC to shift from a major importing economy to a more food self-sufficient and major exporting economy. Again, it is worthy of note that this UK policy complemented objectives of increased production and low food prices.[3]

Under the National Wool Act of 1954, the United States has supported a classic compensatory payments policy, which has continued essentially unchanged since then. This policy was launched in response to sharply declining wool production, and partly in response to vivid memories of a wool shortage during World War II, when submarine warfare reduced imports from New Zealand and Australia. This policy, of particular interest in Congress to senators and representatives of the major wool growing states, was designed to encourage and sustain a domestic wool producing sector.

Under this policy, a target level (actually called a price-support level) for wool is announced for the wool marketing year. Production and marketing proceeds. Then, at the end of the marketing year, if the average wool price received by wool growers is below the target level, a compensatory payment is made. Wool varies widely in quality, however. In order to take this into account, the payment to each grower is not based on total pounds sold, but on the total value of all bona fide market receipts submitted by a grower. For example, if the target level is $1.82 a pound and the average national market price for wool in the marketing year is $1.40 a pound, then the payment to a grower would be 42 cents ÷ $1.40 = 0.30, or 30 percent, of market receipts. For instance, a grower with market receipts of $40,000 would receive a payment of $12,000.

The target levels in the wool program have been related to various standards such as parity, cost-of-production, and ad hoc determinations. The payment levels have ranged from 10 to 160 percent of market prices, depending on production and marketing conditions. But domestic wool consumption has declined as a result of both supply and demand factors, and U.S. production has continued to supply a decreasing proportion of total domestic consumption, now well below 50 percent. However, the domestic industry probably would be substantially smaller without the policy.

Voluntary Production Control and Compensatory Payments Policy Experience

The more general U.S. experience with voluntary production control and various compensation schemes dates back to the AAA of 1933. At that time, voluntary land retirement through participation in acreage allotment programs plus CCC loans for participants were the major commodity programs for farm price and income support. Throughout the rest of the 1930s, these voluntary programs were generally followed until legislative and judicial authority was obtained for mandated programs. Although some production control was achieved, technological advance continued apace, and both domestic and export demand was sluggish. By the time the United States entered World War II in 1941, stocks under CCC loans were burdensome; however, these surpluses unexpectedly became valuable in the war years.

Following the war, the 1950s were, in many respects, a repeat of the experience in the 1930s. Compulsory controls were used for cotton, tobacco, and peanuts, and sometimes for wheat and rice. In 1956, the Soil Bank was adopted as a voluntary program and implemented during the rest of that decade. It offered government payments on a per acre basis—modest indeed—to induce producers to take some of their cropland out of production, that is, "bank it" for future use, under two provisions. The land could be put in an *acreage reserve* on an annual basis or in a *conservation reserve* for a guaranteed annual rental of up to 10 years. Yet the effort was cautious and the payments offered were meager, so the peak of land retired was barely 30 million acres, which was not enough to reduce the growing government "surpluses." Moreover, much of the retired land was marginal for production.

By 1960 price-support programs had reached a crisis stage because the large stocks were unmanageable and economic market conditions for agriculture were deteriorating. Policy makers concluded that bolder action would be required to stabilize the farm sector. The Emergency Feed Grain Act of 1961, which was aimed at taking more land out of production voluntarily, was the first major response. In 1963, with the defeat of the referendum to install mandatory controls on wheat, voluntary land retirement induced by compensations was continued as the primary price and income policy for agriculture.

Thus, in every subsequent year through the 1990 Food, Agriculture, Conservation, and Trade Act, except for 1980 and 1981, an annual voluntary land retirement program, using payments as incentives, has been in effect for one or more of the following crops: feed grains, wheat, cotton, rice, and some peanuts. Tobacco alone continued under a compulsory production control program. The total diverted acreage ranged from a low of one million acres in 1977 to a high of 83 million acres in 1987. Program costs varied greatly with increasingly idled land during the 1960s and up until 1972. This period was followed by more limited programs during the prosperous, heady, export-exploding years of the early 1970s. Increasing amounts of land were again set aside in the late 1970s and most of the 1980s. The fluctuations have been a response to the degree of prosperity or depression in agriculture, and some changes have been related directly to the variability of exports.

The experience of one period is worthy of special note. The USDA administration was very reluctant to institute any substantial set-aside of land during 1981 and 1982, despite some deterioration of the market. The USDA finally reacted by opening up the new national grain reserve program to add large quantities of price-supported commodities, corn and wheat especially. Subsequently, this action was viewed as inconsistent with the stabilization objectives of the reserve program because it resulted in too large a stock of grain. In 1983, following two near-record crops and some slowing down in export growth, the farm financial crisis was so serious that an unprecedented package of incentives was offered. This involved payments-in-kind (PIK) out of the large stock under CCC loan to encourage a set-aside of up to 50 percent of a farm's cropland base. However, instead of just giving farmers a certain amount of grain in lieu of cash, the government issued negotiable certificates to farmers in the program, which they could trade for grain or cash. For a short time, some farmers were also able to gain through arbitrage (buying certificates in one market and selling them in another), which in this case was called "pick and roll."

Implications of Current Policy Approach

The U.S. voluntary production control and compensatory payments policy, continuing under the 1990 Act, was designed to enhance the economic welfare of farmers, generally without regard to their income group or tax bracket. Its provisions of acreage reduction and the accompanying incentives have been applied uniformly to all producers, large and small. Larger farmers have received larger absolute payments, but only proportional to their total land bases and set-asides. Smaller farmers have received smaller payments, but also proportional to their land base or set-aside. In other words, payments have been roughly proportional to market receipts, but slightly more equal. This means, however, that this policy cannot be defended as a program of special aid for small farmers or criticized for not significantly altering the highly skewed income distribution of farmers associated with the market system. Larger payments have been required to get larger farmers to participate in production control programs. This, of course, leaves problems of aid or help for the small farmer, or more equal redistribution of farm incomes, to be solved by other means.

In addition, for several years there have been discussions, and some legislation proposed, to limit deficiency payments to a fixed acreage or quota for a qualifying farm. This would ''decouple'' price or income support from current production, and hence provide less incentive for farmers to increase output beyond the fixed base. However, this idea has been viewed among farmers, or farm organizations, as a movement toward ''welfare'' payments instead of market adjustment, and it generally has been rejected by policy makers.

The ongoing, combined policy has periodically included voluntary land retirement, price supports through nonrecourse loans, target prices, diversion payments, deficiency payments, access to a national grains reserve, export subsidies, and a conservation reserve. In addition, there have been attempts to provide more flexibility in planting options, broader eligibility for disaster assistance, more domestic and foreign food aid, and stricter environmental provisions. As a result of these price and income programs, direct Treasury outlays have risen and fallen with the tides of crisis and prosperity. In real terms, the outlays were relatively high in the late 1950s and 1960s and low in the 1970s. In the next decade they grew again to a high in 1986, followed by a decline of about half in the late 1980s. Government stocks have also gyrated, peaking in the late 1950s and again in 1986.

In summary, through almost half a century of U.S. policy experience, public policy makers appear committed to providing some farm price and income assistance, some food aid to low income consumers, and some resource conservation. Farmers have preferred and the public has concurred in a voluntary ''carrot'' instead of a coercive ''stick'' approach. The financial burden has been carried primarily by the Treasury, or taxpayer, and minimally by the consumer.

National Grains Reserve

Some version of a grains reserve has permeated economic and political discussions since early civilizations. There is a compelling logic for societies to store food during times of plenty for survival during times of scarcity due to bad weather or any other type of catastrophe. An interest in such a policy for the United States mounted during the economic and drought-induced crises of the 1930s, often under the banner of buffer stocks or an ever-normal granary. The idea was widely popularized by Secretary of Agriculture Henry A. Wallace in the 1930s, when he proposed an ever-normal granary, which simply involved increasing stocks in good crop years if prices were low and reducing them in poor crop years if prices were high.

Buffer-stock programs are generally conceived as a means of stabilizing food and agricultural supply in the face of fluctuating crop production, which usually would help to stabilize major crop prices at the farm level and retail food prices as well.[4] By implication, such stabilization programs can provide gains for both producers and consumers within the framework of a market economy. In the case of producers, total revenue derived from a given total production will be larger over several years if the quantity supplied is stabilized, provided demand is more elastic at high prices than at low prices. That is, two crops of average size will sell for more than one large and one small one, other things being equal. Although few domestic food and agricultural demand curves fulfill this condition, some of them do, and this may be especially important for the export market. Stabilizing the export supply curve when crop harvests fluctuate should tend to increase importing countries' demand for these exports over several years, because they will be assured of a stable supply and price from the United States. Buffer stocks will provide insurance against extreme shortfalls in crop production that might otherwise force expensive adjustments on producers, and on domestic consumers and importing countries as well. Less price variability will help grain producers to reduce risk and produce more efficiently, primarily because they can estimate returns from production choices more accurately. More stable grain prices also will help stabilize production and prices in the livestock economy because livestock producers can make more accurate estimates of their costs. But, perhaps most important, more dependable grain supplies and stable prices in the United States will encourage the long-range development of commercial grain exports that both the government and grain producers want.

Finally, reducing the variability in retail food prices, in part by the use of buffer stocks, is a benefit of particular interest to consumers. The utility of a given total quantity supplied over several years should be higher under stable prices. Although consumers may gain from food-price instability from season to season over the course of a year, retail prices tend to move downward slowly, or not at all, and it may be practically impossible for consumers to gain from price instability from one year to the next.

However, regardless of many arguments mustered over many years for a recognized buffer-stocks or grains-reserve policy, the United States has usually left the issue to the market system, to agribusiness strategies, and to private decisions. Since public price and income policy was launched in 1929, the CCC has been involved intermittently in handling stocks of price-supported commodities. Yet the CCC stocks were often perceived as temporary and a sign of policy failure. Price supports and production control programs were thought by many to be incompatible with the realities of supply and demand in domestic and export markets.

The fact is that government-owned stocks, which were largely forfeitures by farmers with nonrecourse loans, grew rapidly to crisis levels during the 1930s and 1950s, subsided in the 1960s, and were wiped out by a massive well-timed Soviet Union purchase at favorable prices in the early 1970s. Since then, they have been moderated somewhat by policy changes. Generally, when these stocks have been low—usually because of strong demand or slow production—price variability has increased, and vice versa.

In 1972 to 1973, during a short-lived panic about world food scarcity, the Soviet Union's startling grain purchase—officially sanctioned by USDA administrators to eliminate the unintended government stocks, even with export subsidies—helped trigger policy debate and eventual inauguration of a new policy. For the first time in the history of U.S. policy, the Food and Agriculture Act of 1977 mandated the establishment of a stabilization-stocks, grains-reserve program, the farmer-owned reserve (FOR), administered by USDA's ASCS and financed by its CCC. Initially, the Act provided for a modest 300 to 700 million bushels of wheat during normal supply times as security for subsequent domestic needs or exports to either food-aid recipient countries or commercial buyers. A fundamental principle of the FOR is that farmers rather than government should own the reserve stocks. When supplies are plentiful and prices low, the FOR program offers incentives to farmers to place their grains in sealed storage for up to five years with an interest-free nonrecourse loan and storage payments. Farmers cannot dispose of those reserve stocks until permitted by USDA-prescribed release prices and procedures. Thus public policy sets the minimum and maximum reserve levels, eligible commodities, appropriate price ranges for entry and exit of stocks, and incentives and disincentives for the voluntary farmer decisions. Incentives have included loans, storage costs, and interest waivers; the disincentives have been the withdrawal of such benefits.

The FOR under the national grains reserve policy was reaffirmed and slightly expanded in the 1981, 1985, and 1990 Acts. It has continued to be controversial as a policy issue, praised by some producers, analysts, and policy makers and criticized by others. Yet it has had significant influence on commodity markets, farmer decisions, and subsequent policy.

A research study has concluded that the U.S. farmer-owned reserve (FOR) program, as an expression of a buffer stock policy, has raised grain prices, increased farm income, reduced liability from nonrecoverable government surpluses, raised total outlays for farm programs, and reduced price variability.[4] These relationships are dependent on a number of factors, some related to the supply–demand situation and some to the nature and stability of the reserve management rules. As was agreed at the 1975 World Food Congress in Rome, major agricultural nations are encouraged to contribute regularly to a world food reserve. Although much food aid has been forthcoming, this kind of buffer food effort is still at the discretion of independent nations, and food security is left primarily to individual country policies.

LOOKING AHEAD

In Chapter 9, we continue this discussion of alternative price and income policies by turning our attention to international trade issues. Chapter 9 (1) sets forth the basic principles of comparative advantage in international trade, which is the general basis for all voluntary trade transactions; (2) shows how to identify the mutual benefits from trade; and (3) considers policy alternatives for dealing with trade.

ENDNOTES

1 See Theodore W. Schultz, *Agriculture in an Unstable Economy,* McGraw-Hill, New York, 1945. Also, William H. Nicholls and D. Gale Johnson, ''The Farm Price Policy Awards, 1945, A Topical Digest of the Winning Essays,'' *Journal of Farm Economics,* 29(1): pp. 267–283, 1946; and D. Gale Johnson, *Forward Prices for Agriculture,* The University of Chicago Press, Chicago, 1947.

2 For a discussion of the Brannan Plan as originally proposed, see articles by D. Gale Johnson, W. E. Hendrix, Harold G. Halcrow, and Roy E. Huffman, *Journal of Farm Economics,* 31(3): pp. 487–519, 1949. For a shorter summary, see Harold G. Halcrow, *Food Policy for America,* McGraw-Hill, New York, 1977, pp. 166–167.

3 *Fifty Years of the National Food Survey,* 1940–1990. The proceedings of a symposium held in London, December 1990, edited by J. M. Slater. Printed in the United Kingdom for HMSO, Crown Copyright 1991.

4 Excerpts taken from Meyers, W. H., et al., *The Farmer-Owned Reserve After Eight Years: A Summary of Research Findings and Implications.* North Central Regional Research Publication No. 315. Iowa State University Agriculture and Home Economics Experiment Station, Ames, 1988.

TERMS AND CONCEPTS

Target return per unit
Voluntary production control
Compensatory payments
Set-aside acreage
Farm base crop acreage
Slippage
Soil Bank
Payment-in-kind (PIK)
Buffer stocks
Ever-normal granary
Farmer owned reserve (FOR)
World food reserve

QUESTIONS AND STUDY SUGGESTIONS

1 Compare a policy of compulsory production and marketing control with voluntary production control and compensatory payments.

2 What are the basic assumptions underlying voluntary production control and compensatory payments policies? What are the necessary conditions for the program to increase farm income? What are the steps for implementing this program?

3 In what ways do the economic and political rationales differ for voluntary production and compensatory payment programs?

4 Give examples of the use of the compensatory payments approach in U.S. policy history. In another country. Explain why the U.S. shifted from price supports to more reliance on compensatory payments.

5 Specifically, how does the U.S. national wool policy function? Explain why the U.S. wool program has failed to increase or even maintain U.S. wool production.

6 What factors contribute to slippage in a voluntary production control program?

7 How is the elasticity of supply and demand of a commodity related to the U.S. Treasury cost of a compensatory payments policy?

8 What economic consequences can be expected of a compensatory payments policy?

9 Why is a voluntary production control program usually combined with a compensatory payments program?

10 What economic consequences can be expected of a combined voluntary production control and compensatory payments policy?

11 How does current U.S. price and income policy affect farmer income distribution?

12 What was the first U.S. grains reserve or buffer stock program? When was it started? Why was it begun?

13 What conditions must be fulfilled in order for a buffer stock program to raise grain prices, increase farm income, and reduce price variability?

14 Show why a compensatory payments program will tend to increase farm output.

CHAPTER 9

INTERNATIONAL TRADE POLICY

From now until well into the twenty-first century, a dominant problem for agricultural prosperity and growth in all major exporting countries will center around the best way to price and sell the increasing volume of food and other agricultural commodities available for exporting. The problem must also deal with producers of major commodities who have been protected to varying degrees from foreign competition. This problem is especially critical for food and agricultural policy because many countries have shielded agricultural producers from foreign competition. The United States is the largest agricultural exporter, and what this country does in trade and trade negotiations affects the entire world.

IN THIS CHAPTER

By studying Chapter 9, you will learn how to answer the following questions:

1 What are the basic problems of agricultural trade policy?
2 How does international trade interact with domestic farm programs?
3 What are possible alternative policies for future agricultural trade?

Since the adoption of the Reciprocal Trade Agreements Act of 1934 and the General Agreement on Tariffs and Trade in 1947 (GATT), the United States has been working diplomatically, in cooperation with other trading nations, to reduce tariffs and eliminate nontariff barriers to trade (NTBs). The NTBs include commodity quotas, licenses, and other government regulations that may be used to restrict trade. More than 100 nations are involved in negotiations under the GATT. Much progress has been made in reducing tariffs and NTBs for most industrial commodities, but trade in many agricultural commodities has continued to be restricted.

In the following sections, we see that agricultural trade has been the most difficult area of trade to liberalize. This is unfortunate because agriculture, with its efficient production potential and limited growth possibilities in the domestic market, has much to gain from freer trade. When certain sectors of U.S. agriculture are protected, the options available to the U.S. government in promoting and attaining greater freedom in international trade are restricted. More freedom, or less restriction on export trade, is of growing importance for the growth and prosperity of U.S. agriculture, and for the United States to earn more foreign exchange through agricultural exports. The policy problem is how to achieve the benefits of freer trade without damaging certain farm sectors that have gained from trade restrictions.

PROBLEMS OF TRADE POLICY

All trade is based on *comparative advantage,* which is defined as the capability of producing goods or services *relatively* more cheaply than some others are willing and able to do. Every national government has to choose whether or not to intervene between its economy and the economies of other countries. This is called *trade policy.* In international trade, someone benefits on each side of the transaction; otherwise, no trade would occur. Within nations, trade occurs because of such benefits, and because of the way resources are allocated, technologies are distributed, and firms are organized. Among nations, trade occurs for similar reasons, except that governments determine the conditions under which all trade takes place. Just as a firm tends to determine what goes on between it and other firms, so do nations. The difference is that, while nations usually prohibit trade restrictions within their borders—for example, the U.S. Constitution prohibits tariffs or other restrictions in interstate commerce—trade among nations may be restricted by any nation or group of nations.

Characteristics of Agricultural Trade

Governments set the conditions for agricultural trade in various ways. Practically all developed countries use tariffs or NTBs such as quotas to protect in varying degrees some of their farm producers against imports from other countries that may have a comparative advantage in producing such commodities. Important examples are tariff or NTBs to restrict imports of sugar, dairy products, and peanuts into the United States; rice and beef into Japan; and wheat and feed grains into the European Community (EC). Many developed countries also subsidize the exports of some commodities as a means of providing aid to their farm sectors. This is done either by supporting prices in the domestic market at levels above prices in other countries and paying export-marketing firms to cover the difference, or by subsidizing farmers while letting prices drop to world levels.

These export policies have been used for many commodities. A few important examples are wheat and most feed grains in the United States, wheat and dairy products in the EC, and wheat in Canada. A few less developed countries have intervened by means of taxes on agricultural exports to try to hold down domestic food prices and to raise revenue. However, they also limit imports of some commodities to protect their producers and save their scarce earnings from exports. India, for example, maintains domestic prices for wheat well above the world market by means of import restrictions. The result is higher prices for India's wheat farmers, which are obtained without direct government subsidy. Agriculture is the largest economic sector in the Indian economy. Wheat is an important part of the agricultural economy. Consequently, it is not surprising that India would increase the incomes of wheat farmers, if this can be done without additional government expenditure. In addition, the government's revenue may be increased without higher tax levies on domestic producers.

The Agricultural Trade Dilemma

The trade dilemma is that, although domestic aid to agriculture may be justified in social, economic, or political terms, restricting imports to help provide this aid alters the use of agricultural resources, which reduces the efficiency of agriculture and raises costs to consumers and governments. The extent of these effects depends on (1) the elasticities of supply and demand, (2) the relative importance of the commodities involved, and (3) the degree to which the terms of trade are altered. This can be illustrated with a series of supply–demand diagrams based on the assumptions of classical economic theory (Figures 9-1–9-4). These diagrams help us to visualize the trade-policy issues and problems.

Supply and Demand in Closed Economies

In Figure 9-1, let us assume that the United States and Germany are closed economies that do not permit trade in wheat. When the two markets are separately in equilibrium, the U.S. wheat price is much lower than the price in Germany. If the U.S. price were supported at the higher price in Germany, there would be an excess supply or surplus

FIGURE 9-1
Supply and demand for an agricultural commodity such as wheat in the United States and Germany. When each country is closed to trade, the equilibrium price is *A* in the United States and *C* in Germany. See Endnote 1 for source and acknowledgment of Figures 9-1 through 9-4.

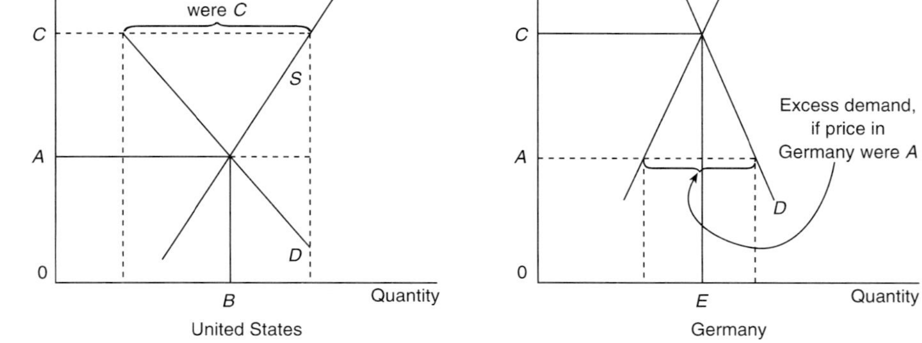

in the United States. If the low U.S. price were established in Germany, by means of price controls for instance, there would be excess demand in Germany. Rationing might have to be used to stabilize the wheat market.

For many years, the price of wheat in the former West Germany was maintained at about double the price in the United States by means of a variable import levy sanctioned by the EC under its Common Agricultural Policy (CAP). The import levy, or tariff, generally was set to fill a major part if not all of the gap between target prices in the EC and lower world prices. In the 1990s, as the EC's agricultural output continued to expand and budget costs grew for export subsidies and for some adjustment payments to farmers, EC countries began maintaining farm incomes with direct payments and a small amount of land set-asides.

Supply and Demand under Free Trade

A supply–demand diagram comparing free trade with no trade is presented in Figure 9-2. Equilibrium is established in the world market by intersection of the U.S. export supply (domestic *S–D*) and Germany's import demand (domestic *D–S*). The U.S. price of wheat rises from *A* to *G*. The price in Germany falls from *C* to *Z*. Thus consumers in the United States pay a higher price, while consumers in Germany gain by lower prices. U.S. farmers gain from higher prices and more exports, while *EC* farmers lose because their prices drop.

This illustration is useful, provided Germany has a comparative advantage in some other commodities that can be exported to the United States or elsewhere. The illustration could apply to any country that functions as a trader in the world economy, or as a trader in a group of nations with a common market. If this is the case, why is it so difficult to get a truly world economy with free trade in agricultural commodities?

Persistence of the Trade Policy Dilemma

The basic problems of agricultural trade policy persist because the world is composed of almost 200 independent, highly nationalistic countries in which producers are concerned about the security of their returns, and consumers about the security of their food supply. Almost universally, farmers and some agribusinesses try to gain an economic advantage by pressuring their governments to restrict imports or subsidize exports. Short-term gains to some economic groups usually are possible and may foster economic growth in those sectors. All analyses suggest that there are net economic losses to societies generally when trade is restricted, however, even though several nationalistic sectorial gains may be perceived otherwise. Such restrictions usually result in protected industries using resources in a less efficient way, and hence a higher cost to consumers.

AGRICULTURAL TRADE POLICY

Agricultural trade policy has a central economic content, despite being affected by national security, social conditions, political interests, and other considerations. Such policy can be divided into two general types, export policy and import policy, each of which in agricultural trade has been dominated in many countries by interests of producers.

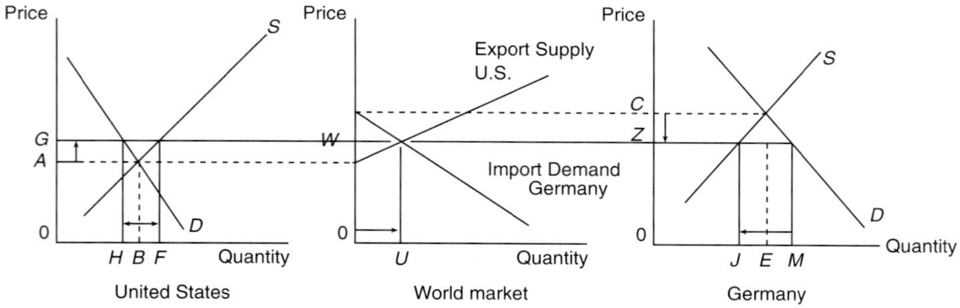

FIGURE 9-2
Free trade between the United States and Germany in wheat compared with no trade. Under free trade, the U.S. price rises from *A* to *G*, and the price in Germany falls from *C* to *Z*. The United States exports *HF*, and Germany imports *JM*.

Agricultural Export Policy

Agricultural export policy usually rests on the economic concept of *market differentiation,* which is based on the general proposition that demands can be differentiated between domestic and foreign markets. That is, demand for a commodity of a given country is usually more inelastic in the domestic than in the foreign market. Hence, if the domestic demand is inelastic and the quantity supplied to the domestic market is restricted, prices will rise relatively more than the supply is restricted, and the total returns to producers from the domestic market will be increased. Then output that is not sold in the domestic market can be exported, with an export subsidy if necessary, to further enhance total gross producer returns. These short-run gains to producers will diminish over time as the producers' returns are capitalized into higher fixed factor costs. The magnitude of producers' gains and their duration will depend on the importance of the commodity in trade, the elasticities of supply and demand in each country, and whether the importing country restricts imports or not.

This policy of market differentiation can be combined with other domestically oriented programs, such as price supports, production controls, consumer subsidies, and food aid. However, such efforts at market differentiation by an important agricultural exporting country, such as the United States, will affect other trading countries and encourage them to set up retaliatory trade policies. Thus the producer gains could be diminished or eliminated as each country in competition takes countervailing measures to protect its own producers against the domestic subsidies or the subsidized exports of other countries.

Agricultural Market Differentiation Trade Model and Consequences

In Figure 9-3, market differentiation as an agricultural export policy is illustrated by the customary supply and demand graph for a major exporter such as the United States. There is, of course, the implicit assumption that the restricted domestic market could be managed effectively and protected by import controls and, further, that the rest of

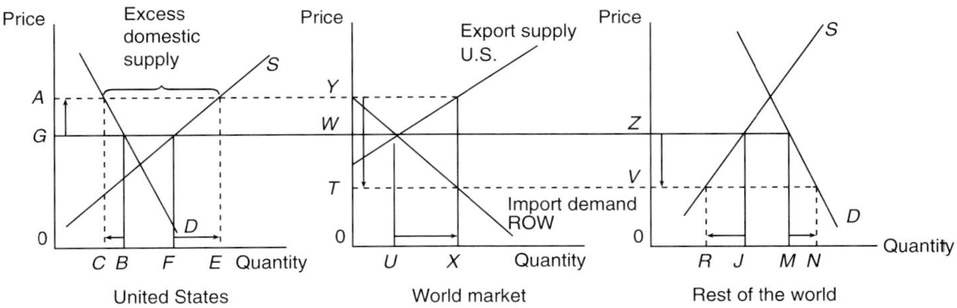

FIGURE 9-3
Effects of an exporter nation, as the United States, differentiating its domestic and foreign markets. The United States restricts the amount of its total production of a commodity available for consumers from *OB* to *OC* in order to maintain a domestic price *A* which is higher than equilibrium price of a free world market of *G*. The U.S. producers respond to the higher potential price by expanding production from *OF* to *OE*. The U.S. government must now purchase *CE*, or otherwise divert it, for export. Now, the U.S. export supply expands from *OU* to *OX* and the rest of the world (ROW) imports *RN* instead of only *JM* of a free world market. The differentiated market prices are now *A* and *T* (same as *V*), with *CE* = *OX* = *RN*.

the world (ROW) would permit the additional imports, thereby depressing its prices. This "pure" market differentiation policy could be combined with other trade policies to increase its effectiveness, to minimize the adverse reaction of the ROW, and to restrict its burden on the U.S. Treasury. These policies could include domestic production control; differential prices to producers, depending on their shares of the two markets; blended producer prices; and various export subsidy schemes.

The magnitude of gains to producers in the U.S. economy depends on two factors: (1) the volume of the commodity or resources involved and (2) the elasticities of supply and demand in each of the markets. The elasticity of the U.S. export supply is determined by the interactions between the U.S. domestic supply and demand, and their respective elasticities. For instance, the elasticity of the U.S. export supply is related directly to the elasticity of the U.S. domestic supply and demand. Likewise, the elasticity of import demand on the world market is related directly to the elasticity of demand and supply in the ROW. Whereas the U.S. economy has its gainers and losers from this export policy of market differentiation, the ROW similarly has its producer losers and its consumer gainers.

Several export subsidy programs, amounting to more than $4 billion annually, have been a mainstay of U.S. agricultural export policy since the 1985 Act. They include (1) the Export Enhancement Program (EEP), which awards bonuses in kind from CCC stocks to private exporters so that they can successfully compete in approved foreign markets already being offered subsidies from other exporters; (2) the CCC Export Credit Guarantee Program, which guarantees that exporters to approved developing countries will be paid within 3 years; (3) a similar export credit program for 3 to 10 years; (4) the Market Promotion Program (MPP), previously the Target Export Assistance Program (TEA), which awards redeemable certificates from CCC stocks, or direct CCC funds, to commodity promotional efforts in markets adversely affected by trade subsidies of other countries; and (5) Foreign Food Aid (PL 480), to be examined more fully in Chapter 10.

Political Effects of Agricultural Export Policy

A U.S. agricultural export policy of market differentiation, paid for by domestic consumers and U.S. taxpayers, benefits U.S. producers such as wheat growers with a higher domestic price. Consumers in the ROW also gain economically, although some governments in the ROW may adopt policies to offset the U.S. trade advantage because their interests often center on encouraging their own producers, rather than on providing gains to consumers. For instance, although Japan can import rice from the United States or some other country for a small fraction of what is paid to Japanese farmers, the interests of producers have been evident for many years in the prevailing Japanese import restrictions. Furthermore, many countries follow self-sufficiency policies for national security reasons. Governments in importing countries can counteract the U.S. agricultural export policy by using tariffs and quotas, or other NTBs to trade. In reacting to the U.S. policy, other exporting countries could (1) let their own producers meet the new lower world market price, or (2) match the U.S. subsidy in whole or part. The greater the production response by domestic producers, the more the world market will be depressed. If other governments take countervailing measures to protect their producers, the ROW import demand could be curtailed, and the cost of the policy to the U.S. government could become exorbitant unless combined with some kind of production control.

AGRICULTURAL IMPORT POLICIES

In addition to being the world's largest exporter of agricultural commodities, the United States is also one of the largest importers. For analytical purposes, such imports may be divided into two classes: *complementary* and *competitive* to U.S. production.

The complementary commodities include those not produced in the United States or produced only in very small volume, such as coffee, crude natural rubber, cocoa beans, bananas, tea, carpet wool, raw silk, and spices. Coffee is the largest imported commodity, generally accounting for about 60 percent of the value of all complementary commodities and at least 25 percent of all imported agricultural commodities. Crude rubber was once the second largest complementary item, but synthetics have slowed the imports of crude rubber as well as wool and raw silk. Most other complementary commodities generally are imported without trade restrictions.

Interest in U.S. import policy making centers on the competitive commodities such as grains, fibers, tobacco, peanuts, meat and meat products, sugar, vegetable oils and other oil-bearing materials, wine, fruits and vegetables, dairy products, and hides and skins. These commodities account for more than 50 percent of the total value of U.S. agricultural imports. The total value would be much higher, however, if it were not for tariffs, quotas, and other restraints to trade imposed by the U.S. government. Most of these commodities compete directly with U.S. products. Historically for these reasons, the U.S. has not been a consistently enthusiastic proponent of completely free trade nor an aggressive negotiator internationally to lower trade barriers.

What effects do import restraints have on U.S. producers, consumers, and the nation's trade? Economic analysis indicates that the restraints are most significant for sugar, peanuts, tobacco, and dairy products. Restraints on beef imports are usually of marginal significance, although important especially in years when U.S. beef prices are at the top of the beef-price cycle. Even though U.S. producers of some other products such as grains are very competitive, imports are essentially prohibited to protect the markets of price-supported products. The effects of import restrictive policies on the individual commodities are primary or immediate, as well as secondary or broader for the strategic position of the United States in total world trade.

Particularly in the short run, import restrictive trade policies generally benefit domestic producers, hurt domestic consumers by higher prices and smaller supply, and inhibit trade. Analysis of the U.S. sugar policy illustrates the general problem.

Example of Import Restrictions: U.S. Sugar Policy

The U.S. domestic sugar industry is the product of a 200-year policy of protection from imports. Before 1934, there was a tariff of 2.5 cents per pound on imported raw sugar. In 1934, industry representatives were successful in getting more protection, which provided for quotas on sugar imports in addition to the tariff. From 1934 to 1974, a system of quotas was used to govern the supply of sugar from all sources, foreign and domestic, according to a formula established by law for distributing the U.S. import quota among sugar exporters. These policy changes were partly stimulated by the memory of stringent sugar rationing during World War I, when imports were hampered by submarine warfare. The secretary of agriculture was mandated to determine the overall size of the quota prior to the beginning of each new year and to allocate quotas according to the formula. Prices received by domestic producers were supplemented by direct payments financed from a tariff of 0.625 cents per pound on imported raw sugar. These payments assured that domestic producers would fulfill their quotas, and holding domestic prices above world prices assured that foreign countries would generally fulfill their quotas. Domestic refiners also continued to be protected by restrictions on imports of refined sugar.

From 1934 to 1974, industry representatives, lobbying for renewal and continuation of the program, argued that it provided an assured supply of sugar to domestic consumers and protected them from higher prices that might occur at times under a policy of free trade. In 1974, however, when world sugar prices began to rise sharply after a few years of poor crops, the domestic prices rose with the world price. The program of quotas and subsidies failed to provide a stable supply to U.S. consumers under the extreme supply fluctuation. Subsequently, the program was allowed to lapse. A price-support program was mandated in the Food and Agriculture Act of 1977 for the years 1977 and 1978. Discretionary authority was left with the secretary of agriculture from 1979 through 1981. In 1976, the tariff on sugar imports was tripled; and in 1982, reimposition of quotas made sugar the most highly protected of all major crops.

In the Agriculture and Food Act of 1981, price supports on raw cane sugar were set at 16.75 cents per pound, and they increased annually to 18 cents per pound for the 1985 to 1986 crop year. This provided price support for domestic cane and beet sugar and a price ''umbrella'' for high-fructose corn syrup (HFCS), which gained favor as a sweetener for some foods and beverages, soft drinks especially. Restrictive sugar import quotas allotted country by country have been in force since 1982 to ensure that domestic prices are maintained at established price-support levels. As a result, since 1982, sugar imports have declined to historic lows, and the real price of sugar to domestic consumers has been about double the average world price. The effects of this policy can be visualized in an economic model, such as the one in Figure 9-4, where the United States is depicted as the high-cost producer relative to other producers.

Economic Model of U.S. Sugar Policy

In Figure 9-4, the imposition of a tariff increases the price to importers and consumers, so less will be demanded. This tariff and import quota reduces the quantity supplied by the ROW to the U.S. market; and as a result, the U.S. price rises and the ROW price falls. The quantity traded declines as indicated.

In the United States, these economic effects tend to be hidden from consumers and producers. The loss may not be apparent to most consumers because much of the increased cost is hidden in processed foods and beverages. Consumers generally have no basis for comparing these prices with what they would be under free trade. The short-run gains accrue to both sugar producers and manufacturers.

Over the years of industry growth, windfall gains due to the U.S. sugar policy have been realized by both producing and manufacturing firms. Restricting imports reduced the supply on the domestic market, and sugar prices increased. Growers of sugar beets and cane could not produce and sell until a processing plant was built in their community. Furthermore, growers of sugar beets did not have a market until they were granted a quota. New areas for growing sugar beets could not be opened up until a new factory could be built, but the capital requirements for this were not always available. Even a relatively small manufacturing plant might cost $100 million or more.

The economic effect has been to create short-run income gains for producers and processors. However, as these short-run gains are capitalized into the value of land and other fixed assets, the opportunity costs of continuing to produce sugar cane or beets increase for the original owners, and costs of entry into sugar production increase for beginning farmers. As the short-run gains increase opportunity costs, the long-term gains to producers are small. Although the marginal cost of the sugar program to the individual consumer is only a few dollars each year, the aggregate cost to U.S. consumers is large, but it varies widely from year to year because it depends on the difference between U.S. sugar prices and the lower international prices for sugar.

Sugar Policy Strategy

The U.S. domestic sugar and sweetener industry now has almost enough capacity to supply fully the total U.S. sweetener market. More restrictive import quotas were

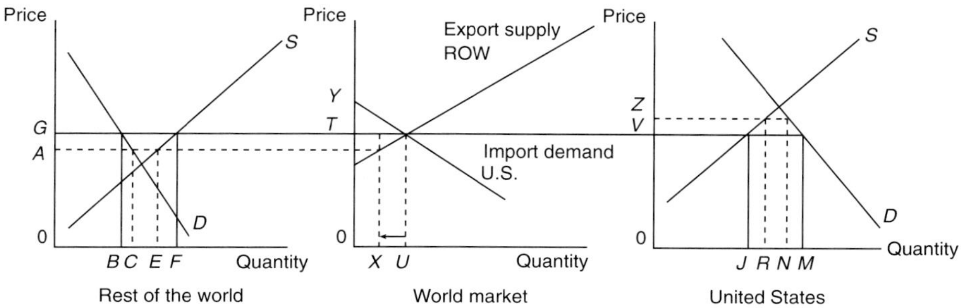

FIGURE 9-4
Effects of a U.S. sugar import tariff on sugar prices, production, and world trade. Under free trade in sugar, the world market would be in equilibrium at the price *T*, with exports from the ROW (rest of the world) of *BF* and U.S. imports of *JM*. ($G = T = V$, and $BF = OU = JM$.)

If a valorem (percentage of value) tariff equal to $Z - V$, is imposed on sugar imported into the United States, the quantity demanded by U.S. importers will decline from *JM* to *RN*. The quantity traded on international markets will fall to *OX* from *OU*. Exports from the ROW will fall from *BF* to *CE*. Since this leaves more sugar to be sold in the ROW, the price in the ROW (or markets outside the United States) drops from *G* to *A*. Then, either the U.S. tariff could be increased more, or an import quota could be applied. Hence, either the tariff or the quota, or both, may be used to protect U.S. producers from lower prices in international markets.

Note that $CE = OX = RN$ and $G - A = Z - V$, however, only if the elasticities of supply and demand are the same in the ROW as in the United States. Note that the size of the U.S. market is an important factor in respect to the practical consequences of U.S. import policy in the ROW. If the United States were a very small country, or if its potential for trade were small, the U.S. policy would make little difference in the ROW. However, this is not the case, and U.S. policy is important to the ROW.

adopted in 1982, and from 1981 to 1988 total sugar imports declined from four million to one million tons. Then, as a result of the 1988 drought, more imports were allowed in 1989 to 1990. If the import restrictions were to completely eliminate imports and if producer prices were to be maintained at prices corresponding to those in the past, the Treasury costs of the program could rise substantially. For instance, the U.S. government might be required to provide and administer a nationwide system of more restrictive producer quotas. One alternative might be to reduce import restrictions, let sugar prices fall, and reimburse producers with compensatory payments or buy them out. Whether this would be cheaper or not depends on several factors, not the least of which may be the continued technological advance in sugar production.

Continuing the protectionist import-restriction policy may have political repercussions on international trade negotiations. For example, at the urging of Australia, GATT negotiators voted to declare the U.S. sugar-import quotas illegal, which might imply further retaliatory measures. Yet, letting sugar prices reach an equilibrium in an open market would be resisted politically by U.S. producers of sugar beets and cane and also by the processing industry. Indeed, an abrupt shift to such a policy without compensation to producers and manufacturers would result in economic loss through fewer jobs and asset depreciation. To compensate for this, a one-time bail-out or buy-out program has been suggested as an economical and perhaps politically acceptable policy.[2] Such a program would help compensate for the fact that U.S. producers' costs have been increased—implicitly or explicitly—by the policy that has been followed.

In the meantime, it is important to keep in mind the strategic economic and political position of U.S. agriculture in respect to the nation's trade. How will this position develop in relation to future trade policy?

AGRICULTURE'S ECONOMIC AND POLITICAL POSITION IN FUTURE TRADE POLICY

Since the end of World War II, U.S. agriculture's economic and political position in international trade has changed greatly.[3] Shortly after the war, U.S. aid to Western Europe brought a boom in U.S. grain exports. For several years following, Western Europe was the largest foreign market for U.S. agricultural commodities. Then the Japanese demand grew more rapidly until Japan became the largest single country importing U.S. agricultural commodities. In the early 1970s, the Soviet Union began to buy U.S. grain in large quantity, not necessarily as a regular or reliable customer, but as a nation with large import demands in some years to offset poor crops.

Other centrally planned countries such as China and most countries in Eastern Europe also became net importers. However, increasing agricultural productivity in Western Europe and expansion of the EC with its strong protectionist policy began to shift the EC from a large importer of agricultural commodities to, on balance, a net exporter. A continuing compensatory-payment and protectionist policy by the EC, with continuing rapid technological advances in European agriculture, may provide little opportunity for growth in EC markets for any exporting country outside the EC. What is to be done?

The U.S. Position in Trade Negotiations

Throughout major rounds of GATT negotiations in the 1960s (Kennedy Round), the 1970s (Tokyo Round), and the 1980s and 1990s (Uruguay Round), only minor agreements were reached in liberalizing trade in agricultural commodities. It has been the general goal of the U.S. government to achieve a world trade position relatively free of tariffs, quotas, and other NTBs. The U.S. position has been weakened by the policy of subsidizing some agricultural exports, especially grain and cotton, and by preventing imports of sugar, dairy products, peanuts, and many minor commodities such as lemons and walnuts. Still, world trade in agricultural commodities has expanded, but the U.S. share of this trade has only slightly increased over what it was in the 1960s, and is much smaller than it was in the 1970s.

For future U.S. agricultural trade policy, this situation suggests that substantial reductions in international trade barriers are likely to come slowly, that both export and import policy interventions may need to be modified, that the United States must continue to be dependent on an export market to maintain a satisfactory supply–demand balance, and that maintaining an efficient domestic agricultural sector and an active, flexible international role are particularly critical. These considerations are important not only because GATT negotiations are never ending, but also because new challenges are being faced and opportunities continue to be explored with the pivotal North American trade talks involving Canada, Mexico, and the United States.

Grain Quality Policy and Exports

For many years, even back to the 1920s and earlier, importers in various countries have complained about the low quality of some U.S. grain exports, especially corn and less frequently soybeans. In the 1970s and 1980s, as U.S. grain exports increased, and in the 1980s, especially as the international market shifted in favor of buyers, such complaints escalated. U.S. research found that most of the complaints arose from quality losses between the U.S. origin certificates and delivery to the final processors' plants in the importing countries. Frequently, grain—especially corn and soybeans—that had received a generally satisfactory grade in U.S. markets arrived overseas moldy or musty or perhaps with an excessive percentage of broken or sprouting kernels. For instance, grain that had passed inspection in the United States but had been dried at high temperatures, tended to crack when handled by the high-speed pneumatic loaders and unloaders used by ocean-shipping grain firms. Broken kernels at standard moisture levels were susceptible to heating and spoilage. The worst problems centered around corn and soybeans, but all grain was involved to some extent.[4]

The Grain Quality Improvement Act of 1986, effective November 10, 1986, established new standards to promote the marketing of high quality grain and to protect the quality of grain marketed in or exported from the United States. The Act required the secretary of agriculture to develop new standards to define dockage and foreign material and establish new regulations for defining broken corn and foreign material. The Act provided that no dockage or foreign material once removed from grain could be recombined with any grain. Dust was prohibited from being reintroduced to the grain stream at any location.

Insect infestations present a serious problem in grain quality, and they are an important cause of international and domestic complaints. The administrator of the Federal Grain Inspection Service (FGIS) is required to set grain inspection standards that will reflect insect infestation more accurately and to evaluate the effects of adopting a new optimal grain grading system. The FGIS is required to adopt, to the extent practicable, the tests that are economically feasible.

In the meantime, according to a broad evaluation by the U.S. Office of Technology Assessment (OTA), improvements in quality can make U.S. grain more competitive in the higher quality, higher priced international grain markets, including markets for both food grains and feed grains. Such improvements must come from a systems approach that encompasses three areas: (1) genetic improvement; (2) government programs, especially discounts and premiums for quality by the Commodity Credit Corporation; and (3) changes in grading and inspection procedures.[5] Quality improvement as well as competitive pricing are important options for the future.

Future Export and Import Policies

Because some five sixths or more of the world's projected population increase over the next generation will occur in the developing countries, and because rising per capita demand in these countries depends on economic growth, national and international policies favoring such growth should be high on the political agenda of U.S. agricultural trade policy. Clearly, it is in the U.S. national interest to encourage such growth. However, this raises a final question (for this chapter) about the role of imports of agricultural commodities that are competitive with U.S. products.

Between the early 1970s and the early 1990s, the value of U.S. agricultural imports increased fourfold, to some $30 billion annually. The increase probably would have been much greater if imports of certain commodities such as sugar and dairy products had not been restricted. Should the U.S. policy allow or encourage more of this import trade, or should the U.S. policy be more restrictive?

Two major issues are involved. One concerns the economic gains from exploiting comparative advantages, an argument for freer trade and for policies that help in the growth of developing countries. These countries are the potential customers for most increases in U.S. agricultural exports. The other issue concerns a major question of safety in food imports. For many years, importing beef from certain countries into the United States has been prohibited because of some known cases of hoof and mouth disease of cattle. This policy is settled and understandable, provided the facts are correct. More recently, however, other food imports have involved increasing episodes of food contamination that the Food and Drug Administration is unable to detect or prevent.[6] Stopping such imports would most likely result in economic losses, retard economic growth in the exporting countries, and inhibit the growth of export demand for other U.S. agricultural commodities. On the other hand, unrestricted imports without regard for quality products may endanger U.S. consumers. A policy compromise, such as expanding the FDA's surveillance of selected food imports could be beneficial for consumers. More careful monitoring of pesticides in the exporting countries may be achieved through international agreement. Finally, however, it is important to remember that unnecessary restrictions of imports due to fears that are not well grounded creates uncompensated losses to U.S. consumers, as well as to farm producers in other countries.

LOOKING AHEAD

Although there is general political support from consumers and some producers for a freer agricultural trade policy, how far the trading nations will proceed in this direction is a matter for continuing policy study, analysis, and debate. Trade policy is linked to other policies that we have explored such as price supports and compensatory payments, U.S. production controls and storage programs, and others to be discussed, such as food aid in Chapter 10, net welfare effects in Chapter 11, and food quality and safety in Chapter 13.

ENDNOTES

1 The authors express their appreciation to Bob F. Jones and Robert L. Thompson for their permission to use Figures 9-1 through 9-4. These figures are adapted from "Interrelationships

of Domestic Agricultural Policies and Trade Policies,'' in *Speaking of Trade: Its Effect on Agriculture,* Agricultural Extension Service, University of Minnesota, Special Report No. 72, November 1978.
2 D. Gale Johnson, *The United States Sugar Program: Large Costs and Small Benefits,* The American Enterprise Institute, Washington, DC, 1974; Harold G. Halcrow, *Food Policy for America,* McGraw-Hill, New York, 1977, pp. 258–266.
3 John Mellor, *The New Global Context for Agricultural Research: Implications for Policy: Food Policy Statement,* No. 6, International Food Policy Research Institute, Washington, DC, 1986.
4 Lowell D. Hill, *Grain Grades and Standards,* University of Illinois Press, Urbana, 1990.
5 U.S. Congress, Office of Technology Assessment, *Enhancing the Quality of U.S. Grain for International Trade,* OTA-F-399, U.S. Government Printing Office, Washington, DC, 1989.
6 Bruce Ingersoll, ''Gone Bad: Tide of Imported Food Outruns FDA Ability to Spot Contamination. Agency Can't Begin to Check All Shipments, Though it Knows Many are Tainted. A Razor Blade in the Banana,'' *The Wall Street Journal,* September 27, 1989, pp. 1, 10.

TERMS AND CONCEPTS

Theory of comparative advantage	GATT
Closed market	Uruguay Round
Free world trade	International market share
Tariff	Agricultural trade dilemma
Nontariff barriers (NTBs)	Market differentiation
Export subsidy	Complementary commodities
Import quota	Competitive commodities

QUESTIONS AND STUDY SUGGESTIONS

1 How can all nations benefit from international agricultural trade?
2 To benefit from comparative advantage, must a nation be the lowest cost producer of every product exported? Why or why not?
3 Why do nations set up barriers to agricultural imports? Why does the United States have an export enhancement program for agriculture? What commodity groups in the United States have been the most resistant to free trade in agricultural commodities? Why?
4 What are the primary factors that determine the comparative advantages among countries in agricultural production? Does the United States have a comparative advantage or comparative disadvantage in production of wheat, dairy products, soybeans, corn, rice, peanuts, tobacco, cotton? Do these comparative advantages hold against all other nations or just a few?
5 What is price discrimination? Will price discrimination be advantageous for agricultural producers of an exported commodity if the demand elasticities for their commodity is the same in all markets? Why or why not?
6 As the major industrial countries have generally increased their share of the expanding world trade in agricultural commodities, why have they restricted agricultural imports more than the imports of industrial goods?
7 What has been the U.S. sugar policy? Do U.S. consumers benefit from this policy? As U.S. sugar producers pay taxes, do other taxpayers benefit? Do sugar producers benefit in the short-run? In the long-run?

8 What has been the effect on agricultural trade of the Uruguay Round of GATT negotiations? What has been the political outcome? Do you expect that negotiations on trade in agricultural commodities will be more successful in the future? Why or why not?

9 Until the beginning of the Uruguay Round of trade negotiations under the GATT in 1986, the U.S. government had favored excluding most of the international trade in agricultural commodities from discussions of free trade. However, the Cairns group of nations (see the Glossary) objected strongly to such exclusion and the U.S. government agreed to include eliminating trade barriers and domestic subsidies on agriculture as the first priority for discussion in the Uruguay Round. Progress on this item was slow, and general blame was placed on the EC for blocking an acceptable agreement. Finally, on September 23, 1992, the U.S. government proposed to put the question to binding arbitration. Suppose such arbitration is successful and holds over several years, what will be the general effect (1) on U.S. agriculture, (2) on the commodities in which U.S. farmers hold a comparative advantage, and (3) on the commodities in which U.S. farmers are at a comparative disadvantage? What will be the effects on EC agriculture, EC farmers, and EC consumers?

CHAPTER **10**

FOOD-AID POLICY: DOMESTIC AND FOREIGN

Throughout the world, the poor have insufficient resources to obtain adequate amounts of nutritious food. Public and private transfers of cash or food help to remedy this situation. The transfers represent an investment in human resources. Moreover, they enhance food demand and serve as outlets for food obtained through some of the commodity programs. This chapter examines the economics of food aid and some policy issues related to the U.S. food assistance programs and foreign aid.

IN THIS CHAPTER

By studying Chapter 10, you will learn how to answer the following questions:

1. What are the economic and political rationales for a food-aid policy?
2. What important economic effects does such a policy have on farm producers, consumers, governmental budgets, taxpayers, and international trade?
3. What has been the general experience with food aid in the United States? What are some of the policy options for the future?
4. What has been the experience with foreign food aid, and what are some of its options?

Food-aid policy may be defined as one or more public programs that supplement the food-buying power of low-income consumers and other targeted groups both at home and abroad to increase their food consumption and, as an associated purpose or indirect result, to raise market prices to producers of farm commodities.

Food-aid policy can have diverse objectives: (1) improving diets and nutrition among poor people in the United States and in developing countries; (2) targeting certain groups such as school children to improve their dietary practices and nutritional levels; (3) alleviating hunger and undernourishment or preventing starvation in the aftermath of war, political catastrophe, or natural disaster; and (4) strengthening the aggregate demand for food, thus helping to improve and stabilize farm incomes.

A "pure" form of this policy has several assumptions: (1) market forces will be left to further allocate resources and distribute food commodities; (2) food consumption of the identified groups will be supported by some means of direct or indirect subsidization; (3) other international trade flows and pricing will be left to market forces; and (4) financing of food subsidies will be supported directly from the federal budget and possibly augmented from state and local budgets.

Two conditions are necessary for the policy to succeed: (1) a means is found to target the food aid so that aggregate demand is increased by enhanced food consumption in the designated groups; and (2) sufficient funding is committed for administration and food subsidies. Even though food aid can be funded either privately or publicly, the magnitude of the task requires that the main programs be public. Legislative bodies and administrative agencies must make many decisions about which segments of the domestic or developing countries' population groups are to be aided, the levels of support, the means of subsidy to be used, which food commodities or nutrient sources will be emphasized, and how unintended obstacles to achieving program objectives can be minimized while maintaining consumer freedom.

ECONOMIC AND POLITICAL RATIONALES

The policy alternative of food aid differs from most other food and agricultural policies because it can be motivated by a public desire to help two considerably different groups: (1) low-income consumers who need better diets, and (2) farm producers, by increasing the aggregate demand for food. Advocates of either group may support this policy. Indeed, political support of both groups has been crucial in establishing a food-aid policy in the United States.

Economic Rationales

The elasticity of demand for food varies inversely with incomes. It is more elastic for low-income consumers than for those with higher incomes. Therefore, simply transferring income from the general tax-paying public to low-income consumers will increase the total demand for food. But, by spending more public funds on food, a successful food-aid policy can exert a proportionately greater effect on improving diets and nutrition. Thus targeting food aid to consumers on the basis of their income is the most effective means for implementing the aid. Furthermore, with the increased demand associated with a food-aid policy, market prices are raised and production increased.

Food aid can be implemented by distributing (1) money or credit; (2) commodities such as flour, dried milk, and cheese that might be obtained from government stocks; and (3) stamps or vouchers that are exchangeable for food. Policy makers and administrators of food-aid policy must decide who is eligible, the form of the aid, its magnitude, and how much choice the recipients will have. From an economic perspective, food aid may not be the most cost-effective way to support farmers' prices or increase poor consumers' food-buying power, yet few policies serve both objectives simultaneously or so directly.

Political Rationales

Food aid becomes politically attractive when there is malnutrition among the poor—particularly among infants, children, the disabled, the handicapped, and the elderly. Food aid becomes even more attractive when unmet food needs persist alongside surpluses of farm commodities. Under such conditions, food aid also seems desirable as a supplement to direct subsidies for increasing farm income. Finally, food aid is sometimes viewed as a safety net by those who see the market system as an answer to the price and income problems of the agricultural sector, as well as by those who look to public policy to answer persistent societal problems. Thus food aid, begun in the United States on a small scale in the 1930s, emerged with broad-based economic and political rationales in the early 1960s, and subsequently grew to absorb half or more of the USDA's annual budget.

What are the economic relationships implicit in a food-aid policy?

ECONOMIC RELATIONSHIPS OF FOOD-AID POLICY

The economic relationships implicit in a policy alternative of food aid are presented in Figure 10-1 for the agricultural industry and the individual farm producer. Without slippage, food aid raises the demand from D to D_{fa}, but with slippage in the effectiveness of the program, demand increases only to D_{fas}. What is slippage in food aid, and what kind of policy problems does it present?

Policy Problems of Slippage

Slippage occurs when some low-income consumers substitute a significant portion of their previous cash expenditure for food, or any cash subsidy, on nonfood purchases. Because of slippage, food aid increases total food demand only to D_{fas}, rather than D_{fa}. If the policy objectives were to increase price to P_{fa} or quantity to Q_{fa}, then the objectives would not be met unless additional food aid is forthcoming to replace the slipped demand.

As a result of slippage, budgetary burdens can increase and become a major problem for this policy. There is a policy trade-off between program costs and discretionary food choices of aid recipients. If recipients were compelled to eat only in one central place or had to use exactly the same previous food expenditures in addition to subsidized food coupons, slippage could be minimized. But such procedures clearly run counter to democratic values and could increase the cost of administering a food-aid program.

These economic relationships and the slippage problem generally apply to foreign food aid as well as to domestic aid. For foreign countries, however, the issues of

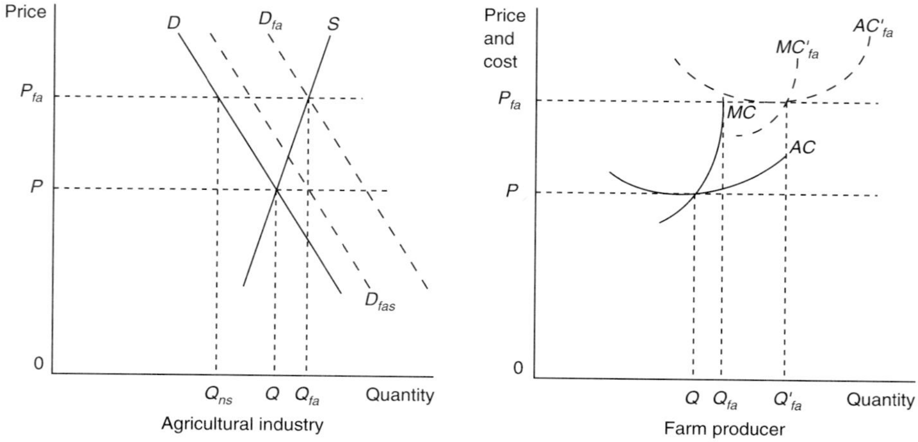

FIGURE 10-1
Effects of a food-aid policy on consumers, the agricultural industry, and farm producers. Without food aid, the industry is in equilibrium at P and Q. Food aid increases the total demand for food to D_{fa}, and the equilibrium price and quantity increase to P_{fa} and Q_{fa}. The increase in demand with slippage is only to D_{fas}, however.

At the farm producer level without food aid, the firm is in equilibrium at P and Q. In the short run, when the price rises to P_{fa}, the firm's output increases to Q_{fa}, and the producer realizes an additional profit equal to $P_{fa} - P$ on the quantity Q_{fa}. However, in the long run, the increased profit attracts more resources and increases the value of other inputs such as land, which shifts average cost and marginal cost to AC'_{fa} and MC'_{fa}, respectively. The farm producer reaches a new long-run equilibrium at P_{fa} and Q'_{fa}, where the profit is normal or is equalized among alternatives, assuming there is no slippage.

eligibility for aid, the distribution of food commodities or money, and the effects on farm producers of other countries complicate the evaluation and implementation of a food-aid policy.

If slippage in the U.S. economy is considered serious, it might be resolved in large part by issuing food stamps as coupons that entitle recipients to purchase food at discounted prices. For example, five different colors of stamps could be issued to eligible recipients, with the values of the stamps varying inversely with income. Those just below the poverty line might have stamps entitling them to 20 percent off the price of food. Those in the next poorer class would be entitled to 40 percent off, and so forth to the nearly destitute, who would be entitled to 100 percent off, or free food. Whether this plan would solve the problem of slippage is of course debatable, and a black market in food stamps could still develop. Indeed, whether eligible recipients should receive free food stamps or pay some minimal purchase price has been a policy issue for many years. Since 1979, however, there has been no purchase requirement for the U.S. food stamp program.

Policy Analysis Using Indifference Curves[1]

The food stamp program increases the demand for food by providing eligible, low-income households with additional purchasing power. Although food stamps can be used legally only to buy food, they may replace income that would have been spent for food in the absence of the program, and that income may be spent on more food or on other commodities. Not all stamps have to be used, but this loss of potential food consumption is small. Food-stamp households, in view of prices, can spend a smaller proportion of real income (money income plus food stamps) on food, and a larger proportion than estimated by policy makers on other commodities. These issues have relevant economic relationships for policy.

A specific-purpose subsidy such as food stamps will likely be viewed as of lower value to the recipient than a general-purpose subsidy of equal amount. Of course, it may not be possible to deliver the benefits in any other way. Figure 10-2 depicts the welfare effects of alternative food-stamp policies. Food expenditures are measured horizontally and expenditures of all other goods and services are measured vertically. Each of the curved lines, called *indifference curves,* connect a series of points representing levels of expenditures jointly of food and nonfood considered equally desirable by the recipient. Successive indifference curves, moving left to right, represent increasingly desirable levels of consumption.

The budget line JD represents what the household can buy at unsubsidized income and the budget line *JLM* represents what the household can buy if its income is subsidized through participation in a food-stamp program, assuming that the price of food in relation to nonfood is the same in each case. The original income determines the height (*J*) of the lower budget line, or its intersection with the vertical axis. The slope of each budget line is determined by the price of food relative to nonfood. At each level of real income, the household, acting to maximize its utility, will plan to buy the quantity of food indicated by the intersection of the price line with the highest indifference curve that it reaches. At the unsubsidized level of income (or assuming that there were no food stamp program), this will result in *OF* expenditures on nonfood

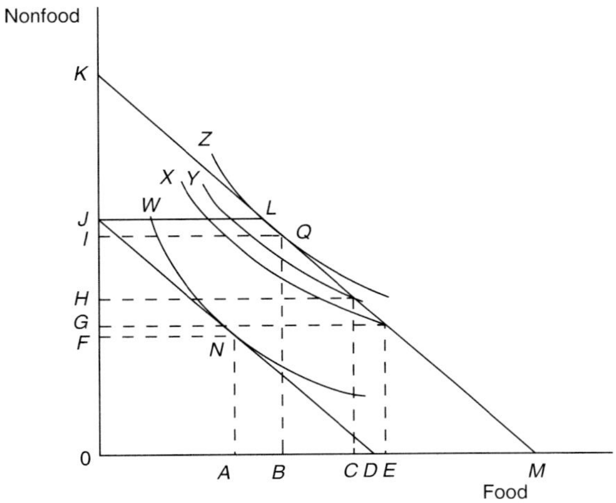

FIGURE 10-2
Welfare effects of alternative food stamp policy options on a representative low-income household. In the absence of a food stamp program, the household is on indifference curve *W*. With the introduction of a food stamp program with a purchase requirement, the household reaches a higher indifference curve (*X*). Eliminating the purchase requirement allows the household to move to indifference curve *Y* and the cash-out alternative would enable the household to reach the highest indifference curve (*Z*) and optimize at point *Q*.

and *OA* on food, when the consumer is on the indifference curve *W*. The consumer optimizes utility at point *N*.

A food-stamp program, with or without a purchase requirement, subsidizes the income of recipient households, and allows them to move to a higher indifference curve. Under either food-stamp policy option, the household faces budget constraint *JLM*. With a purchase requirement in effect, that is, some cash cost to obtain added value of food stamps, the representative food stamp household is on indifference curve *X* and spends the acquired *OE* (food stamp allotment) on food and the remainder of its income *OG* on nonfood—both food and nonfood expenditures exceeding those at the unsubsidized income level. Theoretically, increases in the disposable income of a household using food stamps, which serve to increase disposable real income, will improve nutritional intake if the stamps are spent on healthful food. With the purchase requirement eliminated, households would have more discretionary choice in allocating their real income and might purchase a different combination of food and nonfood. Eliminating the purchase requirement allows households to move to an even higher indifference curve (*Y*) and change their mix of food and nonfood. The household depicted in Figure 10-2 decreases its food expenditures to *OC* and increases its nonfood expenditures to *OH*. If this representative household received cash in place of food stamps, it would be able to allocate all of its real income in accordance with its preferences. The relevant budget line would be *KM* and the household would maximize utility at point *Q*. However, under this option the household reduces its food consumption, relative to the other policy options.

Theoretically, food-stamp households will buy food, not only because the stamps can be used only for food and can give them additional purchasing power, but also because both the *income effect* and the *substitution effect* will be operative. The income effect refers to people's tendency to buy more normal goods, in contrast to inferior goods, when they have higher incomes (income elasticity of demand greater than zero). It is relevant because program participants' incomes are increased. This effect is small for the average consumer because food has an income elasticity of demand much less than one. However, income elasticities vary greatly by income levels; that is, a lower proportion is purchased out of higher incomes and a higher proportion out of lower incomes. *Engel's Law*—a generalization first published in 1857 by Ernst Engel, a German statistician—states that, as income increases, the proportion spent on food decreases. Thus low-income consumers buy proportionally more food than high-income consumers when their incomes are increased.

The substitution effect will also increase the food purchases of recipients because, with food stamps, food becomes cheaper relative to the other commodities they purchase. Therefore recipients will substitute food for other purchases in order to increase utility derived from their total expenditure for a particular period (budget). Thus a food-stamp program should increase the quantity of food demanded and its price.

An increase in the quantity of food demanded affects not only farm producers, but also producers of production inputs such as energy, machinery, and fertilizer. Increased food demand could cause slight increases in the inputs demanded. However, because price elasticities of both the supply and demand for food at retail are relatively low in the short run (as they are for farm inputs), the effects that a food-aid program has on the retail price of food and on farm inputs are probably magnified in the short run, but may have little effect on prices in the long run.

Two conditions would be necessary for the total food-stamp subsidy to be used for additional food. First, the subsidy would have to apply to the entire food budget of the recipient. If the subsidy applied to only a fraction of it, for instance 60 to 80 percent, then the additional food purchases would be at the price listed in the market, and the original food or nonfood preferences would prevail at the margin. Second, recipients would not or could not sell ''surplus'' food to others. Otherwise, recipients might buy food for 80 percent off, for instance, or get it free and sell it to someone else at a profit.

Can these two conditions be met? Probably not entirely, but perhaps in part. On the other hand, is it important to remodel food aid to eliminate slippage? Probably yes, if recipients eat a more nutritious, well-balanced diet. Food-aid is not a panacea for the problems of the poor, or for food and agricultural policy in general, but it can be a beneficial supplement to other policies. What will be its more important economic effects on farm producers, consumers at home and abroad, the federal budget or U.S. taxpayers, and foreign trade and aid?

EXPECTED ECONOMIC CONSEQUENCES OF A FOOD-AID POLICY

The effects of a food-aid policy depend on its magnitude and on how it is targeted, structured, and administered. Until the 1980s, U.S. food-aid programs tended to increase in magnitude, although not necessarily as a percentage of total consumer expenditures. Debate has continued about whether food aid should be continued or increased, and whether its targets, structures, or administration should be changed. What are some of the policy's effects?

Effects on Farm Producers and Agribusinesses

Farm producers are a major beneficiary of a targeted food-aid policy that distributes money, credit, vouchers, or commodities. They gain from both the higher price and the greater volume demanded. Gross farm receipts will increase in the short run, other things being equal, and net income will be as high or higher than it would be otherwise. The extent of change depends on the elasticities of supply and demand and on how rapidly the initial gain in net farm income is eroded by rising costs. Higher fixed costs resulting from the higher product prices and resulting appreciated value of fixed factors, such as land, accrue to owners in any case but, as with any income-enhancing policy, these higher fixed costs become a higher capital threshold for entering farmers. The extent of these effects on producers' welfare will be affected by the elasticities of supply and demand for agricultural commodities. As with the other alternatives considered, most agribusinesses are affected similarly to producers. Producers of course, in countries receiving foreign food aid, can be adversely affected.

Although farm producers gain in the short run from food aid, that gain will be eroded in the long run by increasing average and marginal costs. As shown in Figure 10-1, gross output and income will be higher, but the long-run net returns on investments will be comparable to rates of return on the next best available alternative. The effects of food aid on farm producers will depend on coordinating several policies, not just food-aid policy.

Effects on Consumers

All consumers are affected by this policy, but not in the same way. Recipient low-income consumers benefit from increased consumption at a lower average cost. How they benefit and by how much depends on the magnitude of the aid subsidy and on how it is distributed. They can also purchase more nonfood goods and services because of their enhanced total purchasing power. However, if food-aid programs discourage substitution, then the increase in food consumption will be larger.

Nonsubsidized consumers face the opposite consequences. With the rise in the market price because of increased demand, they pay a higher price, not only for commodities targeted by the policy, but also for other food commodities as well. Their individual losses or costs will be relatively small, however, because such costs are spread over a much broader group, and because the rise in retail food prices is apt to be small, especially if supply is adequate.

In general, these same effects apply to foreign food aid because differential effects are felt by recipient and nonrecipient nations alike. The consequences can of course vary, depending on how other trading nations respond to the aid program of the sponsoring country, and also on whether the donor country augments this policy with other subsidies to its farm sector.

Low-income consumers who are recipients of food aid may become healthier, more productive citizens. Thus, a long-term contribution to economic activity may result from domestic and foreign food aid.

Effects on the Federal Budget and Taxpayers

The major costs of this policy fall on the federal budget and eventually on taxpayers. Under any effective level of food aid, the cost is substantial and is affected by the policy provisions and the demand elasticities of any targeted food groups. Because of this potential budget burden, the pure policy approach needs to be modified by targeting recipients, specifying eligible subsidized food items, varying the level of aid, and limiting recipients' discretionary choice.

Effects on Foreign Trade

Food aid of any kind has consequences for trade, but for a given level of expenditure, U.S. foreign food aid has a relatively greater effect on the recipients than does domestic food aid, due to the lower elasticity of demand in the latter. With domestic aid, the price increase associated with expanded demand tends to restrict U.S. exports of agricultural commodities. Even though more food is produced, more is also demanded domestically. If the previous level of exports is to be maintained, some exports may have to be subsidized. In addition, U.S. producers may not realize a higher price if other exporting countries are free to gain entry to the strengthened U.S. market. This situation might encourage selective U.S. import restrictions. If, on the other hand, the sole objective of the policy is to enhance food consumption of low-income consumers, U.S. food imports at the marginally increased price help to fulfill that goal.

If counteractive measures are not taken, increasing U.S. foreign food aid increases the quantity of food supplied in the recipient countries and tends to lower their food prices. How much these prices will fall depends on the amount of aid, the elasticities of supply and demand for food, and the food-price policies followed by those countries. The slippage problem identified in domestic food-aid policy also exists for foreign aid.

U.S. foreign food aid has just the opposite effect on nonrecipient countries, whether they are commercial importers or competing exporters. Exporters may find that some of their potential market is preempted by U.S. foreign food aid, and they may suffer losses in lower earnings from their own exports. Food importing countries that do not receive U.S. food aid may be relatively unaffected, especially if U.S. commercial export supplies are unchanged. For many other countries, however, U.S. aid policy is almost certain to trigger some programs to protect their own producers' interests. In the long run, it may enhance productivity to the level at which recipient countries import more U.S. products or become trade competitors.

FOREIGN FOOD-AID PROGRAMS

During and following World War I, the United States gave large amounts of food aid to some European countries, thus saving millions of people from acute hunger, perhaps from starvation. Again, only a generation later, U.S. food aid was important during World War II. After the war, food aid was an important part of the massive postwar assistance under the European Recovery Program, often called "The Marshall Plan,"—named after the U.S. Secretary of State, who proposed it—which is credited with greatly accelerating the economic recovery of Western Europe. Also, immediate food aid after that war was critical for Japan and some other Asian countries. Foreign food aid with a continuing commitment first appeared with the Agricultural Trade Development and Assistance Act of 1954 (Public Law 480). It emerged during a period of worsening economic conditions for American agriculture and bulging bins of CCC-owned stocks of forfeited, price-supported grains. Searching for a resolution to the multifaceted problem, impatient producers asked, "Why can't we sell our surplus overseas for foreign currency if they do not have dollars?" After much debate, the policy answer was yes, sell it, and use some of the foreign currencies held in U.S. accounts for mutually agreed-on developmental projects in the recipient countries.

Public Law (P.L.) 480, launched as a new initiative in the evolution of U.S. trade policy, has continued as the nation's primary response to natural disasters and chronic food shortages in developing countries. Through several decades, this law has promoted U.S. exports of agricultural commodities, using programs such as sales for foreign currencies, barter for strategic materials, short- and long-term credit sales, government-to-government assistance, donations through charitable organizations, and mutual assistance.

In general, using P.L. 480 has expedited the exporting of stored commodities from

CCC stockpiles, usually considered surplus. Consequently, it has promoted three objectives: reducing costly public stocks, expanding markets for U.S. producers, and providing food assistance to other nations. It has also served diplomatic, political, and military objectives and has helped particular developing countries during emergencies or food shortages. The general rule was that potential recipient countries had to be financially unable to provide their population with adequate food without jeopardizing their national economic stability, and the food aid could not substantially reduce their commercial purchases. P.L. 480 exports, other than outright donations, have shifted gradually from foreign currency sales and barter to credit sales with terms of varying lengths and financial soundness.

P.L. 480 has been criticized, however. Since 1954, competing exporting countries have at times objected through diplomatic channels that P.L. 480 was subjecting them to unfair competition by restricting their own export markets. Although difficult to prove, such assertions are not entirely unfounded. In general and in some developing countries in particular, large P.L. 480 shipments have in fact depressed agricultural commodity prices and retarded agricultural development. In addition, grain shipments within the importing nations have been wasted or have spoiled, perhaps more so than

TABLE 10-1
VALUE OF U.S. FARM PRODUCTS EXPORTED UNDER P.L. 480, SELECTED YEARS, 1955 TO 1990

Fiscal Year	Title I[a]	Title II[b]	Barter for Strategic Material	Total Current Dollars	Total Constant[c] Dollars
		—Million dollars—			
1955	73	187	125	385	1,415
1960	824	143	145	1,116	3,612
1965	1,300	238	32	1,570	4,645
1970	815	241	—	1,056	2,514
1975	762	339	—	1,101	1,857
1980	865	476	—	1,341	1,565
1981	790	544	—	1,334	1,419
1982	722	385	—	1,107	1,107
1983	810	385	—	1,195	1,150
1984	775	602	—	1,337	1,241
1985	928	699	—	1,627	1,467
1986	766	420	—	1,186	1,042
1987[d]	695	248	—	943	803
1988[d]	693	463	—	1,156	953
1989[d]	722	469	—	1,191	943
1990[d]	658	451	—	1,009	767

[a] Sales for local currency, long-term dollar, and convertible local currency credit sales.
[b] Government-to-government donations, World Food Program, and donations through voluntary relief agencies.
[c] Constant 1982 dollars, using GNP implicit price deflation.
[d] Preliminary.
Source: USDA, Foreign Agricultural Service, *Food for Peace: 1990 Annual Report on Public Law 480,* May 1992.

is usual for unsubsidized exports. Most of the commodities have been wheat and flour, which require efficient handling and well-conditioned storage. Many recipient countries, as a result, face distributional and storage problems.

Although P.L. 480 continues, the programs have been gradually reduced in real terms. As shown in Table 10-1, for many years P.L. 480 exports have been stable in current dollars at $1.0 billion to $1.5 billion a year. When these data are converted to constant (1982) dollars, however, shipments in recent years have been only about one third of what they were in the 1950s and 1960s. Also, foreign currency sales, which were phased out in early 1970, and barter sales, which were stopped even earlier, shifted to long-term credit sales, which grew gradually from the early 1960s to one half to two thirds of the total aid. Finally, donations directly through governments and relief agencies increased slowly in the early years and then generally became stable, accounting for the other one third to one half of all aid.

DOMESTIC FOOD-AID PROGRAMS

Government-supported domestic food aid has been more extensive than foreign food aid and has evolved in a much different way. Originally, in the United States, in the eighteenth and nineteenth centuries, church groups and other local organizations sponsored relief efforts for the poor. In the early years of the Great Depression, they expanded their efforts. Bread lines and soup kitchens became part of the culture of large cities. Then, as policy interest shifted to agriculture, federal aid was initiated as an outlet for surplus agricultural commodities. Over time, emphasis in the government-supported domestic food-aid programs shifted to meeting the nutritional needs of the poor and other targeted populations, particularly infants, children, and the elderly.

Early Experience

To boost farm prices during the Great Depression, the federal government purchased and distributed surplus food commodities to needy families, often in town squares or community centers, in bulk lots as bags of flour, containers of dried milk, or other commodities. Surplus commodities were also donated to schools for lunch programs and milk breaks. In 1935, a new chapter in food-aid policy was opened with the enactment of Section 32 of the amended Agricultural Adjustment Act of 1933 (P.L. 74–320), which made monies available from import tariffs and custom receipts for funding food distribution.

In 1939, an experimental Food Stamp Plan was initiated and tied to agricultural surpluses. Eligible families purchased orange-colored stamps, which they used as currency to buy food in grocery stores or other retail food outlets. In return, they received blue stamps free of charge to buy foods designated as surplus each month by the secretary of the USDA. By the early 1940s, the Food Stamp Plan was available in more than half of the counties, and about 4 million people were participating.[2] The plan was terminated in 1943, however, because there were few unemployed, and food surpluses largely disappeared during World War II. Commodities were distributed directly to the needy and to schools, but at a greatly reduced level.

After World War II, a significant step was made in food-aid policy with the National School Lunch Act of 1946. This Act established a permanent School Lunch Program designed to "safeguard the health and well-being of the nation's children and to encourage the domestic consumption of nutritious agricultural commodities."[3] It authorized the USDA to donate commodities and cash for program operations. When food surpluses reappeared in the 1950s, the National School Lunch Program became an important outlet for them.

The reemergence of surpluses also contributed to renewed interest in a food-stamp program, but political disagreements delayed program development for more than a decade. During the 1950s, Representative Leonor Sullivan (Dem., MO) led the fight for a more effective food-aid program, emphasizing the paradox of huge surpluses while the urban poor and the unemployed were hungry.[4] Apparently in response, the Agricultural Act of 1956 directed the USDA to analyze food-stamp plans. The department concluded that such plans were costly and less effective than direct commodity distribution in reducing agricultural surpluses. However, in 1959, Congress adopted an amendment to the Agricultural Trade Development and Assistance Act of 1954, which authorized demonstration food-stamp programs, but the USDA did not make use of this authority and it was allowed to expire. In contrast, one of the first actions of the new Kennedy Administration in 1961 was to establish a new pilot food-stamp program with Section 32 funds that were obtained from agricultural import duties and custom receipts and designated for domestic market support. By August 1964, 350,000 people in 43 project areas were participating in the program.[5] Evaluations of the program showed that the nutrition of participants improved and that retail food sales rose. Apparently influenced by these evaluations, the Food Stamp Act of 1964 was adopted.[6]

Many analysts cite the trade-offs between members of Congress from urban areas and those from rural districts and states as instrumental in the passage of the Food Stamp Act of 1964 (P.L. 88–525). The national policy objective of this Act was to "raise the levels of nutrition among low income households." Important policy questions—such as who would be eligible, whether the program would contain a purchase requirement, and the appropriate benefit levels—were addressed in the legislation and through administrative decisions. These questions have generated controversy ever since the program was established.

Under the Food Stamp Program, as established by the 1964 Act, states had the option of offering the new program or continuing to offer the old commodity distribution program. The federal government determined the food-stamp benefit levels and bore the cost of funding benefits, federal administrative costs, and about 30 percent of state administrative costs. The states decided who would be eligible and set limits on the income and assets that a household could own and still qualify for the program. Eligible households had to pay a certain amount, called a purchase requirement, and received an allotment of stamps worth more than the purchase price. The difference between the allotment and the purchase price was the subsidy provided by the federal government and was referred to as bonus food stamps. The purchase price was based on a household's income, and all households were required to purchase stamps as a condition of program participation.

During the 1960s, the Food Stamp Program grew slowly. Problems arose with particular provisions of the program and the way in which it and the Commodity Distribution Program were being administered locally, especially in the rural South, where racial discrimination was a lingering problem.[7] In some counties, participation declined when the Food Stamp Program displaced the Commodity Distribution Program, the purchase requirement being cited as a major reason behind the decline. Civil rights activists and the hunger lobby exerted pressure to reform the program.

In the late 1960s, a broad-scale citizens' movement emerged and focused on the plight of the poor, reaching a high point in a so-called "poor people's march" to Washington, DC. Throughout the summer of 1968, thousands camped on the national mall in an area that they declared to be Resurrection City. But as the hot summer continued and the movement's leaders received only a few positive responses from the targeted government offices, they began to concentrate their efforts on the department of agriculture and on more specific reforms in food-aid programs, which indeed had been thwarted by many local and state governments.

The goals of the Poor People's Campaign converged with those of citizens' groups and policy makers who were interested in documenting and solving hunger problems in the United States. Activities that culminated in reform of the Food Stamp Program and authorization of other food-aid programs included the following:

- field hearings by a subcommittee of the Senate Labor and Public Welfare Committee, which resulted in two influential members, Senators Robert Kennedy and Joseph Clark, touring the Mississippi Delta and observing extreme poverty and malnutrition firsthand
- investigations of hunger by citizens' groups such as the Field Foundation and the Citizens' Crusade Against Poverty, which formed an organization called the Citizen's Board of Inquiry into Hunger and Malnutrition in the United States
- publication of a report entitled *Our Daily Bread,* compiled by a national group brought together by several women's organizations, which was critical of the weaknesses of the food programs
- a CBS (Columbia Broadcating System) documentary entitled "Hunger USA," which indicted the food programs for being inadequate in scope and coverage

From the late 1960s to mid-1970s, the Food Stamp Program underwent major revisions, perhaps largely as a result of the reform campaign spearheaded by social activists concerned about poverty and hunger. In addition, new programs such as the School Breakfast, Child Care Food, Summer Food Service, Commodity Food Service, and Special Supplemental Food Program for Women, Infants, and Children (WIC) were established.

More Recent Developments in Food-Aid Policy

On May 6, 1969, President Nixon pledged to "end hunger for all time in America."[8] This commitment signaled new directions for food policy. In late 1969, food-stamp allotments were based through administrative decree on the cost of the Economy Food Plan, which was developed by the USDA. The purchase price for food stamps was lowered so that no household paid more than 30 percent of its income for food stamps, and stamps were provided free for the poorest households (that is, those households with less than $30 monthly net income). The 1971 amendments to the Food Stamp Act made important administrative changes and also established national uniform eligibility standards and annual indexing of food-stamp benefits for inflation.

In 1973, the Food Stamp Program was reformed again. All counties were mandated to operate the program by July 1974, and direct commodity distribution was ended as a major policy instrument for assisting the poor and hungry. The 1973 amendments also called for semiannual adjustment of food-stamp benefits so that their purchasing power would be better maintained. These changes, combined with those implemented in 1969 and 1971, resulted in rapid expansion of the Food Stamp Program.

The preamble to the 1971 amendments indicated that the program's goal was to provide participants with "an opportunity to obtain a nutritionally adequate diet," which the U.S. Court of Appeals ruled in 1975 was a shift in policy from "supplementing diets of low-income households to guaranteeing those households the opportunity for an adequate diet."[9] In response to this decision, the USDA replaced the Economy Food Plan, which had been designed for temporary use in emergency situations, with the Thrifty Food Plan as the basis for food-stamp allotments. Congress revised the preamble to state that the program's goal was to "permit low-income households to obtain a more nutritious diet."[10] The cost of the Thrifty Food Plan was set at the same level as the cost of the Economy Food Plan, and an analytic model was used to select the appropriate mix of nutritious food that deviated the least from food consumption patterns of low-income households. The least costly of the four food plans developed by the USDA, the Thrifty Food Plan includes less meat, poultry, and fish and more dried beans and grains than many households consume on the average.

The Food Stamp Act of 1977 mandated another significant change in food-stamp policy: the elimination of the purchase requirement for all households (see box on page 178). Under the revised program, participating households receive what were originally called bonus stamps as their monthly food-stamp allotment. From the program's beginning, an underlying philosophy held that everyone should pay something, even a minimal 50 cents per month by the poorest, for their food stamps (that "there is no free lunch" adage). Others were concerned that eliminating the purchase requirement would reduce the food expenditures of participating households and weaken the program's effects on the food marketing system. As a result of eliminating the purchase requirement, the Food Stamp Program became more accessible to low-income households and permitted them to reach a higher indifference curve, thereby increasing their utility (see Figure 10-2). The USDA estimated that eliminating the purchase requirement added 3.4 to 3.6 million people to the food-stamp rolls. Revisions of the Food Stamp Program in the 1960s and 1970s transformed it into an important part of the public assistance system that provides a safety net for the poor.

During the 1970s, other food-aid programs grew rapidly (Table 10-2). However, the cost of the food distribution programs has fluctuated. The special Supplemental Food Program for Women, Infants, and Children (WIC), which was created in 1974, represents an innovative approach to food aid. Participation is limited to women who are pregnant, nursing, or postpartum and to infants and children up to age five years who are nutritionally at risk and who have family income below a certain level. Benefits in

TABLE 10-2
FEDERAL COST OF USDA FOOD PROGRAMS, 1970–1990[a]

Year	Food Stamps: Current	Food Stamps: Constant	Puerto Rico Nutrition Assistance	Food Distribution[b]	Child Nutrition[c]	Program for Women, Infants, & Children	Total: Current	Total: Constant
				—Million dollars—				
1970	1,104	(3,145)	—	543	462	—	2,109	(6,008)
1971	1,699	(4,592)	—	609	776	—	3,084	(8,335)
1972	1,980	(5,103)	—	539	947	—	3,466	(8,933)
1973	2,209	(5,349)	—	453	1,097	—	3,784	(9,162)
1974	3,498	(7,791)	—	478	1,381	33	5,390	(12,004)
1975	4,602	(9,354)	—	408	1,684	106	6,800	(13,821)
1976	4,657	(8,904)	—	489	2,010	182	7,337	(14,029)
1977	5,014	(8,970)	—	570	2,267	289	8,139	(14,560)
1978	5,261	(8,725)	—	655	2,453	422	8,790	(14,577)
1979	7,108	(10,852)	—	914	2,778	569	11,369	(17,357)
1980	9,004	(12,558)	—	1,211	3.181	783	14,179	(19,775)
1981	10,968	(13,901)	—	1,096	3,086	863	16,013	(20,295)
1982	10,376	(12,382)	412	1,388	2,950	1,002	16,128	(19,246)
1983	11,673	(12,751)	825	2,318	3,224	1,194	18,680	(21,422)
1984	10,700	(11,728)	825	2,284	3,400	1,417	18,601	(20,441)
1985[d]	10,744	(11,381)	786	2,192	3,471	1,487	18,680	(19,788)
1986[d]	10,605	(10,944)	824	2,173	3,664	1,582	18,848	(19,451)
1987[d]	10,500	(10,500)	856	2,153	3,849	1,678	19,036	(19,036)
1988[d]	11,149	(10,730)	883	1,842	4,049	1,435	19,358	(18,631)
1989[d]	11,676	(10,771)	912	1,417	4,272	1,489	19,766	(18,234)
1990[d]	14,205	(12,582)	940	1,062	4,684	2,116	23,163	(20,516)
1991[d]	17,348	(14,827)	967	1,126	5,182	2,293	27,072	(23,138)

[a] All data are in current dollars except those in parentheses, which are in constant 1987 dollars using the GNP deflator.
[b] Includes Indian reservations, schools, Temporary Emergency Food Assistance Program, and cash in lieu of commodities.
[c] School lunch, school breakfast, child care food, summer food service, and special milk. *Source:* USDA, Economic Research Service, *National Food Review* and *Food Review*, various issues.
[d] Fiscal year.

ELIMINATION OF THE PURCHASE REQUIREMENT AND CASH OUT OF FOOD STAMPS: AN EXAMPLE

Prior to 1979, food-stamp program participants with monthly incomes in excess of the stated minima were required to pay a certain amount, called a purchase requirement, in order to obtain their monthly allotment of food stamps. The allotment was based on the number of people in the household. For example, in July 1975, all participating four-person households received an allotment of stamps worth $162. Household income determined the monthly purchase requirement. Four-person households with monthly net incomes of less than $30 were exempt from the purchase requirement. At the other extreme, those with monthly net income of $480 to $569.99 had a purchase requirement of $138. Four-person households with monthly net income of $570 or more were ineligible for food stamps.

A household with a monthly net income of $150, had a purchase requirement of $41. In exchange for the $41, the household received $162 worth of food stamps (its monthly coupon allotment). The difference between the allotment and the purchase requirement, in this case $121, was a government subsidy or transfer to the household. In effect, the household was compelled to spend the allotment of $162 on food. Under the policy alternative of eliminating the purchase requirement, the household receives $121 in food stamps. If the program is cashed out, the household would receive $121 in cash (which is actually an income subsidy) and is not required to spend any of it on food. However, as incomes increase among the poor, their food expenditures also rise because of their relatively high income elasticity of demand for food.

the form of vouchers redeemable for specified foods are based on age, special dietary needs, and other related factors.

By the end of the 1970s, the federal government was spending approximately $11 billion to provide food aid to some 45 to 50 million people. Studies showed that progress had been made in alleviating hunger and credited the food-aid programs for this progress.[11]

In the early 1980s, a combination of factors contributed to significant changes in food aid. These included the rapid growth of food program expenditures in the 1970s, concern about fraud and abuse in the programs, mounting government stocks of surplus commodities, and the growing federal deficit. Perhaps one of the most significant policy changes was the cashing out of the Food Stamp Program in Puerto Rico. In an effort to control program costs there, the Omnibus Budget Reconciliation Act of 1981 replaced the Food Stamp Program with a block-grant Nutrition Assistance Program, effective July 1982. Under the block grant, Congress established a fixed appropriation level, and Puerto Rico was authorized to design its own program. Commonwealth policy makers decided to provide benefits in cash rather than in-kind such as food stamps and continued this alternative food-aid policy into the 1990s.

Establishing the Temporary Emergency Food Assistance Program (TEFAP) in 1982 was another significant policy. TEFAP reflected the original objectives of food aid, which was to serve as an outlet for distribution of commodities obtained through farm price-support programs. Conditions in the early 1980s were similar to but not as severe or widespread as those in the 1930s, when the first domestic food-aid programs were established. The farm economy was in an economic downturn, unemployment was rising, reports indicated that many people were hungry, and the government had accumulated costly inventories of agricultural commodities. From the initial distribution of processed American cheese, TEFAP expanded to include butter, instant nonfat dry milk, cornmeal, rice, flour, and honey. According to various estimates, 15 to 18 million households received some TEFAP commodities in 1986.[12] By the late 1980s, the huge government stocks had disappeared and questions arose about the future of the program.

In the 1980s, some other important revisions were made that affected participation and costs. For example, the Omnibus Budget Reconciliation Acts of 1981 and 1982 instituted changes in the Food Stamp Program that reduced the allowable income deductions, including the earned income deduction, delayed cost-of-living adjustments, tightened eligibility standards for students, and temporarily lowered benefits so that households with no net income received benefits equal to the cost of the Thrifty Food Plan less one percent. In the National School Lunch Program, reimbursement rates for paid meals were reduced, the maximum charge for a reduced-price lunch doubled and income eligibility requirements for free and reduced-price meals were tightened.[13] However, when the provisions of the 1981 and 1982 Omnibus Budget Reconciliation Acts were implemented, the United States was in a recession that saw the unemployed rate for civilian workers climb from 7.6 percent in 1981 to 9.6 percent in 1983, while the number of people living in poverty increased from 31.8 million to 35.3 million.

During the 1980s new reports of hunger emerged. Many people, including families with young children and the unemployed, sought food aid from soup kitchens, food pantries, and other private volunteer programs, but there were no national scientific surveys or data to document claims that hunger had again become a greater problem. Thus the issue was quite controversial. Two credible organizations issued conflicting reports on hunger in America. The President's Task Force on Food Assistance reported in 1984 that there was no evidence of hunger in America (in the clinical sense), except perhaps among the homeless. Although the report conceded that some people were hungry if the commonly accepted definitions were used, it indicated that the extent of hunger could not be estimated. On the other hand, the Physician Task Force on Hunger in America estimated that 20 million Americans suffered from hunger some time each month in 1985. Furthermore, the report suggested that hunger was "more widespread and serious than at any time in the last 10 to 15 years."[14]

Reports about the reemergence of hunger contributed to revisions in the Food Stamp Program in the mid to late 1980s. In 1984 and 1985, benefits were increased to 100 percent of the Thrifty Food Plan, and the asset limit for nonelderly households was raised. States were required to implement procedures for certifying and issuing food stamps to eligible homeless people and were prohibited from charging sales taxes on food purchased with food stamps. The Hunger Prevention Act of 1988 (P.L. 100–435) mandated additional changes, including raising food-stamp benefits slightly above the Thrifty Food Plan, providing matching funds to states for outreach programs, and making permanent the rules that eased food-stamp eligibility for the homeless. This Act also addressed the problem of reduced commodities for distribution under TEFAP by authorizing the USDA to purchase commodities in the commercial market. This action necessitated a name change—from Temporary Emergency Food Assistance Program to Emergency Food Assistance Program. Also, this was the same solution that was applied to the National School Lunch Program in its early history when quantities of surplus commodities were insufficient to maintain program operations.

During the 1980s, growth in food aid was concentrated in the WIC program (Table

TABLE 10-3
MONTHLY PARTICIPATION IN MAJOR FOOD-AID PROGRAMS, 1980 TO 1988

Year	Food Stamp Program[a]	National School Lunch Program[b]	School Breakfast Program[b]	Program for Women, Infants, and Children
		Million		
1980	19.2	26.6	3.6	1.9
1981	20.6	25.8	3.8	2.1
1982	20.4	22.9	3.3	2.2
1983	21.6	23.0	3.4	2.5
1984	20.9	23.4	3.4	3.0
1985	19.9	23.6	3.4	3.1
1986	19.4	23.7	3.5	3.3
1987	19.1	23.9	3.6	3.4
1988	18.7	24.2	3.7	3.6
1989	18.8	24.2	3.8	4.1
1990	20.1	24.1	4.1	4.5
1991	22.6	24.2	4.4	4.9

[a] Excludes Puerto Rico
[b] Nine-month average
Sources: J. William Levedahl and Masao Matsumoto, "U.S. Domestic Food Assistance Programs: Federal Costs and Public Benefits," in *Agricultural-Food Policy Review: U.S. Agricultural Policies in a Changing World,* Agricultural Economic Report No. 620, USDA, Economic Research Service, Washington, DC, November 1989, p. 193; and *Food Program Update,* Financial Management, Food and Nutrition Service, USDA, FY 90 and FY 92.

10-3) and commodity distribution programs, especially TEFAP. Although WIC funding and participation increased during the decade, it remained a nonentitlement program, which meant that it was subject to appropriations limits and could not serve all eligible people. By the end of the decade, the program served only about half of the eligible people, and program funding remained a contentious issue.

Effects of Food-Aid Programs

The primary beneficiaries of food-aid programs are participants of course, but many people are affected by the programs, including farm producers, suppliers of inputs, and U.S. taxpayers as discussed previously. Let us examine the effects of selected programs.

Food Stamp Program More than three fourths of all participants are children, elderly, disabled, or single parents who are heads of households. Participants use their stamps to purchase food and nonalcoholic beverages, but the types of foods that can be purchased are not stipulated. Thus participants do not necessarily purchase foods that enhance their nutritional status. Further, some households probably use their stamps in place of income that otherwise would have been spent for food. The effects on food expenditures and diet quality could range from no increase to a substantial increase.

Most studies of the Food Stamp Program have used data collected before elimination of the purchase requirement, which reduced the amount of money and stamps that households were required to spend on food.[15] These studies have generally found that the program increases food expenditures, typically in the range of 11 to 30 cents for each dollar of food stamps.[16] Because the remainder of each dollar's worth of stamps (70 to 89 cents) makes more money available to buy nonfood items, the major effect of the program is to supplement the incomes of participating households.

Only a few studies have examined the effects of the Food Stamp Program after elimination of the purchase requirement. These studies found that the program has a positive and significant effect on household food expenditures and has a substantially greater effect on food expenditures than would an equivalent amount of cash transfer.

The effect of the program on nutritional status is less conclusive. Some studies, based on local areas that may not be representative of the national food-stamp population, have found no significant effects. Others, including some national studies, have found evidence that food stamps have a positive influence on nutrient availability.

The effect of the Food Stamp Program on the food and fiber system is generally modest. Sylvia Lane estimated that the program increases total farm cash receipts by about one percent or \$1.5 billion to \$2.0 billion, with some producers and sectors of the food industry such as pork producers receiving a relatively large share of the increase.[17] Because the program increases the demand for food, it leads to an increase in retail food prices, but this increase is very small, perhaps in the lower range of 0.08 to 4.0 percent.

The Food Stamp Program has become valued for its effects during economic downturns when personal income declines and unemployment rises.[18] The USDA estimates that with each one percentage point increase in the unemployment rate, program participation increases by 750,000 to 1 million people (Figure 10-3). Because the program bases participation only on income, assets, and work registration requirements and can quickly add eligible households to its roles, it has become an automatic stabilizer in the business cycle and a safety net for those who become needy because of a changing economy.

National School Lunch and School Breakfast Programs Students from all income classes participate in these programs. They receive subsidized meals and the amount of the subsidy depends on family income. The USDA donates food and cash to help defray meal costs. Students from the poorest families receive free meals while those just above the poverty level receive meals at a reduced price. About half of the participants in the lunch program and 90 percent of the participants in the breakfast program receive free or reduced-price meals.

Evaluations of the National School Lunch Program have generally concluded that it significantly improves the nutritional status of both low-income and high-income participants. The national evaluation of school nutrition programs conducted for the USDA found that participants' lunches contained more of several nutrients such as vitamins A and B_6, calcium, magnesium, and protein than did the lunches of nonparticipants.[19] The positive effects were due to the quality and quantity of the food served.

Benefits accrue to the food and fiber system because the USDA donates surplus agricultural commodities to the National School Lunch Program and also purchases some commodities in the commercial market. The USDA estimates that demand increases for red meats, poultry, and milk because of the program; demand also increases for fruits, vegetables, and eggs, but to smaller extent.[20]

Relatively few studies have been conducted on the effects of the School Breakfast

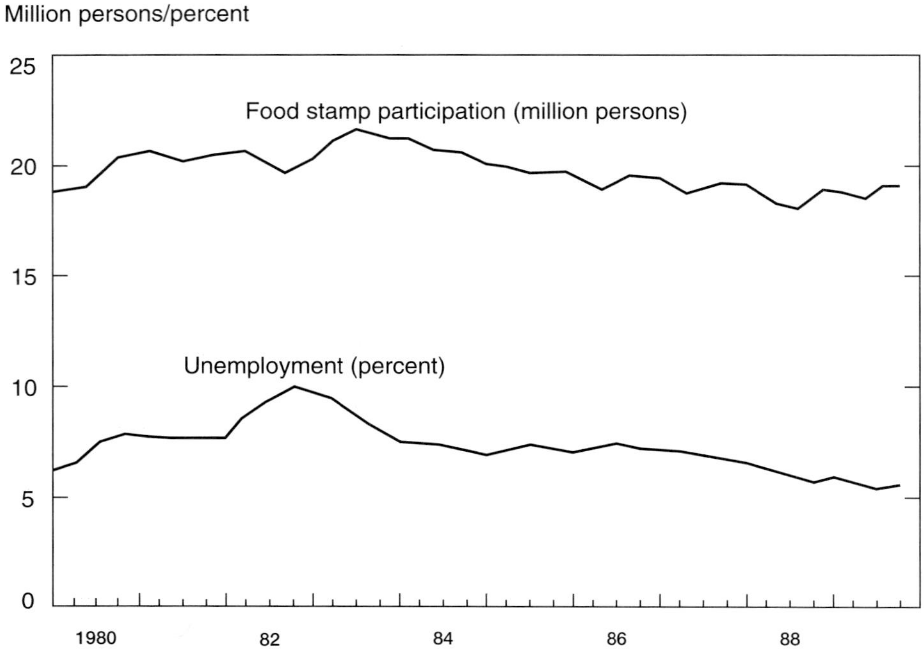

1989 estimated. Seasonally adjusted fiscal year data.

FIGURE 10-3
Unemployment rate and participation in the food stamp program. *Source: 1990 Agricultural Chartbook,* USDA, Agriculture Handbook No. 689, Washington, DC, p. 74.

Program. This fact notwithstanding, the results of the studies are insightful and demonstrate that benefits extend beyond food consumption. According to the national evaluation of school nutrition programs, children enrolled in schools that participate in the breakfast program are more likely to eat breakfast than are children attending nonparticipating schools. However, the breakfast program increased consumption of only a few nutrients relative to nonschool breakfasts. Some evidence indicates that the School Breakfast Program improves academic performance and on-time attendance among young children.[21]

WIC About 80 percent of WIC participants are infants or children. Researchers who evaluate the program have tended to conclude that it improves the diet of participants and increases access to health care, which results in beneficial effects such as maternal weight gain and increased birth weights. Health improvements noted in a 1986 study of WIC included reduction of preterm deliveries, larger head size of children, and perhaps improved brain growth and potential improvement in behavioral and cognitive performance.[22]

Emergency Food Assistance Program This program provides food aid to the hungry and the poor, but its effect on diet quality is probably limited because only a few commodities (primarily cheese and butter) have been available from government stocks. Moreover, the program has displaced some commercial sales, which reduces its benefit for the agricultural sector. The USDA estimates that each 100 pounds of cheese donations under the program displaces about 35 pounds of commercial cheese sales. Although butter sales are unaffected by the program, margarine sales are displaced on a one-for-one basis.

IMPLICATIONS FOR FUTURE POLICY DIRECTIONS

Food-aid programs have been shaped by concerns about hunger and malnutrition and the level of government stocks of surplus agricultural commodities. In recent years, the federal budget deficit has also been of paramount concern. These issues are likely to influence the future direction of food aid. Other issues that probably will be important include the characteristics of the poor and whether society considers them to be deserving of aid. Infants, children, the elderly, and the disabled generally fall in this category.

Eligibility and benefit levels for the food-aid programs have been modified over time to reflect changing social, economic, and political conditions. Because of the trade-offs involved in obtaining a politically acceptable solution, the programs are criticized by those who believe that the standards and benefits are too liberal and by others who believe that they do not provide enough aid to the poor. A recurring issue has been the level of food-stamp benefits and the benchmark for these benefits in the Thrifty Food Plan. Underlying assumptions built into the plan about the nutritional knowledge of household food managers, bulk purchase and storage of food, and level of food waste have raised questions about the adequacy of the Thrifty Food Plan for setting benefit levels.[23]

Studies conducted by interest groups such as Public Voice for Food and Health Policy have shown that food prices vary significantly by location. Consequently, Public Voice has suggested that benefits should reflect the disparities rather than being uniform across the contiguous states. Public Voice found in 1989 that the actual weekly cost of the Thrifty Food Plan marketbasket for a family of four was 8 percent higher, on average, in supermarkets in persistently poor rural areas and 36 percent higher in small- to medium-sized stores in those same areas, than the USDA's defined cost of the marketbasket. Furthermore, because some areas had relatively few supermarkets, participants were compelled to shop mostly at the smaller, more expensive food stores. Inner-city residents face similar problems—higher prices than the suburbs and few, if any, supermarkets.

Congress has gradually raised food-stamp benefits above the Thrifty Food Plan level (100.65 percent of the level for fiscal 1989, 102.05 percent for fiscal 1990, and 103 percent for fiscal 1991 and thereafter). But this increase may not resolve the issue. Some advocates are pushing for the USDA Low Cost Food Plan rather than the Thrifty Food Plan to be used as the basis for benefits and for other reforms such as revising the benefit levels semiannually instead of annually. This latter practice was initiated in the 1970s but eliminated in one of the cost-saving efforts of the early 1980s.

One policy change that economists have often recommended on the basis of efficiency and maximization of utility is cashing out the Food Stamp Program. The USDA has initiated demonstration projects in which participants receive cash rather than stamps. Participation in the early projects was limited to the elderly, but in 1989 the USDA began testing the effects of a cash out on the general population. Cash out of the program will likely depend on the effects of cash, as determined by the demonstration projects, and public acceptance.

The evolution of domestic food policy suggests that incremental changes are likely in the food-aid programs in the coming years. Interest groups will probably continue to lobby for expansion of the WIC program so that more, if not all, eligible people can participate. Other incremental changes are likely to include continued improvement in the nutritive quality of meals served in various programs and perhaps more variety in the foods distributed under the Emergency Food Assistance Program now that commercial purchases are authorized.

In the evolution of U.S. agricultural price and income policy, domestic food aid and foreign aid policies have become an integral, continuing, and important part of the periodic cycles of the nation's policy making. For many decades, U.S. public policy has made a substantial commitment to foreign food aid, through P.L. 480, and also to domestic food aid, which is unprecedented in representative nations of the world. The dominant domestic program of food stamps reaches about 10 percent of the U.S. population. In addition, millions of people have been helped by the various programs targeted to infants, children, and the elderly. Recently, some of these have been given priority. Since 1983, in terms of real value, however, the foreign food aid has declined consistently and substantially, while the domestic programs have declined only slightly.

How much to spend and which programs to emphasize are questions for never-ending debate. In domestic programs for instance, with the public commitment to provide both food assistance and freedom of choice for recipients, the phenomenon of slippage continues to challenge the funding and administration of these programs as they try to achieve the several policy objectives directed at both low-income consumers and farm producers.

Using food as a specific vehicle for foreign aid raises questions about priorities in both the United States and the targeted recipient countries. Therefore, to reach sound judgments concerning the net marginal costs and benefits of the various programs, there is a need for a wide range of knowledge about income levels, nutritional adequacies, and other types of aid available to the targeted domestic groups and foreign countries. This suggests that many valuative questions will have to be answered whenever the pros and cons of future food-aid policy are discussed. It is necessary, for instance, to evaluate what is ethically or morally justified, and what can or cannot be accomplished by public policy. We hope that students of policy will retain a broad vision of the problems in this area and the various alternatives for the future.

LOOKING AHEAD

The next chapter completes this discussion of price and income policy by summarizing some of the basic economic concepts of welfare, and thereby providing a basis for evaluating effects of this policy in terms of welfare concepts. The chapter also provides a foundation for studying the policies discussed in the remainder of this book.

ENDNOTES

1 The discussion and graph draw on an initial analysis by Herman M. Southworth, "The Economics of Public Measures to Subsidize Food Consumption," *Journal of Farm Economics,* 27, February 1945, pp. 38–66. Other economists who have used indifference curves in analysis of food-stamp policy and on whose work we draw include Dennis H. Sullivan, "A Note on Food Stamp Reform," *American Journal of Agricultural Economics,* 58, August 1976, pp. 560–562, and Sylvia Lane, "The Food Stamp Program," in R. G. F. Spitze and Marshall A. Martin (eds.), *Analysis of Food and Agricultural Policies for the Eighties,* North Central Regional Research Publication No. 271, University of Illinois Agricultural Experiment Station, November 1980, chap. 7. The discussion on the income effect, substitution effect, and Engel's Law appeared in Sylvia Lane, *Ibid.*

2 *Legislative History of the Food Stamp Program,* USDA, Food and Nutrition Service, (no date).

3 Public Law 79–396.

4 For further discussion, see Marjorie L. DeVault and James P. Pitts, "Surplus and Scarcity: Hunger and the Origins of the Food Stamp Program," *Social Problems,* June 1984, pp. 545–557.

5 As reported in *Food Stamp Program,* USDA, Food and Nutrition Service, February 1986.

6 Margaret S. Andrews and Katherine L. Clancy, "The Political Economy of the Food Stamp Program," Discussion Paper Series No. RR85–08, National Center for Food and Agricultural Policy, Resources for the Future, Washington, DC, 1985.

7 For further discussion on the operation of food-aid programs in some rural, southern communities during the 1960s civil rights movement, see Marjorie L. DeVault and James P. Pitts, "Surplus and Scarcity: Hunger and the Origins of the Food Stamp Program," *Social Problems,* June 1984, pp. 545–557.

8 Harold G. Halcrow, *Food Policy for America,* McGraw-Hill, New York, 1977, p. 65ff.

9 Margaret S. Andrews and Katherine L. Clancy, "The Political Economy of the Food Stamp Program," Discussion Paper Series No. RR85–08, National Center for Food and Agricultural Policy, Resources for the Future, Washington, DC, 1985.

10 Ibid.

11 For further discussion, see William T. Boehm, Paul E. Nelson, and Kathryn A. Longen, *Progress Toward Eliminating Hunger in America,* Agricultural Economic Report No. 446, USDA, Economics, Statistics, and Cooperatives Service, Washington, DC, 1980; Nick Kotz, *Hunger in America: The Federal Response,* Field Foundation, New York, 1979.

12 J. William Levedahl and Masao Matsumoto, *U.S. Domestic Food Assistance Programs: Lessons from the Past,* Agriculture Information Bulletin Number 570, USDA, Economic Research Service, Washington, DC, January 1990, p. 10.

13 For further discussion on the food program provisions in the 1981 and 1982 Omnibus Budget Reconciliation Acts, see Joyce E. Allen, "Food Programs Grow Slowly From 1979 to 1983," *National Food Review,* NFR–26,1984, pp. 21–25.

14 Physician Task Force on Hunger in America, *Hunger in America: The Growing Epidemic,* Wesleyan University Press, Middletown, CT, 1985, p. xix.

15 These studies include: Joyce E. Allen and Kenneth E. Gadson, *Nutrient Consumption Patterns of Low-Income Households,* Technical Bulletin No. 1685, USDA, Economic Research Ser-

vice, Washington, DC, June 1983; Jean-Paul Chavas and M.L. Young, "Effects of the Food Stamp Program on Food Consumption in the Southern United States," *Southern Journal of Agricultural Economics,* July 1982, pp. 131–139; Stanley R. Johnson, James A. Burt, and Karen J. Morgan, "The Food Stamp Program: Participation, Food Costs, and Diet Quality for Low-Income Households," *Food Technology,* 35: 1981, pp. 58–70; Pamela H. Neenan and Carlton G. Davis, "Impact of the Food Stamp Program on Low-Income Household Food Consumption in Rural Florida," *Southern Journal of Agricultural Economics,* July 1977, pp. 89–97.

16 The major findings of these studies are discussed in Joyce E. Allen and Carlton G. Davis, "Domestic Food Programs: Issues and Alternatives," *Agricultural and Food Policy: Issues and Alternatives for the 1990s,* Robert G. F. Spitze (ed.), Department of Agricultural Economics, University of Illinois, Urbana, pp. 193–210; Sylvia Lane, "Food Stamps," *Choices,* Second quarter 1986, pp. 16–20.

17 Sylvia Lane, "Food Stamps," *Choices,* Second quarter 1986, p. 18.

18 D. Hamermesh and J. Johannes, "Food Stamps as Money: The Macroeconomics of a Transfer Program," *Journal of Political Economy,* 93(1), 1985, pp. 205–213.

19 Jean B. Wellisch and others, *The National Evaluation of School Nutrition Programs: Final Report,* System Development Corporation, Santa Monica, CA, April 1983.

20 Paul E. Nelson, *School Food Service Programs and Commodity Market Support,* Agricultural Economics Report No. 676, USDA, Economic Research Service, Washington, DC, March 1982.

21 Spencer Rich, "School Breakfast Seems to Boost Performance," *The Washington Post,* May 3, 1988, p. A–6.

22 Spencer Rich, "Report on Women's Nutrition Program Faulted," *The Washington Post,* January 19, 1990, p. A–7.

23 Committee on Agriculture, U.S. House of Representatives, *A Review of the Thrifty Food Plan and Its Use in the Food Stamp Program,* Washington, DC, 1985.

TERMS AND CONCEPTS

Food-aid policy	Income effect
Hunger	Substitution effect
Slippage in demand	Cash-out of food stamps
Food stamps	P.L. 480
In-kind assistance	Commodity distribution
Safety net	Section 32
Purchase requirement	Thrifty Food Plan
Engel's Law	Entitlement program

QUESTIONS AND STUDY SUGGESTIONS

1 What is meant by food-aid policy? What are some of its objectives?

2 What are some of the economic and political reasons for the development of a food-aid policy in the United States?

3 What is meant by slippage in food-aid programs? What are some of the problems created by slippage? How can it be reduced and what are some of the constraints to reducing it?

4 How does the Food Stamp Program supplement the income of recipients?

5 Theoretically, would recipients prefer a food-stamp program with a purchase requirement, without a purchase requirement, or cash-out of the program? Why? What are the likely effects of each policy option on food consumption?

6 How are each of the following groups affected by a pure food-aid policy: farm producers, low-income consumers, nonsubsidized consumers, taxpayers?

7 Trace the evolution of U.S. foreign food-aid programs. Why has the United States developed programs such as P.L. 480?

8 What were the major policy changes in the evolution of domestic food-aid programs? How have political, economic, and social factors influenced the development of these programs?

9 Why has the level of food stamp benefits become a volatile issue? What are some possible reforms concerning benefit levels?

CHAPTER 11

WELFARE CONCEPTS AND EFFECTS OF PRICE AND INCOME POLICIES: A REVIEW OF THEORY

The term, *welfare,* as usually used in policy study, means to fare well or do well. In this sense it can refer to good health, happiness, comfort, satisfaction, utility of a good, and prosperity. Such an everyday concept has been widely used in general policy discussions and has permeated the literature of more formal policy analysis. Yet the term often remains inadequately defined, ambiguous like many other common words appropriated by policy researchers, and controversial in the arena of policy making.

However, *welfare economics* has come to be rather specifically defined, with measurement procedures identified in the field of economics, even though its transfer to public policy remains controversial. It is controversial in application to public policy because it is often impossible to accurately measure the parameters involved, such as in price and income policy or food-aid policy. Yet the concepts are important and meaningful to the economics profession, and they are, therefore, worthy of more explicit review.

This confusion regarding the term *welfare* is perhaps understandable when we recall that policy means action taken to achieve a desired end, such as welfare; yet public policy usually deals with controversial ends. It is common to hear and read words to the effect that "the government should provide for the public welfare in its programs" or that "individual welfare should be carefully protected in making public policy." Yet before we can understand policy, analyze public policy problems, and conduct policy research, we need to clarify some of the contributions and limitations of welfare concepts.

The purposes of this chapter are: (1) to summarize the basic welfare concepts that may be used in analyzing the welfare effects of price and income policy; (2) to demonstrate how demand and supply curves are sometimes used to analyze the distribution of costs and benefits of policies; and (3) to outline the kinds of economic analyses used in studying price and income policies, as well as other policies.

IN THIS CHAPTER

By studying Chapter 11, you will learn how to answer the following questions:

1 How are welfare concepts distinguished in private and public policy?
2 How can welfare effects be measured in price and income policies? What tools are used to determine possible gains or losses from price and income policy?
3 What are the effects of a particular program on consumers or producers? What is the concept of net social cost?

WELFARE CONCEPTS IN PRIVATE AND PUBLIC POLICY

In Chapter 1 we made an important distinction between private and public policy, a distinction that pervades every issue of food and agricultural policy. Private policy deals with the goals and decisions of individuals, families, individual firms, and various groups of individuals and firms. By comparison, public policy involves participating representative governments at all levels. No public policy benefits everyone, but trade-offs can be sought between gainers and losers. Hence the policy-making process calls for compromises among the participants as the benefits and costs are evaluated.

Private Policy

In the arena of private policy, individuals and interest groups work for programs that are believed to enhance and protect their own position or welfare, be it economic, social, or political. Their actions, while deliberate and rational, may or may not be good for other individuals, families, firms, groups, or society as a whole.

Most democratic societies jealously guard individuals and private decision makers as the primary evaluators of what is good for their own welfare, provided their actions are legal. Thorough and accurate knowledge is the cornerstone of both a participatory political system and a competitive market system. Hence nations strive to provide this knowledge and to protect the participants' opportunities to obtain it through public education, free speech, and a free press. Private decisions and policies involve what is understood to be true and what is judged to be good. The study of economics and policy is designed to contribute to the knowledge base for private and public decisions.

Public Policy

In the arena of public policy, decisions and actions are taken by participatory governments using welfare concepts as criteria for what is a good action. Perceptions of perceived equity or inequity undergird many such decisions. But policy actions based on welfare concepts raise complicated and potentially troublesome issues. Surely public policy should further the public's economic, social, or political welfare. How, then, can the policy makers recognize the "right" choices to achieve the public's welfare? No public policy is perceived as helping everyone. Important balances must be struck. Usually many trade-offs and compromises must occur. What standards can be appropriated and what evaluative criteria shall be selected?

The accepted role of economic analysis in private decision making is not transferred easily to the public sector. In private policy making, goals and courses of action are chosen by individuals or private interest groups to help themselves, whereas in public policy making they are chosen through participatory compromises. Often it is suggested in the economic realm that "economic theory tells us," or that "economists should have the solutions to the economic problems of society," or even that "economists should determine the answers." Yet economics does not judge; economists are not like medical practitioners prescribing for a patient; and participatory societies choose their own economic systems.

Economic analysis can identify many of the gains and losses, and some conclusions can be stated specifically in terms of dollars and cents. A serious welfare problem emerges, however, whenever the gains of one group or economic sector must be offset by losses to other groups or economic sectors. Societies often do not wish to leave such trade-offs to private interests. For instance, in the arena of food and agricultural policy, how is the continued use of a certain pesticide to be evaluated if 99.4 percent of the U.S. population is believed to be beneficiaries of its use, while the other 0.6 percent is harmed, one out of a million persons is made very ill (250 in the United States), and one out of 10 million dies (25 in the United States)? A problem of this nature emerges in policy analysis of chemicals used in farm and industrial production. What ways do we have to resolve such problems?

Frequently, proposals have been made to delegate policy matters affecting the public welfare to an authoritative body of "technocrats," short for economic managers or technical experts, who are supposedly educated in ways to improve the public welfare. In the United States, agencies in all three branches of government—legislative, executive, and judicial—must make many decisions about how to determine or measure welfare. One of the continuing and important questions is how flexible or how rigid shall the U.S. Supreme Court be in interpreting the Constitution, especially where matters of economic welfare are involved?

Welfare issues permeate most of food and agricultural policy. How can economic concepts be applied to measure welfare and help resolve important policy issues?

ECONOMIC SURPLUS AND WELFARE GAINS AND LOSSES

Economists have persistently sought a method of making policy recommendations firmly rooted in economic theory. One of the pioneers in this quest was an Italian sociologist and economist, Vilfredo Pareto (1848–1923), who expanded a utilitarian principle to mean that a policy or action should be undertaken by society if doing so makes someone better off while making no one worse off. This principle is termed *Pareto optimality,* and is met by the assumptions of a free market, because all trades are voluntary. However, strict adherence to the principle would severely constrain the decisions of economists in judging among alternative policies. That follows because all public policies, or interventions, change the way an economy functions and hurt someone. Thus the principle has been expanded to include the notion that if gainers from a particular policy can adequately compensate losers from that policy and still be better off, then both the direct gainers and the initial policy losers will be better off and there will be a net gain in terms of welfare. If free trade, for example, can improve the welfare of consumers so much that they can compensate the producers who might be hurt, then everyone can be better served. However, this modification raises a question of comparing the utility of money among different groups of people. Welfare economists therefore have tried further refinements. The concepts of *economic surplus* and *net social welfare gain or loss* are now used widely by economists as measures of program performance and economic welfare. On the basis of net social welfare gain or loss, a policy is judged as cardinally ''good'' (valuable intrinsically), or as ordinally ''better or worse'' (valuable in an order), than another in economic terms. From this formulation, economists derive one or more policy prescriptions for the economy and the country's citizenry.

These concepts and measurements are totally grounded in a particular application of the classical economic model and its underlying assumptions of a competitive market equilibrium. Adapted through years of scholarly scrutiny and discourse, the theory has become the foundation of what is generally known as ''welfare analysis,'' which is grounded in the concepts of consumer and producer surplus.

Consumer Surplus

In Figure 11-1, the schematic diagram depicts the industry supply function *S,* demand function *D,* equilibrium price *P,* and equilibrium quantity *Q,* with the industry in equilibrium at *E.* The cross-hatched area above the price line bounded by *PEA* is the consumer surplus at this equilibrium, and therein lies one of the critical adaptions of basic equilibrium theory for welfare analysis. According to this particular interpretation of classical theory, the traditional demand function or curve represents not only the alternative quantities to be demanded for this particular commodity at the various prices, but once equilibrium is reached, these ''values'' still exist for all of the willing consumers at each successive alternative price level. This flows from the familiar concept of decreasing marginal utility. For instance, if the quantity available had been very scarce, such as Q_1, some consumers would willingly have paid the high price P_1, or at Q_2 price P_2, and so forth. Because the market (equilibrium) price is lower at *P,* all those consumers who were willing to buy at the higher prices are in reality getting surplus economic value (welfare gain) bounded by some portion of *PEA*. In food and agricultural policy, this concept of being a consumer applies both to consumers of food and to agricultural producers. Farmers as producers are users (consumers) of various production input goods or commodities.

Note that a fundamental adaptation of classical theory is that the values represented in the traditional demand function persist and are realizable as a welfare gain to consumers who bought at the price below that price at which they were willing to buy. Thus, except for the last (marginal) consumer, a hamburger has a value that is greater

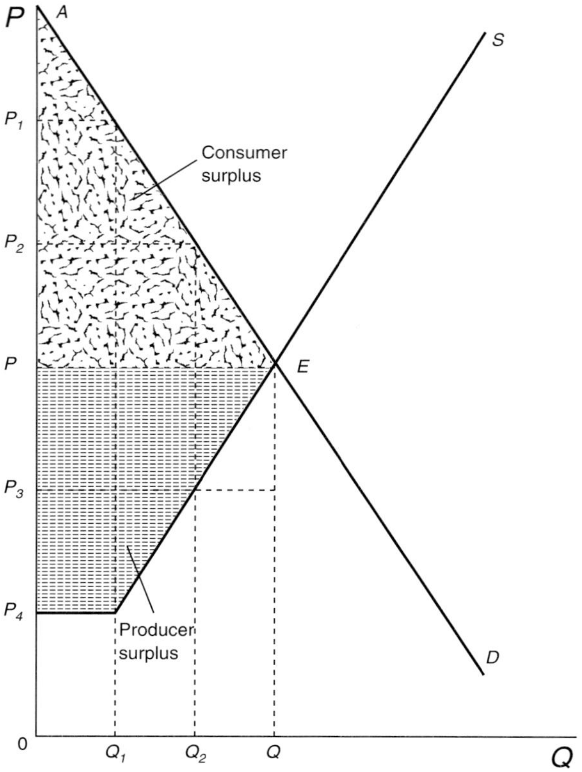

FIGURE 11-1
Selected concepts basic to welfare economic analysis. Consumer surplus is the hatched area above *PE*, bounded by *PEA*, and producer surplus is the area below *PE*, bounded by PEP_4. It is assumed that no firm will produce at a price below P_4.

than the price paid. Once the demand function for any commodity is determined empirically, the magnitude of the consumer surplus can be estimated with quantitative analytical techniques for any specific level of price and quantity. Similarly, units of any commodity used in farm production, except the last marginal unit, have a value to the farmer, who is a consumer of inputs, greater than the market price of the commodity. Rational consumers, including farmers as consumers of inputs, like to buy at the lowest price, which in the formulation maximizes their consumer surplus. Welfare concepts, then, suggest that this economic maximization may be consumers' motivating force in their interest in public policy.

Producer Surplus

In Figure 11-1, producer surplus is the cross-hatched area bounded by PEP_4, assuming that no producer will operate at a price below P_4. The alternative quantities offered to be supplied at alternative prices are Q_1 at P_4, Q_2 at P_3, and Q at P. These values flow from the familiar concept of increasing marginal costs, which in the long run become supply. The surplus values to all except the marginal producer are therefore considered real and can be estimated quantitatively.

Under the basic theory of pure or perfect competition, all costs and resources are variable in long-run equilibrium, and the producer surplus will disappear. That is, the long-run producer supply function is perfectly elastic, because all resources can be shifted to their highest use through the play of competitive market forces. Therefore, although consumers always capture their surplus, which is the ultimate economic prize and the net welfare product or gain of the competitive economic system, producers in pure competition must operate through a continuing or repeating series of short-run equilibria to continue to capture the short-run producer surpluses. Consequently, individual farm producers will tend to lose if they do not keep pace with technological advance.

In imperfect competition, however, some producers can continue to enjoy a producer surplus for long periods of time. But unless they are able to establish a uniquely differentiated market position, such as through patents or a monopoly, the individual firms must advance technologically to be competitive. Given this qualification, some agribusiness input firms and agricultural commodity marketing firms may realize or retain a producer surplus indefinitely in cases in which differentiation of markets is feasible.

Effects of Market Intervention on Producer and Consumer Surplus

In welfare economics, it follows from the configuration of supply and demand under equilibrium assumptions that any market manipulation, such as may occur through private or governmental intervention policy, will change both the consumer and the producer surplus. This intervention will most likely result in a shifting of the economic surpluses, or shift the gains and losses among producers and consumers. According to this analytical approach, any such intervention will result in a higher per unit cost of production, different allocation of resources, and redirected income distributions as compared with pure free market conditions.

This important policy conclusion, which has emanated from welfare analysis formulation, is depicted graphically in Figure 11-2. Suppose that a public policy of reducing the quantity produced or marketed results in a shift of the supply function to S_1. A new equilibrium price at P_1 and quantity at Q_1 are established at the new market equilibrium E_1. If the previous theoretical formulation is followed, consumer surplus shrinks to $P_1\ E_1\ A$, while producer surplus expands to $P_1\ E_1\ C_1\ P_4$ from $PEC_1\ P_4$ where it was before. But now the two segments labeled L, commonly referred to as the ''welfare triangles'' and previously included in the surpluses to either producers or consumers, do not appear as gains to either, but rather as losses. Thus welfare analysts depict these as net welfare losses, or deadweight losses. This concept may be used as the critical conceptual basis for policy analysis to conclude that economic welfare for society declines with any policy of intervention into a perfectly competitive market. It also provides the basis for welfare analysis to make evaluative comparisons among alternative interventions.

Welfare gains, losses, and transfers can be quantified for any identifiable intervention policy and also among policies. An unlimited or infinite number of variations can be

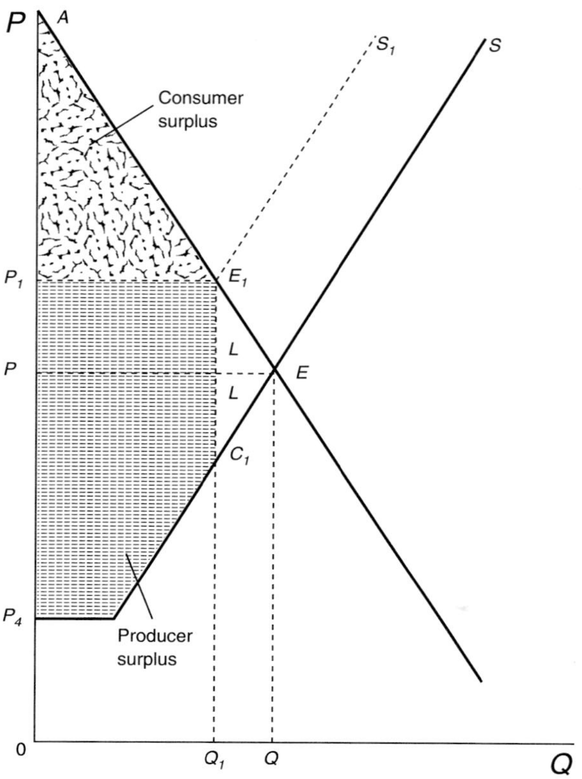

FIGURE 11-2
Effects of a policy of price supporting with effective production control interpreted through welfare analysis concepts. When supply is restricted to S_1 and market price raised to P_1, consumer surplus is decreased and bounded by P_1 E_1 A while producer surplus is increased and bounded by P_1 E_1 C_1 P_4. L and L are the net welfare loss from production control.

made in this basic interventionist policy model. The results depend on the elasticities of supply and demand, the shifts in each of these, the magnitudes of the shifts, the reactions of producers and consumers over time, and the effects of the policy on the structure and performance of the industry and the marketing economy. The elasticities of supply and demand are especially important. The more inelastic the demand for example, the greater will be the net social loss of consumers for any given price increase. Also, if demand is elastic, producers will have a net welfare loss for any given price increase. In welfare comparisons between consumers and producers, the value of any trade-off is often much greater for each producer while a much larger group of consumers are affected.

Based on assumption, one additional theoretical adaptation is made for welfare analysis: When the gains and losses among economic sectors are compared, the marginal utility of a monetary unit of economic gain or loss is the same for everyone. Although such interpersonal comparisons of utility may be difficult, if not impossible, it is assumed that a dollar of loss (gain) to the consumer means the same in economic terms as a dollar of gain (loss) to the producer. This assumption permits the classic welfare policy analysis to be completed with the intellectual power of its analytical rigor and the credibility of its neoclassical foundations intact. The policy analyst, or any other user of these results in policy making, can represent them as policy prescriptions of what is "best" and what should be done for the economic welfare of the economy. Emanating from the implicit assumptions of welfare economics and not from the values of the analyst, these conclusions can then be grounded firmly in economic theory.

Economic Problems of Interpersonal and Intersectorial Welfare Transfers

Many analysts will carry their quantitative policy studies only to the point of determining what would be the likely quantities produced and consumed at likely prices received or paid. This kind of systematic empirical work, based on more limiting theoretical assumptions, leaves to the public policy making process the critical welfare evaluations of what, if any, economic surplus value exists for a sector, and what the trade-offs in values are among sectors and participants. This approach keeps positive-type analysis separated from welfare valuations in public policy making. Yet it still makes an invaluable contribution to policy making in a participatory system.

The methodological integration of fact and value through these normative assumptions, which is inherent in much welfare analysis, achieves an analytical result that can have a persuasive power in policy advocacy. To clarify the picture of the welfare consequences of any policy, however, additional empirical and evaluative types of policy studies are generally needed, because normative economic analysis alone rests on questionable interpersonal comparisons of value through time and between or among various individuals and groups. Furthermore, it may be difficult if not impossible to find a usable, measurable notion of economic value to consumers and producers beyond that associated with the concept of equilibrium market price. Other factors must be taken into account, such as income distribution, the marginal value of money, which is generally not the same to rich and poor, and the relative priorities for various types of public expenditures.

Although welfare analysis is only one of several types of policy inquiry, as discussed later in this chapter, it has become important in the policy analysis of both the input and output markets of agriculture. Its importance stems from several factors, including the strengthened tools of quantitative economic analysis and the well-defined and widely experienced standard types of market interventions that characterize the markets for food and other agricultural commodities throughout the world. For instance, welfare analysis can be applied to the competitive market conditions for many agricultural inputs and for nearly all agricultural outputs. Also, as a result of the competitive conditions in these markets, they are more amenable to this approach than are many of the more highly varied and more complex, imperfectly competitive marketing and pricing conditions in markets of most other sectors.

Strengths and Weaknesses of Welfare Economic Analysis

Welfare economics has both strengths and weaknesses as an analytical tool to improve reliable knowledge and solve public problems. It has the advantages of being rooted in the neoclassical economic concepts of supply, demand, and equilibrium, which are among the best developed facets of economic theory. Welfare analysis is highly amenable to being applied by the rapidly developing field of quantitative economics. It appears to be sensitive to the inevitable trade-offs between different economic groups associated with all public policy. Finally, it is responsive to the historic quest of economists to bring the full power of research and science to bear on this difficult area of public policy; that is, to bring objective, positive economics to the solution of inherently normative, value-laden economic problems.

On the negative side, welfare economics has disadvantages. It rests on questionable interpersonal comparisons of value through time and among people with different values and in various economic situations. It depends on a doubtfully usable measure of economic value for consumers and producers beyond that associated with the concept of equilibrium market price. It implies a superior quality to analytically derived values for making public policy choices, compared with those values arising from participatory compromises in the public policy process. Finally, welfare economics encourages reliance on a policy making authority that may be external to the public participants as a source of policy direction, instead of on an adequately informed public and a respect for the public's ability to make policy decisions.

In summary, various concepts of welfare are prevalent in food and agricultural policy, and they have great emotional and connotational appeal. Consequently, economic policy analysis must involve several different kinds of studies, or several phases of study, to achieve professionally acceptable answers to public problems.

What are the kinds of economic policy analysis? And how are they related to the concepts of welfare?

KINDS OF POLICY ANALYSES

Economic policy analysis involves different kinds of studies, such as estimates of various types of relevant economic trends, relationships among economic variables, characterizations of past and present policies and programs, investigations of policy alternatives, predictions about consequences of policies, and designs of new approaches to resolve policy conflicts. For food and agricultural policy in particular, several disciplines in the social sciences are involved, chief of which is agricultural economics, plus a foundation of some physical and biological sciences.

Estimating the Economic Situation

A study of economic policy usually requires an early evaluation of the situation relating to the public issues under review. Doing so helps the policy maker and the analyst to gain perspective, set the dimensions of a policy problem, and identify some of the major policy options. Thus, in Chapter 4 we presented estimates of the national situation with emphasis on supply and demand and the economic environment for policy relating especially to the effects of technological advance on agriculture. Then, we followed this in Chapter 5 with estimates of the world economic situation affecting food and agricultural policy.

The purpose of such estimates is to broaden our vision and build a foundation for policy study. Although estimates of situations are an important part of policy analysis, no situation can ever be fully described, nor are situations static. Therefore, it is important to establish some of the functional relationships among the variables and provide some examples of these relationships.

Determining Functional Relationships

Another kind of policy analysis seeks to determine functional relationships among variables relevant to policy under clearly stated assumptions and using the rapidly improving methods of quantitative analysis. For example, research has attempted to determine the important effects that various macroeconomic forces, such as monetary policy, may have on the levels of agricultural exports, the differential effects of various levels of cost and return on farm land prices or values, the optimal combinations of resources in agricultural production, and the use of certain policies such as grain stocks as reserves to attain some stated objective. Although researchers generally accept their role in providing reliable knowledge and analysis for policy participants, the line between economic analysis and policy recommendation sometimes becomes blurred. For this reason, professionals must recognize the limitations of analysis, and policy participants must carefully evaluate their sources and the content of information.

Characterizing Selected Policies and Programs

The characterization of past and existing public policies, along with their consequences, continues to be the subject of policy study. For example, researchers have analyzed the provisions of farm credit policy, domestic and overseas programs of food distribution to needy people, and the types and variations of policy in the more than 60 years of U.S. farm price and income policy. The purpose is to inform the policy-making public about some of the costs and returns of the policies and programs so that inferences may be drawn about future trade-offs of gains and losses in policy making.

Designing and Considering Alternative Policies

Designing policy alternatives to solve new or persistent public problems is one of the important contributions that can be made through use of the tools of economic analysis. For example, a specific policy was designed in the 1920s to separate the domestic and foreign markets for agricultural commodities, and those basic policy concepts became the precursors to the export subsidy schemes currently used by the United States and other countries around the world. Generally, the most useful research related to policy experiences also offers new options for policy and comparative quantitative analyses among them. As an example, given the continuing budget constraints, the ability of policy alternatives to achieve specific objectives cost-effectively with a certain amount of governmental expenditures is estimated with welfare concepts.

Such welfare analysis, however, may not solve the problem of what is "best" for the economy or for various individuals and groups. What is "best" has to be determined according to prevailing beliefs and values of the political constituents operating through the given political system. In a democratic system, for example, the general assumption is that the people will participate in deciding relevant economic matters and that they will be well enough informed to make rational, "right" choices.

Predicting Economic Results of Policy Choices

Predicting probable future trends relevant to a public issue or the probable consequences that a specified policy will have on producers, consumers, traders, taxpayers, and other groups provides useful knowledge to citizen participants and policy makers as they evaluate various solutions to economic problems. For example, research identifies the sources or severity of soil erosion and the results of a cropland retirement program on soil loss, the investment return from public expenditures on agricultural research, and the monetary effects of a compensatory payment scheme on grain producers, taxpayers, and consumers. The researchers are professionally responsible for using their best theoretical and empirical knowledge to expand the public's understanding through quantitative analysis of alternative policies. But researchers must be constantly vigilant to avoid misusing their analytical tools to create otherwise unfounded biases, such as to favor or to discredit a given policy, or combinations of policies.

Resolving Conflicts through Policy Analysis

Conflicts in policy making arise out of the differing interests of participants and groups, differing concepts of a situation, and differing beliefs about the outcomes of particular policies or programs. Not all of these differences can be resolved by or through policy analysis. Participants and groups with differing interests presumably will have different policy goals. In a democratic system in which the opportunity for dissent prevails, such differences are resolved through the political process, which calls for many compromises.

Differing concepts of a situation may be clarified by research and economic analysis. Still, participants may disagree because they have different policy goals, or beliefs about the effectiveness of proposed policies. Major reasons for studying public policy analysis include finding ways to do the following: estimate various situations, recognize the differing interests of individuals and groups, and predict or evaluate the outcomes of alternative policies. To recognize the essence of participatory public policy making and the power and limitations of the discipline of economics in policy making is to recognize that policy analysis is indispensable to "good" public policy but that it is not the final source of good policy.

LOOKING AHEAD

The next three chapters—chapters 12, 13, and 14—deal with natural resources policy, food quality and safety policy, farm credit policy, and those aspects of monetary and fiscal policy that are most critical for food and agricultural policy. These areas of policy directly affect the input markets of the food and agricultural economy. Also, they are linked by the fact that natural resources policy affects both agricultural production and the environment, which also has an effect on food quality and safety. Farm credit policy is linked with monetary and fiscal policy, which together also affect the use of resources and the productivity of the food and agricultural economy.

These areas of policy are often referred to as developmental policies, because of their effect on shaping the structure of this nation's food and agricultural economy. They also influence agricultural productivity and growth and the distribution of income. Hence they are basic to understanding both the policy issues in the input markets and the more familiar issues concerning agricultural prices, income, food aid, and trade, which have been covered in Chapters 6 to 11.

TERMS AND CONCEPTS

Welfare
Private economic welfare
Public economic welfare
Pareto optimality
Net welfare losses and gains
Consumer surplus
Producer surplus
Policy gainers and losers
Price and income policy analysis
Cost-effectiveness of price and income programs
Short-run and long-term policy effects
Alternative programs
"Good" policy for society
Interpersonal welfare comparisons

QUESTIONS AND STUDY SUGGESTIONS

1 Why is the study of welfare so important in public policy?
2 What are the differences between the way an individual decides on welfare and the way a whole nation decides on its welfare?
3 What was the rationale for economists in developing the welfare analysis approach to public policy?
4 What are the assumptions on which welfare analyses are based?
5 Identify producer surplus and consumer surplus on the usual supply and demand chart.
6 What are the primary advantages and disadvantages of using welfare analysis to study policy problems?
7 List some of the different kinds of economic analyses that contribute to policy development.
8 How does a participatory public policy making process determine which decisions are in the best public interest?
9 Do economists have any unique capacity to decide the best public policy for a nation?

PART FOUR

DEVELOPMENTAL POLICIES, ISSUES, AND ALTERNATIVES

CHAPTER **12**

NATURAL RESOURCES POLICY

In this chapter, we study the public policies used in acquiring, distributing, developing, and conserving the natural resources used in producing food and other agricultural commodities. These resources include: crop- and grazing land; water used for many purposes, such as irrigating dry land, raising livestock, drinking, recreation; minerals and forest products also used for many purposes in agriculture and other industries. The policy issues include how these resources shall be acquired, distributed, developed, and conserved. Consideration must be given to the supply available, the demands of agricultural producers, and the demands for other uses of these resources.

IN THIS CHAPTER

By studying Chapter 12, you will learn how to answer the following questions:

1. What is the scope and substance of natural resources policy affecting the food and agricultural economy?
2. How was U.S. policy applied to land and water in the land settlement era?
3. How has U.S. policy changed in the post-settlement era?
4. What are the technical and environmental issues that are important for future policy?

OVERVIEW OF NATURAL RESOURCES POLICY

Agriculture's use of natural resources raises many questions: Can increasing agricultural productivity and growth be sustained? How does agricultural production affect the environment? What mix of private management and public policy is required to further the important but often competing goals of efficient agricultural production and protection of the environment? What can a government do to assist in developing and conserving natural resources?

There are at least three distinguishable types of policy involved in answering such questions. One pertains to distribution and ownership of the resources. Another covers the facilities or infrastructures developed to use natural resources. A third type of policy pertains more directly to resource conservation and protection of the environment. Let us briefly sketch an overview of these three types of policy.

Distribution and Ownership Policy

The land distribution and ownership policy followed by government is one of the most important policies in determining how agriculture will be organized and managed. For example, a policy may be designed to favor either private or public ownership of farms and ranches or small or large farms and ranches, or a policy may favor a variety of ownership types, and a broad range of small, medium, or large farms and ranches. A fundamental policy decision in U.S. land settlement was to favor family-sized farms and ranches, and to make land suitable for farming or ranching available at low cost or free to settlers. A large part of the forest land and much of the land valuable for minerals was retained in public ownership. Other land most suitable for recreation, wildlife, public parks, and national monuments was largely retained in or converted to public ownership.

A continuing problem in resource distribution and ownership policy is how best to manage and control natural resource use where the demands compete and conflict. Another continuing problem, which is not solved by resource distribution and ownership policy is how best to develop natural resources for agricultural development and growth. Solving this problem requires what is called developmental infrastructure policy.

Developmental Infrastructure Policy

To reach the highest and best use of natural resources for food and agricultural production and growth requires a variety of publicly developed and supported *physical infrastructures,* the most critical of which are roads, railroads, harbors, airports, canals, and waterways. Such infrastructure is necessary for efficient movement of inputs to the farm and of food and other products from the farm to markets and consumers. As a result, losses can be held to a minimum, production can be specialized in low-cost areas, and products can move to the most profitable market. In the United States, as well as in most other developed nations, transportation infrastructure is generally provided by the government from tax revenues, although it is not uncommon for fees to be charged for use of some infrastructures, toll roads for example.

Sometimes, government indirectly provides a variety of physical infrastructure services by subsidizing the initial investment costs. Examples include government financing or granting resources to aid in developing railroads, communication facilities, and utilities. There are two justifications for governmental support for developing these physical infrastructures. First, for some types of infrastructure, the owner would be in a monopoly position and able to charge monopoly prices. Citizens may choose to have a government help to build a facility, and then regulate the activities of a private firm, as in the case of public utility firms. Second, it is very difficult to charge for the use of some types of physical infrastructure, farm-to-market roads for example, and such infrastructure might not be developed at all by the private sector.

Institutional infrastructure, as contrasted with physical infrastructure, involves the laws, rules, and regulations that define the rights of owners and tenants, regulate markets, and set the penalties for unacceptable or unlawful behavior. Governments are responsible for developing a variety of institutional infrastructures, which affect natural resources, if not the entire economy. Important examples include the infrastructures to promote and maintain law and order, protect freedom of speech and press, support research and education, and establish the rights of individuals to own and manage property. It should be noted, however, that governments can establish and exercise some of the implied powers only at the margin. For example, governments can enforce property rights only if such rights are recognized and learned by individuals, and accepted by the community. The establishment of property rights is more complex than just one of government enforcement.

Natural Resource Conservation and Protection of the Environment

More specifically, in regard to natural resources, governments generally have accepted or must accept the primary responsibility for conserving natural resources and protecting the environment. The reason is that the costs of natural resource degradation and pollution of the environment are only partially reflected in the costs of production incurred by the agricultural or nonagricultural producer. In the case of soil erosion, for example, usually only a very small part of the costs of water pollution and environmental degradation incurred by the public is reflected in the production costs of the farmer, or in the prices paid by the consumer. It is an external cost borne by society, which may not be reflected in producers' costs or consumers' prices. Agriculture is not unique in generating external costs. But it is one of the more important industries in this regard, and the institutional infrastructures that are required to manage these costs are among the more extensive and complex in an economy.

Hence governments generally have to accept the responsibility for implementing soil and water conservation, regulating the harvest of timber in national forests, protecting the quality and purity of ground water, and preserving at least some of the habitats of wild animals, birds, and fish. Governments perform some of these functions by sharing the costs of soil conservation, regulating timber cutting, controlling mining and oil drilling, regulating the sale and use of pesticides, and licensing hunting and fishing. All of these governmental actions involve the institutional infrastructure necessary for the efficient use and conservation of natural resources used by agriculture and other industries. In even the most laissez-faire of economies, this institutional infrastructure is needed.

In summary, governments have many options to apply conservation and environmental policy to the use of natural resources. For example, a government may acquire and maintain ownership of natural resources, or it may grant title to individuals or private firms. The title to land may give the owner unrestricted property rights in its use, or alternatively government may retain the right to regulate the use of a natural resource such as land. For example, a government may ban the use of some pesticides, or adopt specific laws and regulations regarding their use. A government may adopt rules regarding soil conservation practices. It may encourage land settlement for example, or discourage the use of some resources. Hence, such policy may influence or even guide agriculture's use of natural resources.

As agriculture develops under influences of advancing science and technology, however, the role of natural resources used by agriculture—especially agricultural land and water—decreases in relative importance, but the role of government in developing and protecting these resources increases. The effects of agricultural production on the environment may expand also, which increases the role of public policy for conserving soil, managing water under irrigation, improving and regulating pesticides, and controlling pollution. Many of the environmental and pollution-control problems affect rural people and users of recreational resources as well as farmers. The changes in this role can be studied by comparing U.S. policy in the land settlement era with the policy that has come later.

U.S. NATURAL RESOURCES POLICY IN THE LAND SETTLEMENT ERA

From the Articles of Confederation in the 1780s to the end of the homestead era about 1920, U.S. natural resources policy was dominated by settlers' encounter with the frontier, and their efforts to settle the vast public domain. Land was abundant and labor was scarce. The primary roles assigned to the U.S. government were two-fold: (1) to allocate a vast, publicly held domain that farmers and others could exploit for economic development, and (2) to protect the lives of owners and their property rights from outside intrusion. At that time, settlers showed little concern for any adverse effects that private use might have on others. Most settlers established family-sized, partly self-sufficient farms, although larger estates and plantations were allowed, if not encouraged. Until the land frontier ceased to exist, U.S. policy favored farms that were not too large for a family to own and manage, but large enough to provide a good living for a family.

Basic Foundations

Two historic decisions affecting land use were made soon after the founding of the United States. First, property rights granted to owners of land gave them maximum discretionary use of the property and its eventual disposition. This concept of property rights in land is still the focus for much of the public policy related to land use, soil conservation, and environmental protection. Second, eminent domain was adopted as an a priori right of the public over private ownership only when it was deemed, through

TABLE 12-1
METHODS OF LAND DISPOSAL BY LAND ACTS, 1785 TO 1862

Land act	Minimum sale price	Method of disposal	Size of tract	Terms of sale
Ordinances of 1785 and 1787	$1 an acre; discount of one third to companies and acceptance of government paper	Auction	Half of townships offered entire area, other half by 640-acre tracts	Cash until 1787, then one third cash and rest in 3 months
Act of 1796	$2 an acre; evidence of public debt accepted at face value	Auction	Half of tracts 5,760 acres, the rest 640 acres	One twentieth cash, credit of 30 days on balance of first half, and one year on second half
Act of 1800	$2 an acre	Auction	Minimum of 320 acres	8 percent discount for cash; liberal credit system inaugurated
Act of 1820	$1.25 an acre	Auction	160 acres or 80 acres	Cash
Act of 1841 (Preemption)	$1.25 an acre; grants to railroads and canals at $2.50	Selection and settlement, then purchase at minimum price	Not more than 160 acres	Cash
Act of 1854 (Graduation Act)	Price graduated; for example, if on market 10 to 14 years, $1 an acre; 20 to 24 years, 50¢; over 30 years, 12½¢	Offered for sale at stated minimum price		Cash
Act of 1862 (Homestead Act)	Free homesteads of 160 acres; only payment, a fee ranging from $26 to $34	Settle and "prove up"	Not to exceed 160 acres	Only nominal fee, but certain conditions had to be met

Source: Benjamin H. Hibbard, *A History of Public Land Policies,* The Macmillan Press Company, New York, 1924. Roy M. Robbins, *Our Landed Heritage,* Princeton University Press, Princeton, NJ, 1942.

due legal process and just compensation, that invoking their right would serve the public interest.

Evolution of Land Settlement Policy

From the granting of independence in 1780 to adoption of the first Homestead Act of 1862, U.S. land settlement policy evolved through several stages. It changed from initial sales of large tracts of the public domain to smaller tracts at gradually lowered prices, and finally to unrestricted title to 160 acres of free land if the owner would settle on it for at least five years (Table 12-1).

Implementing the homesteading guidelines in semihumid regions of the West created new problems, however. The standard homestead was not large enough to support an average family. Over the next half century, several additional major Acts increased the amount of land from 160 to 960 acres or more that a settler could acquire as a homestead for farming or ranching. By the end of the homestead era in the early 1920s, nearly all of the fertile cropland in the nation's public domain and about a third of the rangeland (nearly 400 million acres) had been granted directly to settlers for farming or ranching. Other major grants were made for building railroads, for education, and for other types of development. Most of the fertile cropland in these other grants was transferred later to farmers by sale or lease. The largest part of the rangeland that was not transferred to private ownership was retained by the federal government, where it is managed by

TABLE 12-2
DISPOSITION OF THE PUBLIC DOMAIN

Area	Million acres
Approximate area of public lands disposed of under the public-land laws, June 30, 1943:[a]	
Granted to states for:	
Support of common schools	77.5
Reclamation of swampland	64.9
Construction of railroads	37.1
Support of miscellaneous institutions	20.6
Purposes not elsewhere classified	16.0
Construction of canals	4.6
Construction of wagon roads	3.3
Improvement of rivers	1.4
Total	225.4
Disposed of by sales and other methods not elsewhere classified	300.0
Granted or sold to homesteaders	285.0
Granted to railroad corporations[b]	88.9
Granted to veterans as military bounties	61.0
Confirmed as private land claims	34.0
Sold under timber and stone laws	13.9
Granted or sold under timber-culture laws	10.9
Sold under desert-land laws	10.0
Total	1,029.1

TABLE 12-2 *(continued)*
DISPOSITION OF THE PUBLIC DOMAIN

Area	Million acres
Area of public and Indian lands in the public-land states, estimated, June 30, 1943:[a]	
Area subject to general disposition under public-land laws:	
Vacant land outside grazing districts	42.8
Vacant land within grazing districts	136.7
Area not subject to general diposition under public-land laws:	
Reclamation withdrawals	14.0
Indian reservations	53.6
National parks and monuments	11.5
Wildlife refuges and game ranges	5.1
Oregon and California revested lands	2.6
Stock driveways	4.7
Power and water reserves	6.1
National forests	137.5
Other withdrawals and reservations	19.2
Area in unperfected entries	0.7
Estimated total area of public and Indian lands (excluding overlap)	413.1
Grand total, United States	1,442.2
Area of public, Indian, and other lands in Alaska:[c]	
Vacant public lands	270.0
Oil and gas reservations (including naval)	48.8
National forests	20.9
Fish and wildlife reservations	7.9
National parks and monuments	6.9
Native reservations	3.9
Military and naval reservations	4.1
Other withdrawals, reservations, and uses (including private)	3.0
Total[d]	365.5
Total public domain, original 48 states and Alaska	1,307.7

[a] *Land Management in the Department of the Interior,* U.S. Department of the Interior, July, 1946.
[b] A deduction of 2.4 million acres has been made to cover recent restoration of certain lands or cancellation of claims of lands by railroads.
[c] Special tabulation of withdrawals and reservations of public land in Alaska, June 30, 1949, by the Bureau of Land Management, U.S. Department of the Interior, subject to revision.
[d] Areas estimated in part.

the Bureau of Land Management in the Department of the Interior (about 160 million acres), and largely leased to livestock ranchers. Also some rangeland (about 34 million acres) is state-owned, and several million acres of the national forests have been leased to ranchers for livestock grazing (Table 12-2).

Irrigation Policy

Most of the fertile agricultural land in the West has been used for farming and ranching without being irrigated. However, early in the homestead era, many families who settled along streams and rivers discovered that irrigation was practical, provided that appropriate facilities could be developed. Some families developed irrigation on a small scale by diverting water out of streams to use on adjacent low-lying land. But the greater potentials for irrigation could not be developed without government help. The federal government first responded to the political requests of settlers by transferring a few million acres of land to the states, implicitly giving the states responsibility for reclamation and development (Table 12-2). Attempts to shift responsibility to the states proved to be impractical, however, and political efforts by farmers to get more federal help continued.

As a result of continuing political support for a family-farm policy, the federal Reclamation Act of 1902 authorized federal funding for large multipurpose dams and for financial help in constructing approved irrigation facilities. It also provided that the sale of irrigation water from federally funded projects be limited to a maximum of 160 acres to any bonafide resident landowner. This provision, which came to be known as the 160-acre limitation, was ineffective without strict and detailed federal supervision however, and such supervision has never been provided. In addition, conflicting laws, such as the Omnibus Adjustment Act of 1926, which provided for the sale of irrigation water to irrigation districts rather than to individual farmers, further weakened the 160-limitation and permitted subsidizing some very large irrigated farms. Differing views about how irrigation water should be allocated have provided the basis for continuing controversy over water policy for irrigation in the Western states.

Policy questions still remain about subsidizing irrigation, especially for large farms, the lowering of water tables of nonrecharging underground aquifers, the dominant use of water by farming, and the increasing demands for water by industry and growing metropolitan areas. The spread of metropolitan areas is slowly absorbing some of the most productive irrigated land. Some marginal irrigated areas are gradually becoming unprofitable for crop production through a process known as ''saltation'' (depositing salts in the soil when water for irrigation is used by plants, evaporates, or leaches into the ground). Hence the area under irrigation has declined by about 10 percent from the peak of 55 million acres, which was reached after World War II. If this trend continues, it can be projected that the area of land under irrigation will drop to about 30 to 35 million acres by the year 2030.

Such projections are tenuous, of course, because the reductions in the land area under irrigation is being slowed down by the spread of sprinkler irrigation instead of ditch and furrow irrigation, and by other new water-saving technologies such as minimum-tillage farming methods. Also, the spread of metropolitan areas could be slowed by increasing costs and time of commuting or by more land-saving city planning. Finally, if the political situation were to favor adopting a national land-use program designed to economize on land used for metropolitan development, such a policy could reduce the quantity of both irrigated and non-irrigated farmland demanded for that purpose. Between 1970 and 1975, members of Congress debated several bills to assist the states in developing such a land-use program, but none was passed. More recent efforts to develop a national land-use program have failed to gain the political support of either major party. Instead, the Soil and Water Resources Conservation Act of 1977 mandated the USDA to make continuing appraisals of the status, condition, and trends of soil, water, and related resources.

In the meantime, all of the 50 states, Puerto Rico, Guam, the Virgin Islands, and the District of Columbia have developed a water resources institute or center, based on the Water Resources Research Act of 1964 and its amendments, and on rules and regulations pursuant thereto published in the *Federal Register.* These institutes, generally located at land-grant college agricultural experiment stations or related facilities, are funded by federal, state, and local sources to carry out broad ranging studies of water resources development, and water purification and conservation for agriculture and other sectors. The basic concept underlying this development is that water is a renewable resource, supplies are limited, and over the long term, water will likely be the most limiting factor in world food and fiber production, exceeding both energy and land in its importance and value. Whether this proves to be the case or not, it is clear that water is a critical resource for sustaining agricultural productivity and growth, and water will become more critical with the passage of time. This is most evident in the regions of relatively low rainfall that have good climates for growing crops, such as a major part of the United States west of the Mississippi River, and it tends to be most critical in areas where these features are combined with rapid industrial and population growth. In these areas, there are strong competing demands for water from both industrial and residential customers, which makes agriculture a comparatively weak demand competitor.

Summary of Land-settlement Policy

The land-settlement era came to a close in the 1920s with more than 1.0 billion acres of the 1.4 billion acres in the original public domain of the first 48 states disposed of under the public-land laws. About one third of this 1.0 billion acres had been transferred directly to settlers under the homestead laws. In addition, most of the good farmland that was originally granted to the states for schools, railroads, reclamation, and other purposes had been sold to settlers at very low prices. However, the railroads kept most of the best forest land for their own purposes, or as an investment. Only a small portion of the original public domain was used for urban development, roads, and railroads (Table 12-2).

Most of the remaining 0.4 billion acres in the original public domain of the 48 states was subject to disposition under the public-land laws but was withheld from private entry under newer laws that superseded the original disposal laws. Major allocations were made for publicly held grazing districts, reservations for Native Americans, national forests, mineral lands, power and water reserves, parks and monuments, wildlife refuges, and game ranges (Table 12-2). Such multiple purposes and uses created many conflicts in making the public-land laws, and many conflicts have continued to the present day.

In the 1920s, when the land-settlement era came to an end, there were about 6.8 million farm and ranch operating units. The majority involved multiple land ownerships, or partial ownership by the operator with some land rented or leased. Very few farms were totally free of debt. About 90 to 95 percent of the operating farm units would have qualified as family-sized farms, but the land distribution pattern was highly skewed, with a relatively small number of very large farms and ranches and many more very small farms. Most farms—both large and small—were deeply in debt, and many were not sufficiently financed to stand much economic adversity. However, the foundation was laid for a highly competitive farm economy, which would soon respond to new conditions for economic development and growth.

NATURAL RESOURCES POLICY IN THE POST-SETTLEMENT ERA

From the beginning of the post-settlement era to the present day, natural resources policy has played a much different role in developing agriculture than it did during the settlement era. Until near the beginning of the 1920s, the growth of agriculture and rural communities was related almost entirely to using more land and water, providing more human capital (labor and management), and raising more livestock. From the mid-1800s to the early 1900s, crop yields had increased very little. The ratio of output to input in the livestock sector had also changed very little. The index of farm productivity was no higher in 1925 than it had been in 1900.

Effects of Science and Technology

After World War I, agricultural growth based on more land and labor gave way to development and growth built on science and technology, with larger scale of firm organization, more capital investment in the new technologies, more and more highly skilled management. The first impetus for this came from farm mechanization, the development of mechanical power for farming, and improvements in the food marketing system. Then came new crops, hybrid seeds, and new techniques in crop and livestock production. Finally came commercial fertilizer and a wide array of chemicals to control crop pests—weeds, insects, and various plant and animal diseases. Indexes of rates of change in factor productivity began to rise rapidly.

Advances in science and technology drastically changed the role of natural resources in developing agriculture and rural communities. The total amount of land and water used for agricultural production and rural communities stabilized. The use of fossil fuels for energy on farms and in agricultural marketing grew rapidly up to about the mid-1960s. Then it leveled off, even though the total output of agriculture continued to grow. Starting in the 1920s and 1930s, new improved farm machinery began to use energy more efficiently. Later the growing use of commercial fertilizer and pesticides encouraged minimum-tillage farming, and much later some no-till crop production. Transportation and marketing facilities improved for handling farm commodities. New methods in processing, storing, and preserving food increased the efficiency of these operations.

However, as the agricultural economy developed, it was recognized that the supply of natural resources was not inexhaustible, that soil resources could be damaged by continued intensity of use, that the off-farm effects of soil erosion were damaging to the environment, and that the growing use of pesticides and fertilizer was a threat to the quality and safety of ground water and food supplies. This brought a new policy emphasis for soil conservation and protection of the environment.

The Changing Role of Soil Conservation Policy[1]

At the beginning of the post-settlement era, after more than a century of rapid land settlement, new fears were being voiced about the possible inadequacy of the nation's natural resources to sustain increases in agricultural productivity and growth. Agricultural land policy issues began to focus on what happens when the free land is exhausted and on what the public's interest is or should be in the private exploitation of soil and water resources.

Such fears had been faced earlier in respect to U.S. timber and minerals policy. However, from the earliest colonial days until near the end of the nineteenth century, the dominant public policy was to encourage ownership and use of all natural resources with practically no restraints imposed by government. In the rapid expansion of the late nineteenth century, however, some influential people began to question and object to this policy. Foresters, playing an early role in the crusade for a change in policy, organized the American Forestry Congress in 1881 and formed the American Forestry Association in 1884. The American Economic Association, founded in 1885, provided a forum for economists to protest the dominant laissez-faire natural resources policy of the federal and state governments. They directed the country's attention to an array of pressing resource problems of public policy, including conservation and general land policy. In 1890, the American Association for the Advancement of Science petitioned Congress to change the nation's natural resources policy in the interest of conservation and more productive long-term use.

The Revision Act of 1891 marked a fundamental shift in natural resources policy away from laissez-faire doctrines and cast the federal government in a new role as the essential conservator of the nation's natural resources. Among other things, the Act incorporated the Forest Reserve Act, which mandated that more than 150 million acres in the public domain would be set aside as forest reserve. This mandate was carried out under Presidents Harrison, Cleveland, McKinley, and Theodore Roosevelt. After the assassination of President McKinley in September 1901, the Roosevelt administration, operating under authority of the 1891 Act, brought a new emphasis to conservation policy. Gifford Pinchot, Roosevelt's close friend and an ardent conservationist, was named head of the new Forest Service. With Pinchot's far-sighted vision and enthusiastic support, Roosevelt vigorously began sequestering additional forest and mining lands still under government ownership. Political opposition to these actions was so strong within the lumber industry, however, that senators and representatives from Western states were able to attach an amendment to an appropriation bill repealing the 1891 Act. But before Roosevelt signed the bill, he issued an order setting up 21 new forest reserves in six Western states, covering 125 million acres. Finally, in March 1908, Roosevelt called the first National Conservation Conference, inviting state governors, Supreme Court justices, cabinet members, congressional leaders, businessmen, and technical experts. This resulted in the National Conservation Commission of 49 members, which brought out a three-volume report supporting the case for conservation.[2] With this support, Roosevelt ordered the secretary of the interior to withdraw from entry some 80 million acres of coal lands, $1^1/_2$ million acres of land adjacent to possible future water-power sites, and nearly 5 million acres of land containing phosphate deposits.

Despite growing concerns about soil erosion, however, public action on soil conservation was not forthcoming until the late 1920s. In 1929, the USDA was mandated to conduct a nationwide study of soil erosion, to estimate the damage being done by erosion, and to recommend policies to reduce such damage. The study concluded in the early 1930s that erosion damage was severe on some 280 million acres used primarily for crops and somewhat less severe but still extensive on another 775 million acres used primarily for grazing.

The case for an effective soil conservation policy was soon made more urgent by widespread drought, extensive crop failures, and great dust storms in parts of the Great Plains. Hence, in the 1930s, in spite of disastrously low prices for agricultural commodities and new government-held surpluses, people widely believed that the future of the nation's food supply was endangered, and that nationwide action would be required if soil losses were to be checked and brought under control. The first action was a conservation program initiated in 1933 under emergency powers granted to President Franklin D. Roosevelt. Organized under the Civilian Conservation Corps in the U.S. Forest Service, a new program provided work in the national forests for formerly unemployed young men, who lived in camps and were given food, lodging, and a small monthly wage. The president soon proposed expanding the program to include erosion control on other types of land. Under emergency powers granted in the National Industrial Recovery Act, the Soil Erosion Service was established in the department of the interior. The new agency was directed to develop conservation programs on public lands and demonstration projects partly or wholly on private lands. Activities included leveling, sloping, and terracing eroding lands; planting trees and grass; building check dams; and developing grass waterways.

The nationwide drought in 1934 created much broader concerns leading to more vigorous action, including the Soil Conservation Act of 1935, which passed without a single dissenting vote in either house of Congress. This Act stated that the wastage of soil and water resources was a menace to the national welfare, and it declared that the permanent control of erosion was a policy of the federal government. The Act directed that "the Secretary of Agriculture shall coordinate and direct all activities relating to soil erosion," and it authorized the secretary: (1) to conduct surveys, researches, and other studies, to distribute results, and to carry on demonstrations; (2) to carry out preventive measures; (3) to cooperate or enter into agreements with any agency or person concerned with soil conservation; and (4) to give aid, subject to the conditions necessary to carry out the purposes of the Act. Finally, the Act directed the secretary of agriculture to establish an agency to be known as the Soil Conservation Service (SCS) to replace the Soil Erosion Service and directed the SCS to exercise the power conferred on the secretary, thus conferring on the SCS considerable status and power. In essence, the new SCS was to provide farmers with plans and technical assistance for farmland conservation on a voluntary basis. As a federal program, it was to function through state and local cooperation, with the SCS serving as the pivotal federal agency for conservation of farmland.

In collaboration with land-grant colleges, the SCS developed a model Standard State Soil Conservation Law. This model law was circulated to governors of the states, together with a letter from President Franklin D. Roosevelt requesting that the states enact legislation authorizing the formation of soil conservation districts. The first state law was passed in 1937, and by 1952 all 48 states plus Alaska, Hawaii, Puerto Rico, and the Virgin Islands had enabling laws. Through these laws, the bulk of the nation's cropland and much of its grazing land, has been included in soil conservation districts. The SCS has offered farmers technical help to design soil conservation programs for their farms, and about 80 percent have received some help. Although the SCS sometimes has a backlog of requests, it has been able to serve farmers on a first-come, first-served basis.

In the meantime, the Soil Conservation and Domestic Allotment Act of 1936, adopted as an amendment to the Soil Conservation Act of 1935, replaced the original Agricultural Adjustment Act of 1933. (The latter had set up a program of acreage allotments for individual farmers to control production. In January 1936, that program was struck down by the U.S. Supreme Court on grounds that it was an infringement of rights to control production, which were rights reserved by the states.)

The 1936 Act set up two classes of crops: *soil depleting* and *soil conserving*. Benefits were offered to induce farmers under contract to shift some of their cropland from soil-depleting crops (such as small grains, corn, and cotton) to soil-conserving crops (such as alfalfa, sweet clover, grass, and pasture crops). This shift would help to achieve both conservation and production control. The Act authorized federal grants to farmers for carrying out designated conservation practices, which were organized under the Agricultural Conservation Program (ACP) with the cooperation of the SCS. The 1936 Act was not satisfactory as a policy to regulate crop production, and it was soon replaced with new and more specific programs for controlling production and conserving soil.

The Agricultural Act of 1956, passed as a new plan combining production control and conservation, initiated the Soil Bank Program. This Act was designed to enable farmers to select which land they would take out of production, while allowing them more freedom to plant on the land left in production. There were two provisions. Under the Acreage Reserve, farmers were paid to reduce the planting of allotment crops—wheat, cotton, corn, tobacco, peanuts, and rice—to or below allotment levels. Under the Conservation Reserve, farmers were paid under long-term contracts to divert cropland to soil-conserving uses. Originally, the Soil Bank was to apply to about 60 million acres. It was funded at less than half this level, however. Consequently, it was ineffective as a measure for controlling production, and was subsequently replaced by more specific voluntary production-control programs.

Succeeding acts, generally passed every three to five years, have continued the policy to pay farmers to use conservation practices on land diverted from crop production while applying more flexible conservation requirements to land being cropped. Very high levels of soil erosion have been experienced over a small percentage of all cropland (such as 15 to 20 percent of the total cropland), and new policy has been aimed at permanently removing this land from crop production. As discussed in previous chapters, this idea was implemented in both the 1985 Food Security Act and the 1990 Food, Agriculture, Conservation, and Trade Act to set aside some 45 million acres of the most highly erodible cropland.

The other main part of conservation policy applies to conservation practices on land that is being cropped. This policy, called Conservation Compliance, introduced in 1985 and continued in 1990, embodies the common sense idea that if the federal government is to shore up the incomes of farmers, the government can expect the farmer–caretakers of the nation's soil resource to do what is necessary to protect it. A farmer who crops land classed as highly erodible by the SCS must take steps to reduce the erosion rate to an acceptable level. Failing this, such a farmer becomes ineligible for government payments for soil conservation or any one of the several other farm commodity price and income programs. The SCS is charged with specifying for each owner of highly erodible land what must be done to comply with the law.

Conservation Compliance holds economic and social interest beyond the confines of agriculture because it deals with the environment and the critical social resource of the nation's topsoil. It also holds political interest because it uses the withholding-of-benefits principle that is widely employed in federal–state relationships to get state and local compliance with federal programs.

Both policies—the permanent retirement of highly erodible land and the encouragement of conservation measures on land in crops—are rational in concept and in economic theory. In practice, however, both encounter very prominent and difficult technological and environmental issues. Unless land retirement can be justified as a policy of production control, retirement is the most expensive way to implement conservation. Also, because of budget limitations in farm programs, the withholding of benefits may be ineffective as a program to implement conservation on land in crops.

Technological and Environmental Issues: Cropland

Continuing appraisals by the USDA of the status, condition, and trends of soil, water, and related resources were mandated by the Soil and Water Resources Conservation Act of 1977 (cited as RCA).[3] By this time, the land base available to agriculture was stabilized. The total supply of crop and grazing land was very inelastic, or slow to change with shifts in agricultural prices. The supply for different cropland uses was more elastic, as farmers shifted from one crop to another, and the total acreage in crops in any year was responsive to changes in government programs. Overall, however, the total land in agricultural use changed slowly. Increases in land used for urban and industrial expansion was balanced by developing more land for agriculture through drainage, land clearing, and land leveling. Major increases in agricultural output were coming from advancing science and technology rather than from using more land or water.

Nevertheless, appraisals of land, water, and other natural resources as mandated by the RCA were becoming increasingly important. Such appraisals, as well as other studies, have found that over nearly all of the fertile cropland and most of the grazing land in the United States, the on-site cost to the farmer of long-term loss of soil productivity from erosion is very small compared with the long-term damage to the environment from eroding soil and pollution of ground water from pesticides and other contaminates. For example, according to estimates by the SCS and the U.S. Water Resources Council, if 1982 management is continued over the next 100 years, the *average loss* of cropland productivity from sheet and rill erosion will be less than 10 percent over more than 90 percent of the nation's cropland. Except for relatively small areas in California, higher losses will be confined to land of generally low agricultural productivity (Figure 12-1).

In comparison, the problem of eroded soil and other pollutants is spread widely over major areas (Figure 12-2), and the total costs in terms of damage to the environment are regarded as several times the value of crop productivity lost from erosion. In fact, in all but the most severe cases of erosion, the loss of soil productivity can be balanced by the natural processes of soil formation, and more than counterbalanced by farmers' use of fertilizer, pesticides, and other technology.

Because of variations in soils, slope, rainfall, and other factors, the location of losses of farm productivity (which the rational farmer would wish to reduce), are not well correlated with the location of damages to the environment (which is largely an externality for the farmer, but a specific cost to those being damaged and a general cost to

FIGURE 12-1

Estimated loss of cropland productivity resulting from 100 years of sheet and rill erosion under 1982 management, by major land resource areas. This map depicts the average loss of productivity in 100 years for major regions. Within each region, erosion in some areas greatly exceeds the average, and some other areas are below the average. Also, this loss applies only to sheet and rill erosion on cropland and does not include the loss from wind erosion on crop or grazing land.

Sheet erosion occurs when soil is dislodged rather uniformly and is carried to another location chiefly by water. Rill erosion occurs when soil is washed away by water in a rivulet or a very small stream or brook. The land may be left with gullies or channels, which in extreme cases are too wide or deep for growing crops. *Source:* Damage categories were compiled by the SCS Inventory and Monitoring Staff using 1977 U.S. Water Resources Council data. The map was prepared by the Cartography and Geographic Information Systems Division, SCS, Washington, DC. Reprinted from *The Second RCA Appraisal,* USDA, 1989, pp. 34–35.

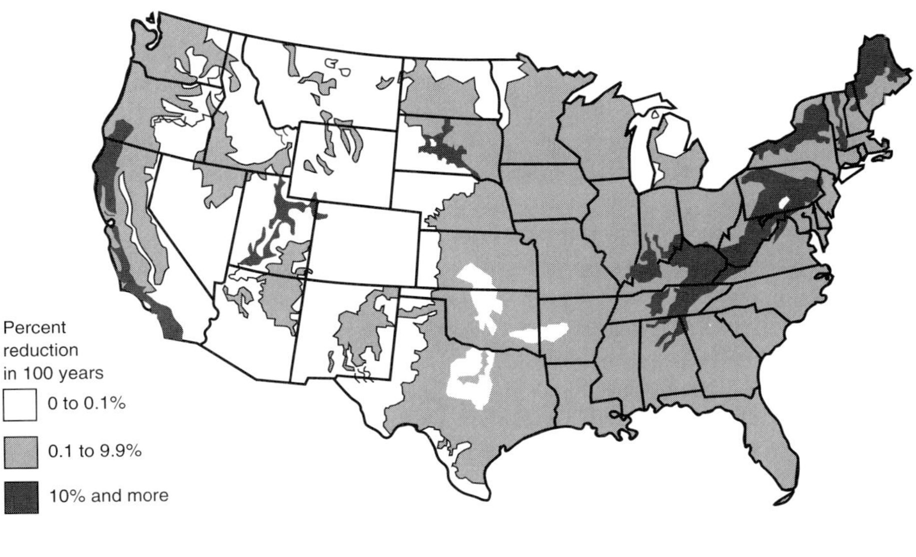

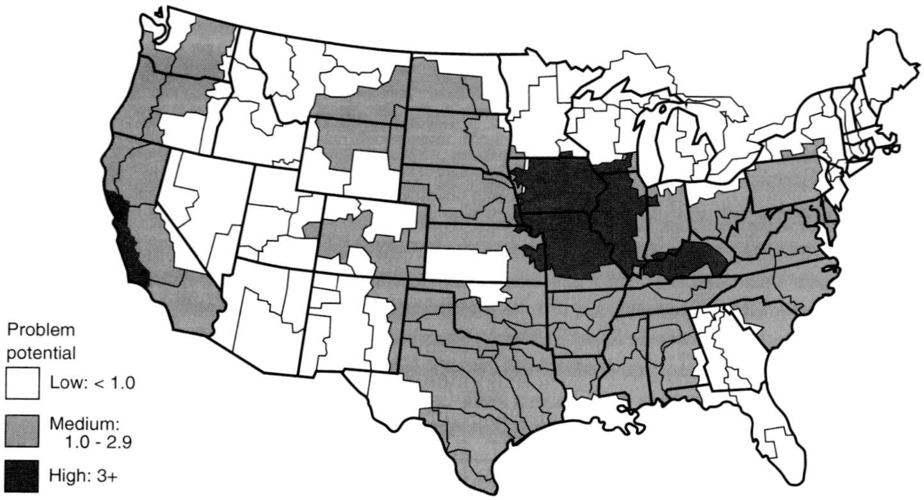

FIGURE 12-2
Estimated sediment yield in tons per acre per year. The potential was determined by estimating the percentage of sheet and rill erosion that becomes sediment in small streams. It was assumed that sediment delivery rates are higher where streams are more numerous and closely spaced and where the surface soils have more fine particles (silt and clay). Also estimated were amounts of pollutants, including nitrogen and phosphorus, organics, and pesticides, absorbed by sediment particles moving into streams. *Sources:* 1982 National Resources Inventory (USDA/SCS, 1984); U.S. Geological Survey Surface Soil Surveys; USDA Soil Survey Laboratory Data State Reports: *The Second RCA Appraisal,* USDA, 1989, pp. 105 and 106.

society). Over most of the best and most heavily cropped farmland in the United States, the current and projected loss of soil productivity from erosion is very low, in many locations almost negligible during the lifetime of the farmer. But the problem potential for damage to the environment is relatively high, in value terms as much as 10 to 20 times as much as the value of loss in farm productivity, for example.

The growing popularity of minimum tillage cropping systems is, of course, reducing the rate of soil erosion on much of the most heavily cropped farmland. However, minimum tillage and no-till cropping systems depend on pesticides to control weeds, insects, and other crop pests. Herbicides, which are the most important sources of pollution among the pesticides, are generally necessary to control weeds. Hence, the erosion problem, while reduced by minimum-till and no-till cropping systems, has not been completely eliminated, and some environmental problems have been increased.

The off-site damage from pesticides has been reduced in some areas by introducing a new system for pesticide management, called "integrated pest management," or IPM. This system involves a more thorough scouting and evaluation of pest problems, evaluation and selection of pesticides if recommended, and programming pesticide use to minimize damage to the environment. IPM may make it profitable for a farmer to reduce the use of pesticides to significantly lower levels. Still, it must be recognized that what is generally most profitable and therefore most acceptable to the farmer is not what is most desired or most acceptable for the environment; this is the essence of the policy problem. Before considering more specific policy alternatives for cropland, let us briefly review the major policy issues relating to rangeland.

Environmental Issues Related to Rangeland[4]

The rangeland of the United States is used primarily for grazing cattle and sheep, with cattle being far more important. Rangeland quality and its continued productivity can be judged by comparing the present vegetation of a site to the climax vegetation. Climax vegetation is generally the most stable and productive natural plant community, or mixture of plants, a given site is capable of supporting; and the percentage composition and plant vigor of the climax species is the best indication of range health or continued productivity. If the climax vegetation is reduced by continuous heavy grazing, drought, or other factors, then the range is impaired and its productivity is reduced.

In the rating system used by the Bureau of Land Management, if 75 percent or more of the present vegetation is of the climax variety, the range is in excellent condition; 51 to 75 percent is good; 26 to 50 percent is fair; and 25 percent or less is poor. Range-condition ratings are not assigned to cropland maintained as permanent pasture, or to pastures rotated with other crops. Also, the rangelands maintained by the U.S. Forest Service are not evaluated according to this system, because conditions vary and different types of appraisals are more appropriate. About 600 million acres of the 770 million acres of rangeland is appraised according to this system, however, and the general results of such appraisals appear as in Figure 12-3.

The data presented in Figure 12-3 are useful for appraising the quality and productivity of rangeland and give some indication of the efficiency of its management. However, these data alone do not provide an adequate basis for conservation policy. The carrying capacity of rangeland varies widely, and the public benefits from more restrictive controls over grazing also vary widely. Arguments persist between ranchers and public officials concerning the relative costs and benefits of public versus private ownership, governmental regulations for protection of wildlife, and fees charged for grazing rights on public lands.

Environmental policy issues are dominated by competition and conflicts relating to multiple uses. Shall the interests of livestock ranchers and increasing agricultural output be given top priority? Or shall livestock ranching be restricted by policy measures, such as higher grazing fees on public lands, to conserve soil and water? Or shall protecting wildlife and improving the environment for wildlife be the guiding light for policy?

Since about the beginning of the twentieth century, the nation has struggled with these issues. During the period of land settlement, most of the public land listed in Figure 12-3 remained in public ownership because ranchers were able to use it as open range and some land that was homesteaded reverted to county or state ownership when settlers failed to pay taxes on it. Until the early 1930s, much of this public land was used by ranchers largely according to who was able to put livestock on it and provide winter feed and other care. However, much of the land was being grazed too heavily and the quality of the vegetation was deteriorating. Serious problems of excess runoff and of wind erosion were becoming apparent. The lack of clearly defined rights to the

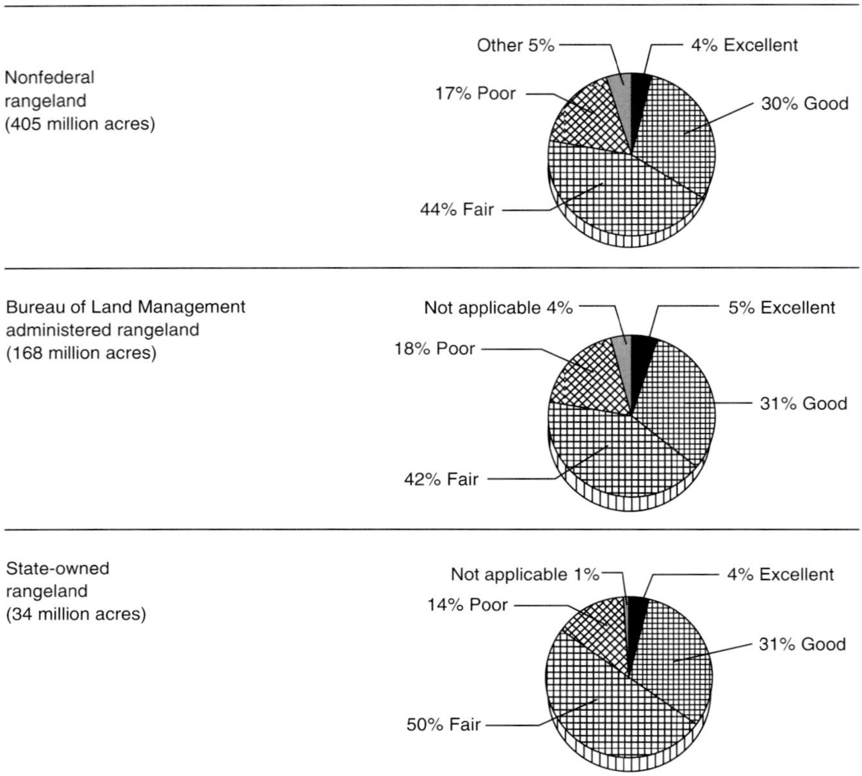

FIGURE 12-3
Condition of nonfederal rangeland (mostly privately owned), federally owned rangeland administered by the Bureau of Land Management, and state-owned rangeland. *The Second RCA Appraisal,* Public Review Draft, 1987, pp. 3–6.

public land made it unprofitable for ranchers to conserve the soils, build dams and reservoirs, control the grazing population, or make long-term plans for land improvements.

The Grazing Act of 1934, popularly called "the Taylor Grazing Act," was adopted to control the use of publicly owned grazing land and to integrate its use with that of privately owned land. This Act authorized grazing districts in which local ranchers would have rights to graze specified numbers of animals at a fixed fee per head. Local advisory boards elected by the ranchers were given legal recognition and permanent status by a 1939 amendment. A National Advisory Board Council was established in 1940, through an amendment to the Federal Range Code for Grazing Districts.

When the Grazing Service was first established, about 166 million acres of grazing lands were under federal ownership. The original act limited to 80 million acres, the land that might be organized in grazing districts. The specified area was raised to 142 million acres in 1936. In 1954, the limit was removed entirely. Thus some 160 million acres of public grazing land were brought under control of the Division of Range Management, Bureau of Land Management, Department of the Interior. This land, added to the 160 million acres of national forest under the U.S. Forest Service on which some grazing is also permitted, constitutes the bulk of federally owned land that is used for food production in the 48 contiguous states. In addition, states and counties own some land used for grazing, and the federal government owns about 98 percent of the 365 million acres of Alaska (Table 12-2).

As livestock ranchers normally try to maximize net returns from their operations and therefore tend to graze up to the maximum sustainable yield and as the government agencies are charged with protecting wildlife, and regulating the use of the grazing lands to prevent excessive water runoff and wind erosion, there is a natural conflict in policy. Public hearings and studies have stabilized the levels of grazing permitted, with revisions in the levels a matter for periodic evaluation.

Policy conflict continues, however, about whether this land could be managed more efficiently under private ownership, and about how much protection shall be given to protecting wildlife and certain endangered species. These issues go well beyond the supply and demand for food, to issues concerning the total environment, and the quality and safety of the nation's food and water supply (discussed in the next chapter).

PRODUCTION AND CONSERVATION POLICY ALTERNATIVES

Theoretically, the general problem for natural resources policy on land that is profitable for farming or grazing is to develop conservation programs that will encourage farmers and ranchers to use land and water in ways that will maximize *net social income.* Net social income is defined as the net value of crop and livestock production after all production costs have been deducted, including the internal on-site damage costs of soil loss and the external off-site damage costs. Since it is not practical to actually measure all of these costs, a generalized method has been developed to estimate them.

The universal soil loss equation (USLE) has been developed by soil scientists to estimate soil loss from an acre of land. The USLE uses a rainfall factor for a type-of-farming area or crop region; a soil erodibility factor; a slope and length factor; a cropping management factor, including crop rotations and tillage systems; and an erosion control-practice factor. The equation is written as follows:

where

$A = RKLSCP.$

A is the computed soil loss in tons per acre per year

R is the rainfall factor for the region

K is the soil erodibility factor

LS is the slope and slope length factor

C is the cropping (or livestock) management factor

P is the erosion control-practice factor

Measuring the Off-site Damage Done by Soil Erosion

In 1993, the USDA adopted a new revised formula (RUSLE), which retains the basic USLE equation, but uses newer, more specific data on rainfall, seasonal changes such as freezing and thawing, and adds such factors as crop growth, crop residue cover and decomposition, surface decomposition, surface roughness, and soil moisture. The new RUSLE is more accurate in estimating soil loss, applies to more diverse situations, and is more helpful to conservation agencies and farmers in controlling soil erosion. However, it may be replaced or supplemented in the future by other new and more sophisticated programs, such as the water erosion prediction program (WEPP), currently under development by the Agricultural Research Service (ARS).

To estimate the damage that sediment causes as it is delivered off-site by the drainage system, a sediment damage equation can be used to express aggregate sediment damage in dollars per year: where $D_s = A_N E D_r P_r$. Then, D_S is sediment damage (dollars per year), A_N is the net drainage area, E is soil loss (tons per acre), D_r is the sediment delivery ratio, and P_r is the cost of sediment damage per ton of sediment.

In this equation, the net drainage area is defined to include a watershed or subwatershed. The soil loss (tons per acre) is taken from the previously used universal soil loss equation. The sediment delivery ratio is expressed as the percentage of total sediment that is actually delivered to the place where it causes damage, as in polluting ground water, filling a reservoir, or clogging a drainage system. The cost of sediment damage per ton is estimated from costs of treating ground water, dredging to remove a ton of sediment from a reservoir, the average cost of building a new reservoir, or dredging a new drainage system. Alternatively, if a new system is not to be built, the off-site damage cost of sediment may be estimated in terms of ecological damage, including crop losses downstream, pollution and losses of lakes and reservoirs, and possible flood-loss damages.

For purposes of illustration, an application of these equations to a particular situation is given in Table 12-3, with the external damage costs of sediment set at $4 per ton. Net farm income is maximized with a cropping plan of continuous corn, conventional tillage, and cultivation straight across the field identified as up-and-down cultivation; but when the cost of off-site damage is added, the net social income is more than $6 per acre lower than it would be with plow–plant tillage, contouring, and terracing. If the farmer were to shift to this system, however, net farm income would be reduced by more than $2 per acre. Then, if a policy were adopted to pay the farmer $2 per acre for using plow–plant tillage, contouring, and terracing, the net gain to society would be $4 per acre.[5]

The example in Table 12-3 is only one from many situations that could be cited (from this or other studies). The possible examples are virtually endless. They will vary from situations in which few farmers without government help could afford to adopt recommended soil conservation-farming systems, to situations in which both net farm income and net social income are maximized, or nearly maximized, under recommended conservation systems.[6]

To define efficient soil conservation programs, it is necessary to (1) identify the value of the net loss of farm productivity and the net external costs of erosion under various conditions, (2) determine the relative efficiencies of alternative strategies in

TABLE 12-3
NET INCOME AND SEDIMENT DAMAGE PER ACRE PER YEAR
(Polecat Creek Subwatershed Cropland in Illinois)

Crop rotation[a]	Tillage system[b] (1)	Conservation practice[c] (2)	Net farm income[d] (3)	Sediment damage[e] (4)	Net social income[f] (5)	Sediment damage percent of net income (6)
Continuous corn	Con	UD	$82.55	$10.20	$72.35	12.4
		CT	81.80	4.96	76.84	6.1
		CT/T	81.51	4.50	77.01	5.5
	PP	UD	81.50	3.84	77.66	4.7
		CT	80.75	2.31	78.44	2.9
		CT/T	80.46	1.88	78.58	2.3
	CP	UD	68.99	0.77	68.22	1.1
		CT	68.24	0.30	67.94	0.4
		CT/T	67.94	0.29	67.65	0.4
C-Sb	Con	UD	78.40	12.68	65.72	16.1
		CT	77.65	6.28	71.37	8.1
		CT/T	77.36	5.70	71.66	7.3
	PP	UD	77.35	5.94	71.41	7.7
		CT	76.60	2.76	73.84	3.6
		CT/T	76.31	2.56	73.75	3.4
	CP	UD	66.57	1.28	65.29	1.9
		CT	65.82	0.56	65.26	0.9
		CT/T	65.53	0.31	65.22	0.5
C-Sb-C-Wx	Con	UD	67.87	8.55	59.32	12.6
		CT	67.12	4.08	63.04	6.1
		CT/T	66.83	3.63	63.20	5.4
	PP	UD	66.82	3.88	62.94	5.8
		CT	66.07	1.77	64.30	2.7
		CT/T	65.78	1.44	64.34	2.2
	CP	UD	57.14	0.65	56.49	1.1
		CT	56.39	0.31	56.08	0.5
		CT/T	56.11	0.28	55.83	0.5
W-M-M-M	Con	UD	47.58	0.13	47.45	0.3
		CT	47.39	0.06	47.33	0.1
		CT/T	47.32	0.05	47.27	0.1
	PP	UD	47.32	0.12	47.20	0.3
		CT	47.13	0.05	47.08	0.1
		CT/T	47.06	0.04	47.02	0.1
	CP	UD	46.16	0.08	46.08	0.2
		CT	45.98	0.04	45.94	0.1
		CT/T	45.91	0.03	45.88	0.1

[a] Where C is corn, Sb is soybeans, Wx is wheat with a catch crop, W is wheat, and M is meadow.
[b] Con is conventional tillage, PP is plow–plant tillage, and CP is chisel–plow tillage.
[c] UD is up-and-down cultivation, CT is contouring, and CT/T is contouring and terracing.
[d] Net of nonland costs, weighted average for the cropland.
[e] As discussed in *Economic Analysis of Erosion and Sedimentation,* University of Illinois at Urbana–Champaign, Department of Agricultural Economics, in cooperation with State of Illinois, Institute for Environmental Quality, AERR 135.
[f] Column 3 minus column 4.
Source: M. T. Lee, A. S. Narayanan, and E. R. Swanson, *Economic Analysis of Erosion and Sedimentation: Upper Embarras River Basin,* University of Illinois at Urbana–Champaign in cooperation with State of Illinois, Institute for Environmental Quality, IIEQ document no. 74–41, AERR 135, April 1975, pp. 24–25.

reducing erosion, and (3) compare some of the alternative incentives that can be used to implement programs, such as education, new technology, payments or taxes, zoning, land purchase or set-aside program.

As long as the U.S. government is attempting to control crop output by setting aside or retiring land, there may be a net benefit in retiring the most erosive cropland, or land that contributes to the largest off-site damages. However, if land retirement for purposes of output control is not a major objective, then taking productive land out of production may be the most expensive way to control environmental damage costs. Programming the use of farmland in a way that will maximize net social income may be the optimum economic policy for combining production efficiency and conservation goals. The value of the benefits in these goals and the costs of achieving the goals must be recognized and faced for more effective policy making.

Political Problems and Constraints

From the beginning of federal soil conservation programs in 1935, many programs and much of the total expenditures have not been well targeted. On-farm planning and technical assistance by the SCS has been provided on request to farmers, on a first-come-first-serve basis, which may satisfy most farmers' preferences, but frequently does not achieve the best results for conservation. More importantly, SCS plans have often been too expensive for farmers to carry out, especially for farmers who use the more erosive soils. Also, a major portion of the soil conservation payments under the Agricultural Conservation Program (ACP) has not been well targeted on conservation problems. A 1977 study by the General Accounting Office (GAO), for instance, concluded that in 1975 some 55 percent of cost-sharing funds under the ACP went to increase production, rather than to conserve soil or water. A 1980 study by the Agricultural Stabilization and Conservation Service (ASCS) found that more than half of the ACP's cost-share funds were being applied on land that had already met the SCS conservation target of annual soil loss from erosion of less than 5 tons per acre, and only 21 percent of the funds were being spent on land with an annual soil loss of 14 tons per acre or more.

The Acts of 1985 and 1990 established the Conservation Reserve Program (CRP), and mandated retiring some 45 million acres of the most erosive farmland, but less emphasis has been placed on funding the necessary programs to achieve higher levels of conservation on land currently being cropped. A broad reordering of priorities will be required to allocate federal funds to encourage farmers to adopt practices that will maximize net social income on land being cropped or grazed. Achieving the economic optimum for conservation funding will remain as much a political problem—that is, both a legislative problem and an administrative problem—as it will an economic problem.

LOOKING AHEAD

The next chapter examines policy alternatives relating to quality and safety of food and water supplies. It deals with some of the effects of modern technology on quality and safety, and the benefits and costs related to policies designed to improve quality and increase safety of food and water supplies.

ENDNOTES

1 Much of the discussion in this section draws on Harold G. Halcrow, *Food Policy for America,* McGraw-Hill, New York, 1977, pp. 333–385; and Harold G. Halcrow, Earl O. Heady, and Melvin L. Cotner, eds., *Soil Conservation Policies, Institutions, and Incentives,* published for North Central Research Committee III: Natural Resource Use and Environmental Policy, by the Soil Conservation Society of America, 7515 Northeast Ankeny Road, Ankeny, Iowa 50021, 1982.

2 U.S. Congress, *Senate Report of the National Conservation Commission,* Senate Document 676, 60th Congress, 2nd Session, 1909.

3 *The Second RCA Appraisal: Soil, Water and Related Resources on Nonfederal Land in the United States, Analysis of Conditions and Trends,* U.S. Department of Agriculture, Public Review Draft, July–August 1987.

4 For extensive background discussion, see especially Wesley Calef, *Private Grazing and Public Lands, Studies of the Local Management of the Taylor Grazing Act,* The University of Chicago Press, Chicago, 1960; Phillip O. Foss, *Politics and Grass, The Administration of Grazing on the Public Domain,* University of Washington Press, Seattle, 1960; Marion Clawson, *The Western Range Livestock Industry,* McGraw-Hill, New York, 1950; Marion Clawson, *Uncle Sam's Acres,* Dodd, Mead, New York, 1951; J. R. Penny and Marion Clawson, ''Administration of the Grazing Districts,'' *Land Economics,* February 1953, pp. 23–34; U.S. Bureau of Land Management, *The Federal Range Code for Grazing Districts,* 1949 and later revisions; and Public Law 375, 83 Cong., May 28, 1954.

5 This point is demonstrated more fully in a series of research publications developed under the leadership of Earl R. Swanson. For a general summary and presentation of conclusions, see Karl Gunterman, M. T. Lee, A. S. Narayanan, and E. R. Swanson, *Soil Loss from Illinois Farms, Economic Analysis of Productivity Loss and Sediment Damage,* Illinois Institute for Environmental Quality, Chicago, IIEQ Document No. 74–62, December 1974.

6 The latter situation is demonstrated in C. Arden Pope III, Shashanda Bhide, and Earl O. Heady, *The Economics of Soil and Water Conservation Practices in Iowa: Model and Data Documentation,* CARD Report 108, SWCP Series I, The Center for Agricultural and Rural Development, Iowa State University, Ames, IA, August 1982.

IMPORTANT TERMS AND CONCEPTS

Natural resources
Natural resources policy
Property rights
Distribution of the public domain
Family farm
Fee simple ownership of land
Eminent domain
Soil erosion
Environmental protection
Homestead Act of 1862
Revision Act of 1891
Reclamation Act of 1902
Civilian Conservation Corps
U.S. Forest Service
Agricultural Conservation Program (ACP)
Soil Conservation Act of 1935
Soil Conservation Service (SCS)
Soil Conservation and Domestic Allotment Act of 1936
Grazing Act of 1934 (Taylor Grazing Act)
Federal Range Code for Grazing Districts
Agricultural Act of 1956, Soil Bank Program
Climax vegetation on grazing land
Efficiency of alternative soil conservation programs

Loss of farm productivity due to sheet and rill erosion
Composite natural resources policy
Soil and Water Resources Conservation Act of 1977 (RCA)
RCA projections of cropland required
U.S. farm use of energy and fertilizer
Conservation Reserve Program (CRP)
Integrated pest management (IPM)
Integrating soil conservation and pesticide policy
Location and comparison of damages done by pollutants

QUESTIONS AND STUDY SUGGESTIONS

1 What do we mean by a natural resources policy?
2 How did the natural resources policy of the United States change in the twentieth century?
3 In what way did the U.S. land distribution policy change during the first 75 years of the nation's history? What important things did the homestead policy accomplish? Which of the policy goals did the homestead policy fail to accomplish?
4 What were the major provisions of the Revision Act of 1891? What natural resources were affected? Why were some of these provisions opposed? Who opposed them?
5 What were the major provisions of the Reclamation Act of 1902? What changes in national policy were established? What limits were set for the sale of irrigation water? Why? Was this provision generally enforced? What has been the result in terms of farm size?
6 What are the respective roles in land-use policy of local, state, and federal governments? Why might agricultural groups oppose the involvement of the federal government in land-use planning and in controlling land use? Why do some other groups support it?
7 What is the Conservation Reserve Program (CRP)? When did it begin? What are its purposes and scope?
8 The off-site costs of soil erosion appear to be much greater than the on-site costs to the farmer. What general policy issues arise from this? What is the general inference for conservation policy?
9 How does the off-site damage done by soil erosion tend to vary among major farming regions in the United States? What are the inferences for natural resources conservation policy?
10 What are the major environmental issues for natural resources policy applied to rangeland? How have these issues been approached in federal policy? Why is there a continuing policy conflict over agriculture's use of rangeland?
11 Define *net social income.* In terms of natural resources policy, what are the necessary conditions to maximize net social income? What are some of its limitations as a guide for natural resources policy relating to cropland? Relating to rangeland? Why do profit-maximizing farmers frequently fail to maximize net social income?
12 What is included in the universal soil loss equation? How may it be used or applied in planning conservation programs? What is in the *sediment damage equation*? How may it be used? What does it measure?
13 In general terms, how much of the inherent erosion of soils can be controlled by using recommended crop rotations, tillage systems, and conservation practices? Supposing a farmer has perfect knowledge concerning this, why might the recommendations not be used?
14 How will an effective soil conservation policy differ from location to location if (a) the off-site damages from soil erosion are generally much larger than the on-site damages, and (b) the off-site damages are relatively much greater in some situations than in others?

CHAPTER **13**

FOOD QUALITY AND SAFETY POLICY

In addition to seeking an adequate food supply, people have always tried to obtain food free of substances that might endanger their health. Concern has deepened, however, with the modernization of agriculture, particularly with the use of pesticides, animal growth hormones, and food and color additives. The scope of food and agricultural policy has expanded in attempts to improve the quality and safety of food and water sources.

IN THIS CHAPTER

By studying Chapter 13, you can learn how to answer the following questions:

1 How have food quality and safety policies evolved?
2 What are the major government agencies involved in food quality and safety and what are their responsibilities?
3 What are the most critical food quality and safety problems?
4 What are the major policy issues in food quality and safety?

The U.S. food supply is generally of high quality and one of the safest in the world. A myriad of government agencies and private organizations are involved in food quality and safety activities such as: approving pesticides for use in food production, inspecting meat and poultry, grading fruits and vegetables, regulating food advertising, certifying organic food, and distributing information on handling, preparation, and dietary guidelines. Yet outbreaks of contamination occur periodically and raise concerns about the safety of the food supply. For example, the presence of salmonella bacteria, probably from mixing unpasteurized and pasteurized milk in a processing plant, resulted in more than 16,000 confirmed cases of food poisoning in the Chicago area in 1985, the largest recorded outbreak in the United States.[1]

Food quality programs involve establishing standards of cleanliness, grading food products according to the standards, certifying that products meet the standards, and inspecting products for compliance with the standards. *Food safety* programs ensure that food is healthful, wholesome, and properly labeled. In recent years, the term has been expanded to encompass the link between a nutritious, well-balanced diet and health. For example, food safety now refers to several factors such as reducing cholesterol levels and fat intake to lower the risk of heart disease.

In this chapter, we will first discuss how food quality and safety policies evolved. We will then examine problems and policy options in the use of pesticides, explore alternative food quality and safety policies, review the role of regulatory agencies responsible for food safety and quality, examine the major safety problems, and address policy considerations for the future.

EVOLUTION OF FOOD QUALITY AND SAFETY POLICIES

The evolution of food quality and safety policies has been shaped by advances in technology and changes in food consumption patterns and life styles. Scientists can now detect minute amounts of potentially harmful substances in food, which raises questions about the scope of current food quality and safety policies. Furthermore, U.S. consumers now eat more fresh fruits and vegetables than formerly, which raises questions about the amount of pesticide residues ingested. Fish consumption has expanded greatly over the past two decades, and more Americans are eating uncooked fish or fish cooked rare, which has created new food safety concerns. Moreover, there is no mandatory inspection program for seafood as there is for red meat and poultry moving in interstate and foreign commerce. These and similar trends are bringing new challenges to food safety policy making.

Development of Food Quality and Safety Policies

Policies for food quality and safety have developed piecemeal, responding to farm organizations, consumer organizations, and technology. During the nineteenth century, agricultural producers were at the forefront of efforts to establish public policy in this regard. Economics was the primary impetus for their efforts. Because of trichinosis, European countries banned imports of U.S. salted pork and bacon, and the loss of foreign markets led to reduced prices. Producers, pushing for a federal meat inspection program for exports saw their efforts rewarded in 1890, when an inspection program was established. To protect their markets, producers also worked through farm organizations to obtain regulations for the sanitation of milk, eggs, and other products and to discourage the production of oleomargarine and other substitutes.[2]

The twentieth century marked the beginning of consumer lobbying for effective food quality and safety policies. Concerns about food processing and meat packing formed the rallying point. *The Jungle* by Upton Sinclair exposed unsanitary conditions in the meat packing industry.[3] Journalists advanced this idea in popular magazines, thus helping to galvanize public concerns. The chief chemist of the USDA, Harvey Wiley, provided information to policy makers and the public on the dangers of some additives that his staff found in manufactured foods. In response to questions about the safety of the food supply, the federal government adopted the first comprehensive food quality and safety law, the Pure Food and Drug Act of 1906. This Act controlled the amount and type of certain substances permitted in food.

Also in 1906, the Federal Meat Inspection Act was adopted. Amended a year later, it required that meat and meat products moving in interstate and foreign commerce be federally inspected to ensure that the products were safe, wholesome, and correctly marked, labeled, and packaged. The legislation was amended 60 years later to deal with intrastate commerce. The Wholesome Meat Act of 1967 established the federal–state cooperative program under which the federal government helps to fund state inspection programs for meat sold in intrastate commerce. The state programs are probably at least as thorough and competent as the federal inspection program. In 1968, the Wholesome Poultry Products Act extended the same provisions to poultry inspection.

Another piece of legislation, the Federal Insecticide Act of 1910, was significant because it laid the foundation for "truth in packaging." It required that pesticides be labeled to show product ingredients to protect agricultural producers from unscrupulous manufacturers who made false claims about insecticidal action. Insecticides, herbicides, and fungicides are critical to producers in controlling unwanted insects, diseases, and plants. The use of these pesticides results in enormous benefits such as higher crop yield, improved animal health, energy savings, land conservation, and reduced labor.

During the 1930s, statutes were enacted to deal with marketing and the use of food and color additives. Modified over the years, these statutes are critical components of food policy. The Agricultural Marketing Agreement Act of 1937 mandated that programs and agreements be established to regulate the quality, quantity, or pack requirements for fruits, vegetables, and some specialty crops. The Act also set quality standards for commodities being imported into the United States when there were marketing orders for the domestic production of those commodities.

Grass-roots support from several national women's clubs was instrumental in enacting the Federal Food, Drug, and Cosmetic Act of 1938. This Act regulated the use of food and color additives. *Food additives* are those that preserve or flavor food; *color additives* are those that color food to make it more appealing to consumers. Under this Act, pesticides are considered food additives. Controversial from the beginning, the Act was opposed on the one hand by the food industry, presumably for going too far in setting regulations, and it was opposed on the other hand by the major consumer organization, Consumers Research, for not going far enough.

After World War II, food quality regulations were expanded with the Agricultural Marketing Act of 1946. This was a broad comprehensive Act that, among other things, authorized the secretary of agriculture to inspect, certify, and identify the class, quality, and condition of agricultural products destined for interstate sale. The Act covers many products, including livestock, meat, poultry, shell eggs, dairy products, fruits, vegetables, rice, processed grain products, fish, and shellfish.

In 1947, with the advent of new pesticides, the 1910 Federal Insecticide Act was amended and updated. The amended Act, called the Federal Insecticide, Fungicide, and Rodenticide Act (FIFRA) continued the provisions of the earlier Act but added two important regulatory reforms: (1) new pesticides must be reviewed by the USDA before marketing, and (2) the manufacturer must provide pesticide safety and efficacy.[4] Eventually, responsibility for enforcing the Act was transferred to the Environmental Protection Agency (EPA), which was organized in 1970. The EPA was charged with registering pesticides, specifying the terms and conditions of their use before marketing, and removing unreasonably hazardous pesticides from the market. EPA is guided in its decisions by making estimates of economic, social, and environmental costs and benefits.

The provisions in the 1938 Federal Food, Drug, and Cosmetic Act (FFDCA) proved to be inadequate and were strengthened with two amendments—the Food Additives Amendment in 1958 and the Color Additives Amendment in 1960—which set federal policy regarding carcinogenic food additives (the Delaney clause) and carcinogenic color additives. The amendments, however, do not apply to food or color additives approved for use before 1958.

The Delaney clause provides that no additive found to induce cancer when ingested by humans or animals shall be added to processed food. Literally, it bans such additives regardless of how small the dose and thus establishes a zero tolerance level. *Tolerances* refer to the maximum legally acceptable levels of a substance allowed in commercially sold food. These levels apply to imported food as well as to domestically produced food and animal feed.

When first enacted, the Delaney clause could seldom be enforced because little scientific information was available to link low levels of pesticide residues to tumor formation or to show the degree to which residues concentrated during food processing. Improved technology has allowed scientists to identify contaminants and carcinogens in foods in amounts as low as one part per trillion (that is, one gram of the substance per one trillion grams of food), thus making it possible to estimate very small risks. Because of the zero tolerance level and recent scienific developments, the EPA has not approved some new pesticides with very low cancer risks, while older pesticides approved before advances in detection of potentially hazardous substances remain on the market. This situation is often referred to as the Delaney paradox.[5]

The Delaney paradox also arises because the clause applies to processed food, but not to raw food such as fresh fruits and vegetables, which fall under the risk–benefit standard of section 408 of the FFDCA. Under this standard, tolerances are set for pesticide residues. The benefits of pesticide use are weighed against the human dietary carcinogenic risk. Thus two different standards are in place, one based on zero tolerance and the other on risk–benefit, and the decision about which standard to apply rests on where the risk occurs, either before or during food processing. There is no public health reason for this distinction.

Modern Era of Policies Affecting Food Quality and Safety

New policies were developed in the 1970s and 1980s in response to improved technologies and procedures to identify risk more accurately. More rigorous methods were developed for testing chemical residues in food and determining the maximum levels that would be safe, and policy makers responded to this development. A 1978 amendment to FIFRA mandated stricter requirements for approving pesticides for use on specific crops and called for reregistration of those currently in use. By the mid-1980s, EPA had analyzed 74 of the 289 registered pesticides and concluded that 53 of them caused either benign or malignant tumors. These 53 pesticides accounted for 90 percent of fungicide use, 40 percent of insecticide use, and 38 percent of herbicide use.[6]

Because reregistration of pesticides is both time-consuming and costly, many that were approved before 1972 remain in use. The 1988 amendments to FIFRA deal with these concerns. They require that the testing, reregistration, and possible cancellation of older chemicals be completed by 1997. Before the 1988 amendments, EPA bore all the expenses for toxicology, chemistry, and other tests needed for reregistration, but the new law allows the agency to pass some of these costs to the manufacturer. Recognizing that Americans are eating more imported foods, the 1988 amendments require the Food and Drug Administration (FDA) to enter into cooperative agreements with foreign countries to obtain pesticide usage data on products imported from these countries.

A 1987 report by the National Academy of Sciences (NAS) has influenced recent policy making.[7] This study pointed out that even though some new pesticides have a lower health risk than older ones, they cannot be introduced into the marketplace because of the Delaney clause. The drawbacks to the Delaney clause as a food safety tool were noted, particularly its application to processed foods but not to raw foods. According to the NAS report, this double standard results in a greater health risk than an alternative policy that would allow tolerances, set at a low level, for pesticide residue on both processed and raw foods. This alternative policy would be based on negligible risk, such as one additional incidence of cancer per one million people over a 70-year life span. NAS estimated that a negligible risk policy would eliminate 98 percent of the total dietary risk from the carcinogenic pesticides, compared with eliminating 100 percent of the risk if the Delaney clause with its zero tolerance level for carcinogenic compounds were applied to both processed and raw foods.

In October 1988, EPA announced that it was moving to a policy of negligible risk in pesticide registration. This decision allows many new pesticides to be approved while facilitating the cancellation of older, more hazardous ones. Some consumer and environmental groups have filed a lawsuit to stop the shift from zero tolerance to negligible risk, which means that this issue may not be resolved.

Policy makers have begun to address the lack of scientific data concerning the long-term health effects of pesticides and the risks associated with their use. Such information is needed to develop effective food safety policy. The Pesticide Monitoring Improvements Act of 1988 mandates that the FDA develop data management systems to track, summarize, and evaluate pesticide monitoring data and to implement a plan to guide the development of methods to improve the efficiency of food monitoring. A Pesticide Data Initiative Program esablished by the USDA in fiscal 1991 will provide funding for data collection and analysis of pesticide use, residue levels, and potential exposure levels.

USDA has traditionally inspected meat and poultry visually for signs of disease or abnormalities. In recent years, however, it has paid more attention to reducing microbial contamination, which is a major food safety problem. This new focus should help to improve the safety of meat and poultry. USDA, industry, and consumer organizations continue to disagree about what constitutes adequate inspection procedures, particularly for processing plants, which are not continuously inspected. Discretionary inspection authority, granted to the USDA's Food Safety and Inspection Service (FSIS) in 1986, allows the agency to vary the type of inspection on the basis of a processing plant's compliance record, the commitment of plant management to control its operation, and other factors.

As evidence mounts that ground water, on which more than 117 million people rely for drinking water,[8] has been contaminated by pesticides and nitrates in some areas, policies have been introduced to deal with water quality. Concerned about the safety of drinking water, some Americans have begun to use bottled water. Between 1976 and 1988, sales of bottled water rose threefold, reaching about $2 billion annually.[9] Under the Safe Drinking Water Act, the FDA in consultation with the EPA sets standards for bottled drinking water.

The Water Quality Act of 1987, which amended the Federal Water Pollution Control Act, is designed to control pollution from agricultural production. Runoff from agricultural land contains pesticides, nitrogen, phosphorous, and other substances that account for a significant portion of the nonpoint-source water pollution. Referring to pollutants that cannot be traced to a specific source, this type of pollution is often more difficult to control because the polluters do not directly bear the costs for cleanup of the contaminated water. Unlike previous laws, the 1987 Water Quality Act requires states to consider nonpoint as well as point pollution in water quality programs. The Act also expands the scope of water quality regulations to include ground water, a major source of drinking water in many rural communities.

The comprehensive Food, Agriculture, Conservation, and Trade Act of 1990 also contained provisions affecting water quality. The Act (1) added cropland that adversely affected ground water or surface water quality to the Conservation Reserve Program and (2) provided incentives for the enrollment of land near wellheads, areas inhibited by threatened or endangered species, or land on which agricultural production poses a threat to water quality. If a farmer enrolls in the Conservation Reserve Program to protect land near a wellhead, or land from which runoff or other factors can contaminate water quality, then that farmer can receive up to $3,500 annually and cost-share assistance up to $1,500.

In recent years, some states have set new policies for food quality and safety. California has emerged as a leader in regulating the use of pesticides. This is significant because California produces more than 40 percent of the fresh and processed fruits in the United States, nearly half of the fresh vegetables, and more than half of the processed vegetables.[10] Under California's monitoring program, chemical manufacturers must provide information on pesticide chemistry, effectiveness, and residues that concentrate on food before a pesticide can be registered with the California Department of Food and Agriculture. Moreover, the state collects samples of food from various outlets, including wholesale and retail stores and ports of entry, and tests them for pesticide residues.

California has also been a leader in addressing the safety of water. The California proposition 65, also known as the Safe Drinking Water and Toxic Enforcement Act of 1986, mandated that people be informed about exposure to chemicals that are carcinogenic or have reproductive toxic effects and that significant amounts of such chemicals not be discharged into drinking water. Because it requires consumer notification through labeling, this Act marks a new direction in food quality and safety policy. Implementing the Act is difficult, however, because few data are available about the hazards of chemicals and the allowable levels of exposure are not clearly outlined.

Some consumers concerned about pesticide residues in fruits and vegetables are now demanding organically grown food, that is, food produced without the use of chemicals. Annual sales of organic foods exceed $1 billion.[11] Typically, these foods cost more than those grown with the use of chemicals. Before the Food, Agriculture, Conservation, and Trade Act of 1990 was enacted, there were no national standards specifying the conditions under which an agricultural product could be labeled as organically produced. To protect consumers, some states started their own organic certfication programs. For example, in 1988, the state of Washington launched a program that includes on-site inspection of food production and sampling of foods labeled as organic to test them for chemical residues.

Title XXI (the Organic Foods Production Act) of the 1990 Food, Agriculture, Conservation, and Trade Act responded to the need for a consistent set of standards to protect consumers and to facilitate interstate trade. The Act required the USDA to establish regulations for production, materials, handling, and testing of products labeled organic. Establishing national organic food standards may benefit consumers as well as agricultural producers. An important issue is whether the benefits exceed the costs. For example, standards can restrict competition among producers, increase record keeping, and lead to increased prices for organic food.

PROBLEMS AND POLICY RELATED TO PESTICIDE USE

Pesticides are mostly poisons that are critical for maintaining and increasing the productivity of agriculture. A few pesticides have been used for several decades, and one or two for centuries. For example, formaldehyde (formic aldehyde) has been used in solution as a disinfectant since about 1873. It was in common use for treating seed wheat just before planting to prevent wheat smut. Paris green (a vivid light green pigment composed of acetoarsenite of copper) has been used for several decades as a dry powder, or in solution as a spray, to control potato bugs and other pests. The rapid growth of pesticides in U.S. agriculture is marked from three dates: (1) the early 1940s, when the chlorinated hydrocarbons, such as DDT, began to be used in the control of insects; (2) 1957, when other insecticides began to be used more widely; and (3) 1961, when use of herbicides for control of weeds, which began in the 1940s, entered a rapid growth stage. The increases in use of pesticides and their effect on productivity is unprecedented, but problems of safety in pesticide use persist. What are the problems? How have they been approached? What are the policy options for the future?

Problems of Safety in Pesticide Use

Pesticides are part of the larger problem of environmental pollution that is spreading around the world, leaving no country untouched and practically no major part of the environment unharmed. The problems related to pesticide use are: (1) protecting the health of workers who apply pesticides or who work in fields where pesticides have been applied, a problem most acute on large-scale fruit and vegetable farms; (2) reducing and controlling the off-farm dangers that come from careless or excessive use of pesticides; and (3) protecting the food and water supply from pesticide residue.

These problems are both unique and pervasive. The residues from pesticides and fertilizers have contaminated to varying degrees the drinking water of some 50 million people in the United States.[12] Some 19 million of these people get their water from private wells, often shallow wells, and these people are the most vulnerable. Residue from pesticides and nitrates from fertilizers do not always occur together in the infected areas, however, which suggests that it is important to target both pesticide and fertilizer control strategies. Over broader areas of agricultural production, chemicals in the food chain are a more specific threat to health.[13] For instance, according to an estimate of the EPA, some 6,000 cases of cancer a year are caused by just one third of the approved pesticides in use today. In some cases, the spraying of crops is a serious health threat to people living near the sprayed fields even if approved methods are followed.[14] And, although the EPA can ban the use of pesticides that are the most lethal to bird or animal life,[15] this does not solve the broader problems of pesticide safety and use.

More energetic and strict enforcement of pesticide laws must be considered an important part of the policy solution. But a still broader and more selective discriminating policy response may be required to maximize the social welfare.

Pesticide Types and Uses

Five principal types of chemical pesticides have been developed, based on the target species to which they are applied.

1 *insecticides,* for direct control or eradication of various insects or insect hosts
2 *herbicides,* as preemergence selective weed controls or as selective poisons applied to growing weeds
3 *fungicides,* for control of rusts, mildews, and molds in grains, fruits, and vegetables
4 *rodenticides,* for control of gophers, rats, mice, coyotes, and sometimes birds, by highly toxic chemicals applied as bait materials
5 *acaricides* and *miticides,* for control of ticks and mites, which generally are not killed by most insecticides

Other chemicals are sometimes used as attractants to lure pests to a particular location and as repellents to force pests away. Chemicals are sometimes used as growth regulators, such as to advance or retard the harvest date of a crop; or they may be used as an aid to harvest, such as *defoliants,* which induce a plant to drop its leaves but do not harm it otherwise; and they may be *desiccants,* which draw moisture from a plant, causing it to wither and die.

Although people may be sickened or even killed by inhaling a pesticide vapor or from a severe skin exposure, nearly all pesticide deaths in both humans and animals are caused by eating or drinking the product or its residue. Nearly half the accidental deaths in the United States from pesticide poisoning are those of children under 12 years of age.

The Social Optimum[16]

The policy problem, then, is to decide how much to limit and regulate specific pesticides; what means to use, such as state or federal regulation, education, or publicity; and what system or agencies to use for implementing the policy. Limiting or constraining the risks to the health of humans, domesticated animals, and wildlife is of fundamental concern in pesticide policy. A solution must be found that will yield the best results in terms of society as a whole, taking full account of the values inherent in both benefits and costs. This means counting not just the costs incurred by the user, including cash outlay and possible adverse side effects of direct personal consequences, but also those costs external to the firm, including such things as loss of wildlife, possible poisoning of the food chain, contamination, and damage to the environment.

The general problem of the social optimum may be visualized as in Figure 13-1.[17] Suppose that positive benefits equal to OQ_0 or Q_0 can be produced with no pesticide. As pesticide inputs are added to a level of P_0, with other inputs held constant, total positive benefits are equal to Q_1. At this point, the distance between the "total positive benefits" curve and "total cost of pesticides at market prices" is a maximum, equal to $Q_1 - N_0$. However, if there are external costs associated with this level of pesticide use, the total cost is N_1 and the net social gain is diminished to $Q_1 - N_1$. If the external costs can be accounted for and brought into the decision framework, pesticide inputs will be reduced to P_1, which maximizes the net social gain.

Because of the magnitude of external costs and the difficulty of controlling them, biologists generally agree that good pest control does not and should not begin with a toxic chemical and work backward to the pest. Rather, first emphasis should be placed on nonchemical factors, such as breeding of pest-resistant crops and livestock, development of biological balances in nature to contain pests and crop diseases, with due consideration for social, economic, and biological limitations. Given this effort, pesticide policy then may concentrate on management of the chemical pesticides, in which the guiding rule is equalizing total costs and benefits at the margin. But this problem is complicated by the fact that some of the costs are not immediately apparent and experience must be accumulated.

Lessons in Using Pesticides

There are important lessons for policy in past efforts to control various insects and other pests. In some cases, considerable damage has been done to the environment without much, if any, long-term gain. In others, more beneficial results have been achieved. What are some of the lessons?

In the late 1950s and early 1960s, when the powerful effectiveness of DDT began to be recognized, the federal government was strongly urged and supported to undertake large-scale campaigns to eradicate the spruce budworm, gypsy moth, Japanese beetle,

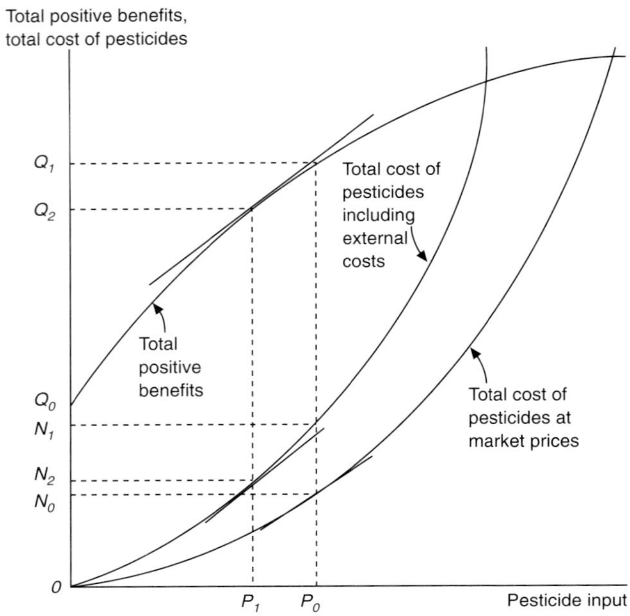

FIGURE 13-1
Optimum use of pesticides considering external cost. *Source:* J.C. Headley and J.N. Lewis, *The Pesticide Problem,* Resources for the Future, 1967.

Dutch elm disease, and imported fire ant, each of which was causing widespread damage. These campaigns, except that against Dutch elm disease, essentially involved aerial spraying over wide areas with light applications of DDT or a derivative. Some of these campaigns poisoned the food chain and were discontinued. For example, it was found that the quantities of DDT necessary to kill the bark beetles that transmitted the Dutch elm disease far exceeded the quantities sufficient to kill birds in or about sprayed trees. As sprayed leaves of elm trees fell to the ground and were eaten by worms that normally lived on them, the poisoned worms would be eaten by birds, chiefly robins. Millions of birds are believed to have been killed by this sequence.[18] In the end, although the spray programs are thought to have delayed the spread of the disease, these programs—like the attempts to use chemicals to eradicate the spruce budworm, the gypsy moth, the Japanese beetle, and the imported fire ant—ended in costly failure. Except for isolated elms, those in the affected areas all died and with this the enthusiasm for large-scale attempts to control most pests also died, at least for a while.

Lessons Derived from the Poisoning of the Food Chain

Other lessons emerged from poisoning the food chain. The food chain can be poisoned by use of pesticides when plants absorb the poison in the pesticide. Then the plants are eaten by livestock, poultry, birds or fish, and then these foods are eaten by people or other animals. Any agricultural crop that absorbs a pesticide may constitute a hazard. Consequently, pesticides should be continually studied and evaluated before permission is given for their manufacture and sale. Even with a supposedly safe pesticide, however, poisoning of the food chain can occur as a result of concentration. There are several incidences that have important lessons for food and agricultural policy.

One episode pertinent for farm pesticide policy occurred in the Klamath Basin refuges in Northern California, which are administered by the U.S. Fish and Wildlife Service.[19] In the summer of 1960, the refuge staff began to pick up dead and dying birds, and soon 307 birds were identified, all fish eaters. They were found to contain lethal doses of taxaphene, DDD, and DDE, concentrated in liver, kidneys, and subcutaneous fat, ranging from as little as 2.2 parts per million (ppm) of taxaphene to over 6,000 ppm of DDT and DDD. This condition was traced to spraying, some two years before, of crops that were irrigated from the Upper Klamath Lake, just over the border in Oregon. The runoff from these sprayed fields, spreading out over the swampy refuge, carried the lethal concentration.

Although the risks to the food supply from pesticides are imposing and perhaps alarming, they are very low compared with the hazards to health arising out of the use of drugs and tobacco, or heavy use of alcohol. However, individuals often have divergent levels of willingness to accept voluntary and involuntary risk. For example, individuals who smoke cigarettes (voluntary risk) may be opposed to a risk–benefit standard because they perceive that such a standard would expose them to involuntary risks. The risks from pesticides and other forms of pollution are important and amenable to further policy analysis.

A reliable appraisal of the damage done by agricultural pollutants must involve a wide range of observations, an assessment of individual pollutants, and a system of judging the severity and extent of pollution. In an appraisal mandated in the Resources Conservation Act (RCA), the analysts weighted the major agricultural pollutant categories by rating their effects on human life and health above their economic importance, as follows: pesticides 4.0; animal waste 3.0; nutrients, including nitrogen and phosphorus from fertilizer 2.5; salts 2.0; and soil sediment 1.0. Then, to suggest possible priorities for policy, a composite rating for combined potential nonpoint source problems was determined.

Pesticide Pollution Figure 13-2 presents RCA estimates of the potential for pesticide pollution by crop areas in the 48 coterminous states and the method for determining the potential. The amount of pesticides applied in each area was determined from various sources and divided by the total acres of agricultural land in each area. Then a composite index of runoff and application rate for each area was developed to establish the potential.

Clear and definite trends toward either improvement or further deterioration in the quality of the nation's water supply as affected by pesticides have been difficult to establish. The chlorinated hydrocarbons, such as DDT, have been banned for agricultural use because of their persistence in the environment. The *National Water Summary* for 1984, concluded that chlorinated hydrocarbon insecticide levels have decreased in

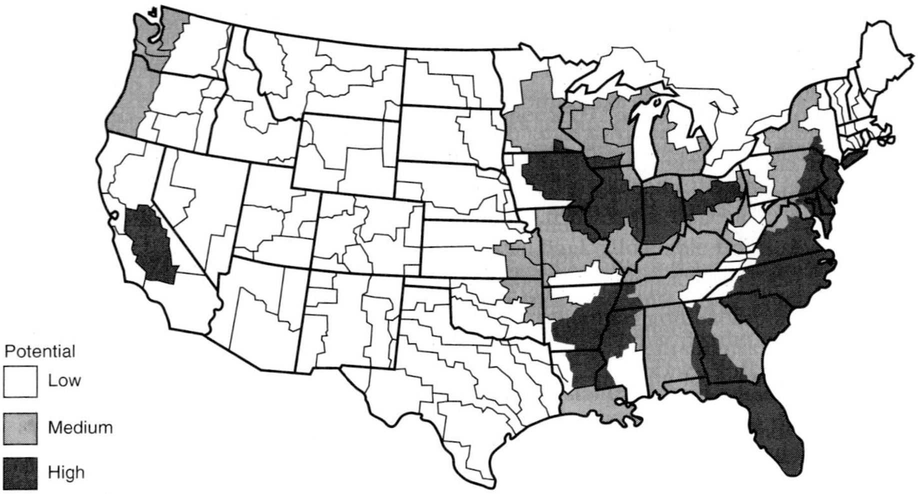

FIGURE 13-2
Estimated potential for pesticide problem. This map identifies estimated potential for surface water pollution by pesticides. The pollution potential was determined by using estimated crop acreages for different parts of the country, multiplied by an availability factor which estimates the percentage of an application leaving a field and then adjusted that figure by a runoff value for the growing season. This process estimates pesticide delivery for each area. To confirm the existence of pesticide pollution, instream and lake monitoring is necessary. *Source: The Second RCA Appraisal,* USDA, 1989, p. 107.

most of the major rivers, but no clear trends were evident in concentrations of the organophosphate insecticides and herbicides that were monitored. The use of more target-specific insecticides, better management, and improved application equipment may be reducing the total quantity of insecticides used by farmers. Furthermore, the use of pesticides is enabling farmers to use more minimum tillage, which reduces erosion and the total pesticide runoff. But some of this effect may be offset by continuing buildup of salt or salinity in arid and semiarid regions and the off-site damage from erosion also tends to be cumulative over many years.

Nutrient Contamination The nutrient forms that most often affect water quality are particulate and dissolved phosphorus and nitrogen, which are ammonia, nitrite, and nitrate. Both nitrogen and phosphorus have a toxic effect in water making it unfit to drink.

Nutrient sources may be multiple and diffuse. In the agricultural environment, nitrogen and phosphorus may enter water supplies in runoff and infiltration associated with commercial fertilizer, animal wastes, crop residues, and erosion from cropland. The potentials for contamination generally increase in proportion to intensive agricultural land use, significant cropland erosion, and (to some extent) population density. Such areas generally are heavy producers of sediment, as some nutrient forms are associated with sediment particles (Figure 13-3).

Health hazards caused by nutrient contamination of private and public water supplies

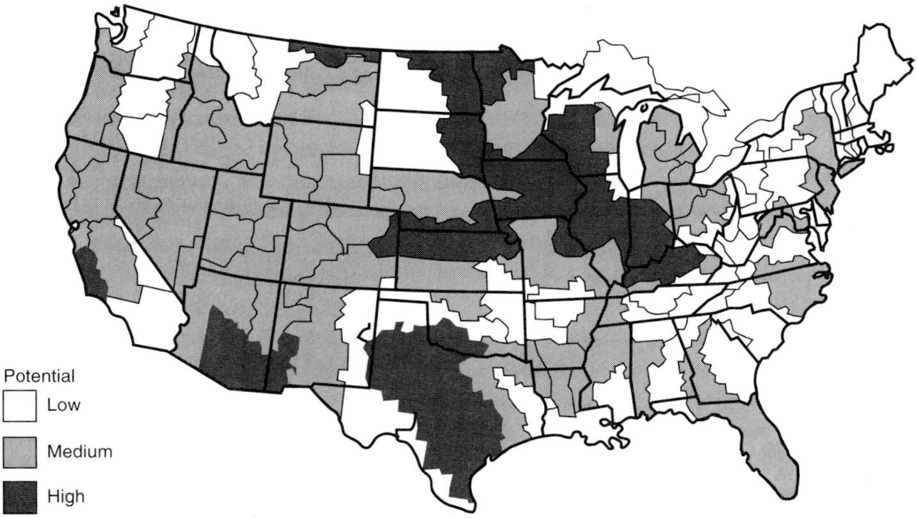

FIGURE 13-3
Estimated potential for nutrient contamination. The potential for impairment of water quality was estimated by determining nutrient concentrations by form and comparing them with the respective threshold levels at which they threaten desired water uses. *Source:* WATSTORE (U.S. Geological Survey data from water quality stations). *Source: The Second RCA Appraisal,* USDA, 1989, p. 108.

are commonly due to nitrite–nitrogen pollution. Eutrophication is detrimental to fisheries and wildlife, water-based recreation, and aesthetic values. In addition to the costs of nutrient pollution that farmers share with the general public, there are on-site farm costs caused by waste of nutrients lost from fields and by lowered crop yields. Individual farm wells and surface water supplies may be harmed by nutrient pollution that makes the water unfit for use.

Animal Waste Animal wastes may cause water contamination. The primary pollutants in animal wastes are nutrients (primarily nitrogen and phosphorus), pathogenic bacteria, and organic material. Areas in which livestock is concentrated, as well as fields and pastures where manure is spread, can be a source of pollutants.

The potential for animal waste pollution is high in areas with very large populations of livestock, notably in the southeastern and southwestern parts of the United States. The large quantities of manure produced in these areas require proper use or disposal, often in areas in which cropland and grassland are in short supply. In relatively dry regions where crops are irrigated, the application of irrigation water and manure on cropland and grassland increases the potential for pollution. These warmer, dryer areas have open feedlots, in which rainfall concentrates nutrients and organic runoff. Pollution is also pronounced in areas along the Gulf of Mexico and the Atlantic seaboard, where there is more rainfall.

In the Corn Belt, large livestock enterprises are generally found in areas of highly productive cropland. Livestock producers grow much of their own feed and cycle the

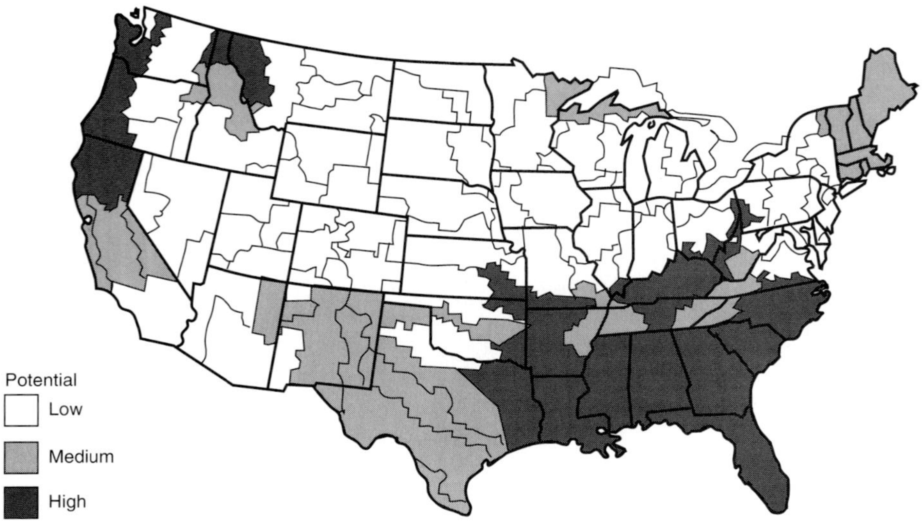

FIGURE 13-4
Potential for animal waste pollution. *Source: The Second RCA Appraisal,* USDA, 1989, p. 109.

manure, but pollution potential is low to moderate because many new and more efficient manure management systems have been implemented in the last several years (Figure 13-4).

Conclusions and Inferences for Policy

Major problems of environmental pollution, water and food safety, are a matter of continuing policy concern, and much of the public concern centers on pesticides. Pesticides are of great benefit in saving resources and increasing the productivity of agriculture, and it is not appropriate to infer that consumers would be better off without them. However, the problem is complex and requires broader knowledge of the possible injurious effects of pesticides, improvements through creation of more efficient and safer pesticides, more strict and comprehensive monitoring of their use, and a broader testing of foods for concentration of pesticide residue in foods.

These conclusions may apply aptly to domestic production and consumption of agricultural commodities, to exports of some crops and animal products, and to food imports. The use of pesticides is universal and many countries—perhaps most countries—lag far behind the United States in setting pesticide standards and controls. In fact, one of the most worrisome parts of the entire pesticide problem arises out of concerns about the levels of pesticide residues in winter imports of fresh fruits and vegetables from Mexico and other tropical countries, and from other temperate countries in the Southern Hemisphere. This has continued to be a policy issue in trade and trade agreements between the United States and Mexico, and in debates over the North

American Free Trade Agreement. The issue continues because pesticides are even more important for farm producers in tropical countries than in temperate zones, and in the last few decades their use has increased rapidly in the tropics. Advances in transportation have increased the risk of contaminants being spread across countries in a short period of time. But detecting pesticide residues in food imports is slow and costly and generally too late to assure protection of U.S. consumers. More monitoring of food imports, plus agreements with exporting countries over pesticide standards and controls may be a basic requirement. More generally, it may be concluded that pesticide policy, especially in implementation, has lagged behind the fast pace of development of modern agriculture. If the major benefits of pesticides are to be realized in sustaining increases in agricultural productivity, while creating a safe food and water supply with less damage to the environment, then important changes must be made in pesticide policy.

What can be suggested as needed changes in food and agricultural policy? Possible features of an appropriate pesticide policy for the long term might be: (1) a continuing public commitment to education; (2) lowering of government grades on some fruits and vegetables to make some pesticide treatments unnecessary; and (3) reorganization of government licensing, monitoring, and inspection services, with a federal agency for enforcement of standards independent of specific consumer or producer interests.

GOVERNMENT REGULATORY AGENCIES

Effective policies to protect consumers from harmful contaminants are important, but how well these policies work depends on how they are implemented and enforced. Thus the regulatory agencies are prime actors in safeguarding the food supply. Twelve federal agencies in four cabinet departments and two independent agencies share responsibility for ensuring food quality and safety.[20] The principal agencies are the Food and Drug Administration (FDA) of the Department of Health and Human Services, the USDA's Food Safety and Inspection Service (FSIS), and the Environmental Protection Agency (EPA). According to the General Accounting Office (GAO), each of these agencies had fewer staff members available in fiscal 1989 than in 1980 to conduct food quality and safety activities and fewer funds in constant dollars.[21] Hence federal resources decreased at a time at which workloads were rising. The GAO did not evaluate the effect of reduced staff and funding on the effectiveness of food quality and safety, but according to other studies the regulatory system deteriorated during the 1980s relative to the task at hand.[22]

Food and Drug Administration

The FDA has the major responsibility for the safety and wholesomeness of domestic and imported food except meat and poultry and it shares authority with the USDA for regulating the safety of eggs. Its mandate comes primarily from the Federal Food, Drug, and Cosmetic Act, the Pesticide Monitoring Improvements Act, and the Safe Drinking Water Act. The agency carries out many food quality and safety activities, notably (1) monitoring the production and manufacturing of food, (2) reviewing and registering food and color additives and animal drugs, (3) monitoring and enforcing the pesticide residue tolerances established by EPA, and (4) ensuring that food is properly labeled and free of substances that endanger human health.

FDA regulates pesticide residues through the Regulatory Monitoring Program and the Total Diet Study Program. Under the regulatory program, FDA collects samples of food—typically in states that are the largest producers of fruits and vegetables and from major importers—and analyzes them for chemical residues. Producers or distributors with a history of exceeding the limits of pesticide residues are sometimes targeted under this program. FDA has the authority to prohibit food from being marketed if the samples reveal excessive residue levels. Since the program was started more than a quarter of a century ago, only a very small percentage of the sampled food has contained residues exceeding tolerances. For example, 63 percent of the 3,020 domestic vegetable samples analyzed in 1987 contained no residues. Less than 1.0 percent contained residues above the established tolerance levels. About 2.0 percent contained some residues with no set tolerance levels.[23] Tolerances have been set only for crops for which a particular pesticide was approved. Thus some producers were using approved pesticides, but on crops for which their use had not been approved.

The Total Diet Study Program, also known as the Market Basket Study, facilitates analysis of dietary pesticide intake. The FDA monitors intake by collecting food samples, preparing them for consumption, and then determining the level of pesticide residue. Like the Regulatory Monitoring Program, this program has found very few incidences of excessive residues. In 1987, 936 food samples were tested for 200 compounds; residues were detected for 53 of them. The dietary intake for nearly all the tested pesticides was less than 1.0 percent of the acceptable daily intakes established by the United Nation's Food and Agriculture Organization (FAO) and the World Health Organization.[24]

Food Safety and Inspection Service

The policy tools that USDA's FSIS adopted for promoting the safety of meat and poultry products include inspection of products moving in interstate and foreign commerce and consumer education. As a part of its mandate to ensure plant sanitation, FSIS requires that facilities, equipment, and procedures be approved before a plant can begin operating as a federally inspected business. Approval is mandatory if the meat and poultry products will be sold interstate or in other countries. The agency then monitors plant sanitation after the business opens. The FSIS's procedures also include examining all animals at federally inspected slaughtering plants for symptoms of disease and other abnormal conditions on animals to be slaughtered and on their carcasses and internal organs. More than 119 million livestock and 5.6 billion poultry were checked at slaughtering plants in fiscal 1988.[25] At processing plants, inspectors use sampling techniques and testing to check for residues of animal drugs, pesticides, or other chemicals. Products with excessive levels are condemned and cannot be used for human consumption. Few products have been condemned—less than one percent of meat and about one percent of poultry.[26]

FSIS also monitors imported meat and poultry to assure that they are as safe as domestic products. As a part of its monitoring process, FSIS reviews the food safety policies of exporting countries, plant operations, and enforcement activities. Under agreements with other countries, the FSIS staff can conduct on-site reviews of foreign plants and laboratories. As with domestically produced meat and poultry, the agency does statistical sampling and testing for residues.

Environmental Protection Agency

The EPA sets the maximum allowable pesticide tolerances for both domestic and imported foods. If it determines that a particular pesticide will cause no unreasonable adverse effects to human health or to the environment, the agency registers the pesticide and specifies the terms and conditions for its use. It also has the authority to remove unreasonably hazardous pesticides from the marketplace. In making decisions about food safety, the EPA uses benefit–cost analyses, summing the benefits associated with a change and comparing them with the economic, social, and environmental costs, plus any potentially negative effects of the change.

The EPA regulates the safety of drinking water produced by municipal systems of a certain size. In addition, the agency has developed programs to address water quality problems related to agriculture, such as the concentration of nitrates or pesticides in some drinking water. The EPA has granted the primary responsibility for ground water protection to the states and requires them to develop public policies to protect water quality.

SOME CONTINUING FOOD SAFETY PROBLEMS

Even if scientists and advanced technology could technically eliminate all risks to human health from the food supply, the costs would be prohibitive. Furthermore, food prices would rise substantially. Policy tools such as tolerance levels, government inspection, and food safety inspection are used to protect consumers at affordable costs. This approach has resulted in a food supply that is one of the safest in the world, although some illnesses from pathogenic microbes and chemical contaminants are still tolerated, and death from contaminated food is still possible.

Food scientists rank food carrying pathogens (bacteria and viruses capable of causing human disease) as a major food safety problem, far out-ranking chemicals in terms of the number of people affected. Other problems include naturally occurring food toxins such as aflatoxins, which are produced by molds that grow readily on corn, soybeans, nuts, and certain other foods; chemical residue contamination; and the poor choice of foods that cause or exacerbate health problems. Consumers, on the other hand, have tended to rank pesticides and chemical residue contamination as the top problem, which suggests that they may underestimate the risk from foodborne pathogens and not take proper precautions against them. Indeed, many incidences of foodborne diseases are due to contributory negligence. Poor food storage and preparation practices at home are the causes of many food safety problems. At issue is whether food can be made safe for irresponsible or uninformed consumers, and whether the expense in trying to do this is justified or not. Education and information programs could be critical in addressing foodborne diseases and other problems such as diets that are too high in sodium, fat, cholesterol, sugar, and calories. Let us look at some of the major food safety problems in the United States.

Foodborne Disease

Studies indicate that 6.5 million to 33 million Americans, or 3 to 14 percent become ill each year from foodborne pathogens.[27] Moreover, about 9,000 of these people die as a result of their illness. Because some foodborne diseases can lead to loss of work time and even hospitalization and death, the costs in terms of medical expenses and lost productivity may be in the billions of dollars. In addition, the costs to the food industry can be high if a product is recalled or consumers stop buying it because of real or perceived health hazards.

Livestock, dairy, poultry, and seafood (especially seafood from tropical areas) are the major carriers of microorganisms that contaminate food and cause outbreaks of illnesses. Meat and poultry inspection and consumer education are the typical policy instruments used to alleviate dangers from the food supply. However, as seafood inspection is voluntary, many seafood products are not government inspected. A continuing policy issue is whether or not such inspection should be mandated.

Salmonellosis One of the most common foodborne diseases is salmonellosis, an intestinal illness with symptoms mimicking those of the flu. In fact, some people with salmonellosis may even think that they have the flu rather than a foodborne disease. Approximately 40,000 cases have been reported each year in the United States, and researchers have estimated that the total number of reported and unreported cases may be near 2.0 million. Furthermore, since 1955, the number of cases has increased dramatically. An estimated 2,000 Americans have died each year from salmonellosis.[28]

One of the main causes of salmonella food poisoning has been contaminated beef, which has accounted for about one third of the cases. The FSIS sampling data from chickens has showed that about one third of the carcasses were contaminated with salmonella when they left the slaughterhouse.[29] Chicken, which has become increasingly popular in the American diet, has accounted for about 10 percent of reported outbreaks of salmonellosis. Freezing, refrigerated storage, and adequate cooking is essential to control salmonellosis.

Only recently have economists conducted research on the costs of foodborne illness. In some cases, the estimates are based on limited data and extrapolation. According to economist Tanya Roberts, the average cost of salmonellosis per case has ranged from \$500 to \$700, and total U.S. costs from \$983 million to \$1.4 billion.[30] These estimates are for medical costs and productivity losses only. Salmonellosis and other foodborne illness may also entail household costs for pain and suffering, lost leisure time, food industry costs for product recall and plant closing, public health costs for disease surveillance, and public costs of investigating the outbreak.

Campylobacteriosis Researchers have suggested that campylobacteriosis (an intestinal disease) has been slightly more prevalent than salmonellosis. An estimated 2.1 million cases have occurred annually in the United States, with perhaps 2,000 resulting in death.[31] About half of the cases probably have resulted from eating contaminated chicken. Using available data, researchers have estimated that each case has cost $700 on average. Most cases are mild, involving only abdominal pain and diarrhea; while a few are more severe and have involved higher medical costs, and income or productivity losses. Estimated medical and productivity costs for some outbreaks of campylobacteriosis have run as high as $1.5 billion.[32]

Pesticide Residues in Food and Water

Although consumers rank pesticide residue as a top food and water safety concern, pesticides pose much lower hazards to the food supply than do foodborne pathogens.[33] However, concentrated pesticide residues affect not only food, but also drinking water. In addition, water quality has been adversely affected by nitrates from fertilizer. Nitrates pose a particular health problem for infants who can develop methemoglobinemia, a condition in which nitrates are converted into nitrites in the digestive system, thus affecting an infant's ability to carry oxygen. Also, some evidence suggests that nitrites may be carcinogenic.[34]

In rural areas especially, where people rely heavily on ground water for drinking water and water for livestock and irrigation, there is more reason for concern about ground-water contamination because such water provides some 97 percent of the drinking water, 55 percent of the livestock water, and 40 percent of the irrigation water. However, the scope of the ground-water quality issue extends beyond rural America. In 1980, 40 percent of the population (nearly 74 million people) served by public water supplies used ground water.[35] The major regions with potential ground-water contamination from pesticide use are the Eastern Seaboard, the Gulf Coast, and the Upper Midwest, whereas the major regions with potential ground-water contamination from nitrates are the Great Plains and some areas in the Northwest, Southwest, and the Corn Belt. Parts of the Corn Belt, the Lake States, and the Northeast also have potential for contamination from both pesticides and nitrates. Moreover, water contamination can pose a serious problem for years to come because it is difficult if not impossible to purify ground water once it has become contaminated. Alternatives such as treating the water at the wellhead or providing other sources of drinking water are expensive. Treatment is difficult to justify economically because many agricultural chemicals often take years to reach the water supply and in some cases they never do.

Although high levels of pesticides and nitrates in drinking water can impair health, the relationship between traces of the contaminants and risks to human health are not well understood. Continued exposure can result in reproductive problems, cancer, and other adverse effects on health.[36] It is expensive and sometimes almost impossible to determine the actual scope of the problem in potentially contaminated areas. For instance, in the late 1980s, USDA economists estimated that first-time monitoring costs for private wells could total $1.4 billion, while first-time monitoring costs for public ground-water systems could total $14 million.[37]

Public policy is responding to concerns about ground-water contamination. For instance, the Water Quality Initiative, started during the administration of President George Bush, helped to determine the magnitude of the problem and to develop strategies to protect ground water from agricultural chemicals and other contaminants. Educational programs and technical assistance to encourage farmers to use production practices that protect ground water are a part of this strategy. Research and extension programs continue to develop farming techniques that will reduce pesticide pollution, and data collection on the use of agricultural chemicals can further clarify and limit economical approaches to the problem.

Diet Quality and Nutrition Policy[38]

The American diet has changed over the past two decades, with some of the changes enhancing diet quality while others have had an adverse effect. Increased consumption of poultry, fish, grains, and cereal products has generally contributed to increases in diet adequacy. On the other hand, higher consumption of fats and oils has lowered diet quality and contributed to obesity. Hunger or malnutrition in the United States is generally limited to economic factors, poor food choices, lack of knowledge concerning nutrition, and various physiological and sociological factors. For example, some people have poor diets because of inadequate resources and limited education while others prefer foods that are high in fat, sugar, or sodium. Various policy tools are used to combat the problems of poor diets. Food-aid programs help the poor to obtain nutritious food. Educational programs, such as the Expanded Food and Nutrition Education Program (EFNEP), provide nutrition information. Dietary guidelines are developed and widely distributed jointly by the USDA and the Department of Health and Human Services.

Because the average American diet has been high in fat, this nutrient has been targeted as the chief villain in many cases of obesity and heart trouble. In most cases, however, fat in the diet would not be harmful if the amount were reduced to less than 35 percent of total calories. People with high percentages of fat in their diets would enhance their health if they lowered the amount of dietary cholesterol and shifted to polyunsaturated fat and away from saturated fat.

Americans also consume high levels of sucrose, which is the most common form of sugar and is obtained from either sugarcane or sugar beets. At high levels, sugar and fat are converted to body fat unless intake is balanced by exercise to burn the extra calories. These extra calories contribute to obesity and poor nutrition because more nourishing foods are displaced in the diet. A number of health problems are associated with obesity including diabetes, high blood pressure, and heart disease.

The American diet is also high in sodium, much of it obtained from eating processed food. In addition, Americans add table salt (sodium chloride) or flavor enhancers such as monosodium glutamate (MSG) to food. Excessive sodium is often the culprit in high blood pressure, which, if uncontrolled, can lead to stroke and heart disease. The food industry has responded to concerns about sodium levels by offering consumers a choice between low sodium products and the regular products. However, the products with less sodium are often not as palatable and may cost more than the regular products.

CONTINUING POLICY ISSUES FOR FOOD QUALITY AND SAFETY

Other ongoing issues related to food quality are how to objectively measure quality, and whether the grading system offers incentives to promote the production and marketing of quality grain and other food. Legislation, such as the Grain Quality Improvement Act of 1986, which addresses improvement in the quality of grain exports, corn and soybeans especially, mandates a study of incentives for high quality grain. Over the years various schemes have been proposed, such as having the Commodity Credit Corporation pay premiums to producers and exporters for quality above base levels or for exceptionally clean, high quality grain.[39] Research on objective measures of quality and the potential effects of alternative incentive plans is likely to lead to further improvements in grain grading and grain quality.

Unresolved concerns about foodborne illnesses, pesticide residues, water quality, and diet quality will be debated for many years to come. Public policies to address these and emerging issues are likely to be influenced by a host of factors, including research and education. For example, information is needed on the full costs of foodborne illnesses. Consumers need information on the effects of irradiation and other methods that could be used to protect the food supply. More education is required to help consumers assess risks accurately. The task of research is to provide the necessary information and to ascertain the actual or potential health hazards in food production, handling, and preparation. Because the dangers are not constant across population subgroups, new information must be obtained periodically on how particular subgroups such as infants, young children, the elderly, and pregnant women might be affected.

The GAO, in its publication *Food Safety and Quality: Who Does What in the Federal Government,* lists the critical issues that were identified by regulatory agencies.[40] These issues, as reported by the FDA, include the following:

1 the safety of foods produced by biotechnology,

2 the microbial safety of foods, including the implications of increasing the ease and sensitivity of pathogen detection,

3 monitoring the food supply, including the proper balance of surveillance and inspection activities,

4 educating the public about food quality and safety, including the issues of food labeling, nutrition labeling, and dietary advice to high-risk groups,

5 use of appropriate regulatory tools, including the application of food standards and food labeling with warnings.

The FSIS identified three critical issues: foodborne pathogens, chemical residues, and modernization of meat and poultry inspection. Of these three, foodborne pathogens were considered the most critical.

When policies for food quality and safety are established or revised, a number of questions must be addressed. Chief among them are: (1) how to determine whether a food is safe or not, (2) what standards should be set, and (3) how to implement the standards. For example, to determine if a particular substance is safe for human consumption, scientists usually conduct tests on rats that are fed large amounts of that substance. Important questions about this approach include the extent to which results from rats are applicable to humans, and whether the substance is hazardous or not if consumed in small amounts.

As government policy shifts toward negligible risk, new questions will arise. What risk standards should be set for suspected carcinogens? Should the costs of hazards in the food supply be balanced against the benefits or should the zero tolerance requirement as mandated by the Delaney clause be the standard? Some consumer groups argue that the zero tolerance policy should not be modified, but many of the benefits and costs are unclear. This issue is likely to be settled by the judicial branch of government, which is often influential in making food and agricultural policy.

Most people believe that the government should bear the chief responsibility in cooperation with the food industry for determining that food and drinking water are safe. To do so, research must be applied in assessing the risks, inspecting meat and poultry, monitoring the safety of the food supply, grading various foods, distributing information on food handling and preparation, and preparing and publishing dietary guidelines. In 1992, new guidelines were set for food labeling. However, the proper mix of responsibilities between government and the food industry are sometimes the subject of debate, and no policy can succeed if consumers are not knowledgeable and also responsible. In the 1980s, the emphasis on deregulation resulted in a policy shift from banning to using more information strategies. More reliance was placed on the private market in areas involving food quality and safety.[41] In the 1990s, some people questioned deregulation and claimed that it had gone too far, or that the emphasis is misplaced. Thus the balance between the roles of the private and the public sectors is likely to continue to be an important issue for food and agricultural policy for years to come.

LOOKING AHEAD

This chapter concludes the current discussion of food quality and safety policy. Chapter 14 opens up an entirely new area, which involves monetary policy, agricultural credit policy, and fiscal policies of federal, state, and local governments. Emphasis is on the effects of these policies on agriculture and the food economy.

ENDNOTES

1 Julie A. Caswell, "Food Safety Policy Fights: A U.S. Perspective," *Northeastern Journal of Agricultural and Resource Economics,* October 1990, pp. 59–66.

2 Don F. Hadwiger, "Nutrition, Food Safety, and Farm Policy," in Don F. Hadwiger and Ross B. Talbot (eds.), *Food Policy and Farm Programs,* Proceedings of The Academy of Political Science, Vol. 34, No. 3, New York, 1982, pp. 79–88.

3 Upton Sinclair, *The Jungle,* Doubleday, Page & Company, New York, 1905.

4 Katherine Reichelderfer and Maureen Kuwano Hinkle, "The Evolution of Pesticide Policy: Environmental Interests and Agriculture," in Carol S. Kramer (ed.), *The Political Economy of U.S. Agriculture: Challenges for the 1990s,* Resources for the Future, National Center for Food and Agricultural Policy, Washington, DC, Annual Policy Review, 1989, pp. 147–177.

5 Tanya Roberts and Eileen van Ravenswaay, *The Economics of Safeguarding the U.S. Food Supply,* Agriculture Information Bulletin No. 566. Economic Research Service, U.S. Department of Agriculture, Washington, DC, July 1989.

6 Roberts and van Ravenswaay, *Ibid.*

7 National Academy of Sciences, *Regulating Pesticides in Food: The Delaney Paradox,* National Academy Press, Washington, DC, 1987.

8 Stephen R. Crutchfield, ''Agriculture's Effects on Water Quality,'' *Agricultural-Food Policy Review: U.S. Agricultural Policies in a Changing World,* Agricultural Economic Report No. 620, Economic Research, U.S. Department of Agriculture, Washington, DC, November 1989.

9 Roy Carriker and Amy Purvis, ''Agriculture and Water Quality: Old Problem, New Urgency,'' *Southern Journal of Agricultural Economics,* 22(1), 1990, pp. 27–32.

10 Catherine Greene and Glenn Zepp, ''Changing Pesticide Regulations: A Promise for Safer Produce,'' *National Food Review,* July–Sept. 1989, pp. 12–16.

11 Lori Lynch, ''Congress Mandates National Organic Food Standards,'' *Food Review,* January–March 1991, pp. 12–15.

12 *The Magnitude and Costs of Groundwater Contamination from Agricultural Chemicals,* Ag. Ec. Report 576, ERS/USDA, October, 1987.

13 See *Chemicals in the Food Chain* (report of a conference proceedings), UC Agricultural Issues Center, University of California, Davis, 1988; especially ''Chemicals in the Human Food Chain: Sources, Options, and Public Policy,'' by Carole and Frank Nuckton.

14 Anne Cook, ''Complaints Surround Chemical Use,'' *The Champaign–Urbana News Gazette,* January 8, 1989, p. C–1.

15 The Associated Press and The News-Gazette, ''EPA to Ban Version of Corn Pesticide,'' *The Champaign–Urbana News Gazette,* January 7, 1989, p. C–8.

16 This discussion is adopted from Harold G. Halcrow, *Food Policy for America,* McGraw-Hill, New York, 1977, pp. 114–116.

17 J. C. Headley and J. N. Lewis, *The Pesticide Problem: An Economic Approach to Public Policy,* Resources for the Future, Inc., Washington, DC, distributed by The Johns Hopkins Press, Baltimore, 1967.

18 R. J. Barker, ''Notes on Some Ecological Effects of DDT Sprayed on Elms,'' *Journal of Wildlife Management,* 22(3), 1958, pp. 269–279.

19 Robert J. Rudd, *Pesticides and the Living Landscape,* The University of Wisconsin Press, Madison, 1964, pp. 255–259.

20 For additional information on the responsibilities of the various agencies see General Accounting Office, *Food Safety and Quality: Who Does What in the Federal Government,* GAO/RCED-91-19B, U.S. Government Printing Office, Washington, DC, December 1990.

21 *Ibid.*

22 Caswell, ''Food Safety Policy Fights,'' *Op. cit.* (endnote 1).

23 Greene and Zepp, ''Changing Pesticide Regulations,'' *Op. cit.* (endnote 10).

24 Greene and Zepp, ''Changing Pesticide Regulations'' *Ibid.*

25 Sharin Sachs, ''Safe Meat and Poultry,'' *National Food Review,* July–Sept. 1989, pp. 31–33.

26 Sachs, ''Safe Meat and Poultry,'' *Ibid.*

27 Tanya Roberts and Eileen van Ravenswaay, ''The Economics of Food Safety,'' *National Food Review,* July–Sept. 1989, pp. 1–8.

28 Roberts and van Ravenswaay, ''The Economics of Food Safety,'' *Ibid.*

29 Tanya Roberts, ''Microbial Pathogens in Raw Pork, Chicken, and Beef: Benefit Estimates for Control Using Irradiation,'' *American Journal of Agricultural Economics,* 67(5), 1985, pp. 957–965.

30 Tanya Roberts, ''Human Illness Costs of Foodborne Bacteria,'' *American Journal of Agricultural Economics,* 71(2), 1989, pp. 468–474.

31 Roberts, ''Microbial Pathogens,'' *Op. cit.* (endnote 29).

32 Roberts, ''Human Illness,'' *Op. cit.* (endnote 30).

33 Roberts and van Ravenswaay, ''The Economics of Food Safety,'' *Op. cit.* (endnote 27).

34 Stephen R. Crutchfield, ''Agriculture and Water Quality Conflicts,'' *Food Review,* April–June 1991, pp. 12–14.
35 Elizabeth G. Nielsen and Linda K. Lee, *The Magnitude and Costs of Groundwater Contamination from Agricultural Chemicals: A National Perspective,* Staff Report AGES870318, U.S. Department of Agriculture, Economic Research Service, Washington, DC, June 1987.
36 Crutchfield, ''Agriculture and Water Quality,'' *Op. cit.* (endnote 34).
37 Nielsen and Lee, *The Magnitude and Costs, Op. cit.* (endnote 35).
38 For further discussion on the foundations of current dietary and nutrition policy, see Harold G. Halcrow, *Food Policy for America,* McGraw-Hill, New York, 1977, pp. 492–545.
39 Lowell D. Hill, *Grain Grades and Standards: Historical Issues Shaping the Future,* University of Illinois Press, Urbana and Chicago, 1990.
40 General Accounting Office, *Food Safety and Quality, Op. cit.* (endnote 20).
41 Caswell, ''Food Safety Policy Fights,'' *Op. cit.* (endnote 1).

TERMS AND CONCEPTS

Food quality
Food safety
Pesticides
Federal Food, Drug, and Cosmetic Act (FFDCA)
Federal Insecticide, Fungicide, and Rodenticide Act (FIFRA)
Delaney clause
Tolerances
Delaney paradox
Zero tolerance level
Food additives
Color additives
Risk benefit standard
Negligible risk policy
Poisoning of the food chain
Resources Conservation Act (RCA)
Meat and poultry inspection
Water quality
Proposition 65
Regulatory Monitoring Program
Total Diet Study Program
Foodborne disease

QUESTIONS AND STUDY SUGGESTIONS

1 How did the modernization of agriculture affect the safety of the food supply and drinking water?
2 What factors have influenced the evolution of policies for food quality and safety?
3 What are the major Acts that set policy for safeguarding the U.S. food and water supply? How have consumers affected policy making in food quality and safety?
4 What are the implications of the Delaney clause? Explain the Delaney paradox.
5 What role do the states play in setting policies for food quality and safety? What are some state initiatives in this area?
6 What are the principal federal regulatory agencies responsible for ensuring that the U.S. food supply is safe? List the major responsibilities of each of these agencies.
7 What are the major food safety problems as identified by experts? By consumers? What are the implications in the divergence of the rankings?
8 Why is foodborne disease a major food safety problem? What are some of the costs associated with outbreaks of foodborne disease? What are some policy alternatives for dealing with the problem?
9 What are the major policy issues for food quality and safety in the 1990s? What role can research and technology play in effective policy making?

CHAPTER **14**

MONETARY POLICY, AGRICULTURAL CREDIT, AND FISCAL POLICY

The U.S. food and agricultural economy is affected in important ways by changes and interactions among federal monetary policy, agricultural credit policy, and the fiscal policies of federal, state, and local governments. This chapter explores the role of these policies and presents policy alternatives for the future.

IN THIS CHAPTER

By studying Chapter 14, you will learn how to answer the following questions:

1. What are the effects of monetary policy on agriculture?
2. What is the scope and substance of agricultural credit policy and what effect does it have on agriculture and rural communities?
3. How is federal crop insurance linked to federal emergency farm credit, and what are the alternatives for crop insurance?
4. How are agricultural producers and rural communities affected by federal, state, and local fiscal policies?

MONETARY POLICY

The link between the nation's monetary policy and the prosperity of the agricultural economy—the farm sector especially—is strong and enduring. Generally, agriculture prospers in times of easy money or inflation, and it is depressed when money is tight or deflation occurs. For example, the 30 years following the Civil War, from the mid-1860s to the mid-1890s, were a time of most extreme hardship. During this time, farmers joined with many nonfarm rural people and others in a broad political protest, called "The Populist Revolt."[1] Farm prices were depressed, money was tight, farmers and other businesses in rural areas suffered as well.[2]

In their revolt, the populists campaigned for a wide range of reforms including a larger or more liberal money supply, control of monopoly, and regulation of banks and railroads. Some of the policy issues were resolved by adoption of the Sherman (antitrust) Act of 1890, and by establishing the Interstate Commerce Commission, which was granted specific powers to regulate railroad-freight rates, and prevent certain other restraints on interstate commerce. Some of the populist's monetary issues declined in urgency in the Golden Age of American Agriculture from the mid-1890s to 1914. During this time, farm credit became more readily available, but the market for long-term farm mortgage credit was not developed, and most farmers and rural businesses

FIGURE 14-1
Capital gains and losses in farm real estate and physical assets. Changes in farm real estate values less net capital investment were mostly unrealized. Other physical assets include machinery and motor vehicles, livestock and poultry, and crops stored on farms. *Source: 1990 Agricultural Chartbook,* USDA, Agriculture Handbook No. 689, Washington, DC, p. 36.

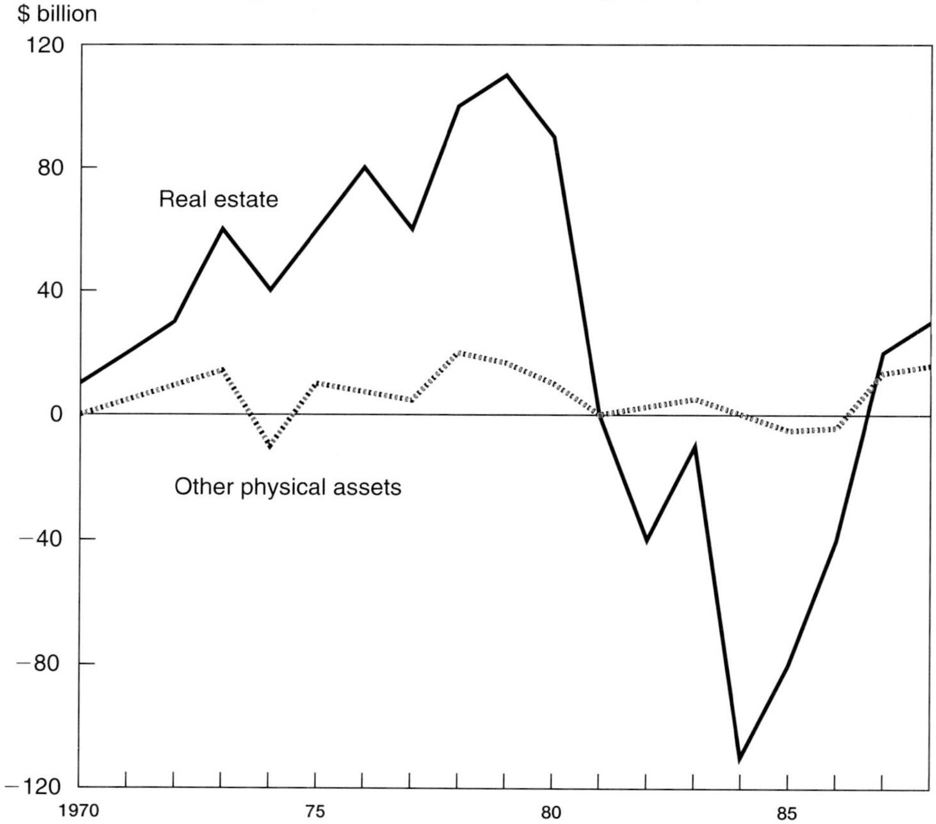

faced high-cost credit and uncertainty. Such conditions and their lingering effects helped prompt a broad reorganization of the nation's monetary system.

The founding of the Federal Reserve System in 1914 provided the nation with an agency, somewhat shielded from direct political influence, that could regulate the quantity of money and credit in the economy (see the appendix to Chapter 14: How the Fed Manages Money and Credit). Since then "the Fed," as it is often called, has functioned as the federal agency for controlling the nation's supply of money and credit. However, the Fed alone has not been able to stabilize the nation's economy against fluctuations of the business cycle and inflation, or shield agriculture and rural communities from the effects of the cycle or other sources of instability.

Agriculture and rural communities—the farm sector especially—have been destabilized by the interactions of the nation's monetary policy and agricultural credit policy. For example, the great farm boom of the 1970s, and the subsequent financial collapse of the farm economy in the early 1980s, can be traced directly to the interactions of monetary and agricultural credit policy. In the 1970s, weakness of the dollar in international exchanges contributed to a large increase in farm exports by making them cheaper in terms of other countries' currency, or the prices paid by importing countries. In the early 1980s, a much stronger dollar depressed exports. Added to these effects, in the 1970s a more liberal farm credit policy magnified the farm boom, stimulated the market for farmland, and helped to escalate farmland prices to record levels. In the late 1970s and early 1980s, the tightening of money and credit by the Fed to control inflation restricted the money supply, drove up interest rates and intensified the farm financial crisis. These events were reflected in huge capital gains and losses in farm real estate and physical assets between 1970 and 1988 (Figure 14-1).

During the 1980s and early 1990s, large federal deficits, financed by the sale of government securities, expanded the money supply, which soon drove down interest rates, greatly increased the federal debt, and eventually set the stage for a possible return of inflation. In the meantime, reform and refinancing of the federal farm credit agencies added financial strength to agriculture and many rural communities, but important questions remain concerning the nation's monetary and fiscal policies, and the links of these policies to agricultural credit policy.

Agricultural Credit Policy

The supply of agricultural credit comes from two general sources. On the average, for more than half a century, about three-fourths of the total has been supplied by private sources (commercial banks, insurance companies, individuals, and other private lenders). The other one-fourth has been supplied by two public sources: (1) the federally sponsored, privately organized, and cooperatively owned Farm Credit System (FCS), and (2) the direct government lending and loan guarantees of the Farmers Home Administration (FmHA).

All private sources that supply credit to agriculture operate in open competition with other lenders. Nearly all of the private lenders operate under federal and state credit laws, and rules established by federal and state regulatory agencies. For example, all commercial banks must have a federal or state charter, and strong regulations are provided through various federal and state regulatory agencies. The interest rates they charge and other terms they offer are strongly influenced by the rules and regulations of the Fed, and its control of the nation's money and credit supply.

The interest rates and other terms offered to agriculture by commercial banks strongly influence the agricultural credit market and other agricultural lenders, as well as other credit markets. Most empirical evidence indicates that the administrative costs per dollar of lending to farmers and agribusinesses is essentially the same as the administrative costs per dollar of other commercial and business loans. Therefore, the decision by private lenders of whether to loan to agriculture or not generally rests on their judgment based on comparative yields, costs, risks, and uncertainties.

The case for federal sponsorship of the FCS, and federal support of the FmHA rests on the unique economic structure of the agricultural credit market, possible gaps in private credit sources available to agriculture, the political influence of farmers and agribusinesses and their organizations to solicit support of the federal government, and the government's responsibilities to promote the general welfare. Most of the farm credit market is not perfectly competitive because there are large numbers of borrowers and often very few private lenders in most farming communities. Many agribusinesses, such as agricultural cooperatives, have very specialized and highly seasonal demands for short-term credit.

Perhaps most important is that farmers cannot always receive satisfactory long-term credit from local banks or other private lenders, and short-term credit also may be unavailable in sufficient volume or on terms that are satisfactory to them. In addition, it has been demonstrated that in times of farm financial crisis, such as occurred during the Great Depression of the 1930s, and again in the mid-1980s, the magnitude of the problem necessitates a public response. The federal government provides credit and financial support for agriculture through the FCS and the FmHA.

The Farm Credit System

Federally sponsored, long-term, amortized farm mortgage credit was authorized by the Federal Farm Loan Act of 1916. This Act was based on recommendations of two national commissions, which had been delegated to study the long-standing agricultural credit cooperatives of Germany and Scandinavia. Differences among the commissions' members resulted in two sets of banks to supply long-term mortgage credit to farmers. One set included 83 joint-stock land banks, which were placed under federal supervision and were almost entirely privately financed. Most of these banks were inadequately financed, however, and soon many of them failed. Consequently, in 1931, provisions were made for their liquidation.

The other set of banks founded by the 1916 Act consisted of 12 federal land banks (FLBs), one in each of 12 newly created farm credit districts. The initial capital was loaned by the federal government—$9 million, or $750,000 for each bank. These funds, plus much greater amounts for expansion in the 1930s, were loaned to the banks at rates of interest generally based on the federal government's cost of funds, which was below the commercial market rates. All of these loans were later entirely repaid to the

Treasury. A policy of requiring borrowers to invest five percent of each new loan in stock of the bank assured that the banks eventually would be owned by the borrowers.

The policy goals of farm people and their organizations also included cheaper and more reliable short-term credit. The Agricultural Credits Act of 1923 authorized $60 million to establish 12 federal intermediate credit banks (FICBs) to supply short-term production credit at uniform competitive rates of interest in each of the 12 farm credit districts. Finally, in 1929, as a result of continuing hard times for farmers, The Agricultural Marketing Act provided for a system of 12 district banks for cooperatives (BCs) and a central bank to serve the district banks and provide credit for large loans. These banks were financed with an appropriation of $500 million to carry out price-stabilization through storage of farm commodities. This failed, but the BCs were continued to provide both long-term and short-term credit at competitive rates to agricultural cooperatives.

The early years of the Great Depression of the 1930s were a time of extreme financial hardship for agriculture. Following the great stock-market crash of 1929,[3] prices of farm commodities dropped sharply. Incomes of farmers and rural businesses were depressed further by a sharp contraction in the money supply.[4] A high percentage of the nation's 6.8 million farm families were in deep financial distress. Most agribusinesses and rural communities were suffering as well.

The Farm Credit Act of 1933 placed the three sets of banks (FLBs, FICBs, and BCs) under policy control of the new Farm Credit Administration (FCA). It also restructured the program for production credit by authorizing $120 million more for purchase of capital stock in some 500 local production credit associations (PCAs), which became local lending agencies for the FICBs.

The Federal Farm Mortgage Corporation Act of 1934 provided for restructuring and greatly expanding the programs of the FLBs carried out through about 500 federal land bank associations (FLBAs). Farm mortgages held by the FLBs as well as by many other lenders were restructured and refinanced. Leadership of the FLBs in this program apparently encouraged some other lenders to carry out a similar program. In summary, the primary implements and techniques of the federal government's agricultural credit policy were to provide temporary subsidized capital, as "seed money," plus innovative credit terms such as long-term amortized mortgages at interest rates geared to general money-market costs. All of the government's original capital investments were paid back to the Treasury, and the basic structure of the FCS was kept intact until reorganized by the Agricultural Credit Act of 1987.

Experience of the FCS[5] The experience of the FCS has important lessons for current and future monetary and credit policy. The role of banks and their scale of operations changed substantially in response to changing economic conditions and political efforts to solve economic problems. Before 1933, the FLBs held only a small part of the total farm mortgage debt (3.5 percent of the total in 1920, 13.5 percent in 1932). At the end of 1932, the FICBs held only 6.0 percent of all nonreal estate farm debt outstanding.

With the deepening of the Great Depression and the start of the Roosevelt New Deal, this situation changed dramatically. From 1933 through 1936, the FLBs were financed by the federal government to engage in an emergency farm refinancing program that far surpassed previous lending activities. During the period, from the end of 1933 to the end of 1935, the combined farm mortgage holdings of the FLBs and the new Federal Farm Mortgage Corporation (formed for emergency financing to supplement the FLBs), increased from less than $1.3 billion to more than $2.9 billion. By the end of 1936, when emergency refinancing was largely completed, the FLBs held more than 40 percent of the total farm mortgage debt. In addition, this action encouraged other lenders to refinance many loans, rather than foreclose on delinquent borrowers.

During the next 35 years (1936 to 1971), the FLB's share of the farm mortgage market declined relative to the share of other lenders, and by 1971 the FLBs held only 24 percent of the farm mortgage debt. In the meantime, the FICBs and PCAs had been directed to follow a similarly conservative policy so that, at the beginning of the 1970s, they also played a minor role in financing farm production and marketing. Such a conservative policy was unsatisfactory to many farm families and most farm organizations, and a 27-member commission, established by the Federal Farm Credit Board, recommended a much more aggressive lending policy.

The Farm Credit Act of 1971, which was based largely on recommendations of the 27-member commission, directed the FLBs to raise their lending limits on farm real estate from 65 percent of *normal agricultural value,* which generally was well below current land values, to 85 percent of *appraised current market value.* The maximum loan to a single borrower, without prior FCA approval, was raised from $100,000 to $400,000. Additional steps were taken to increase the lending powers of the banks. Consequently, from 1972 to 1983, when nearly all farm lending agencies were adding capital to the farm sector, the FLBs rode the crest of the farm credit expansion, leading all farm mortgage lenders, as the total farm mortgage debt mounted to record-breaking levels (Figure 14-2).

The 1971 Act also greatly increased the roles of the FICBs and PCAs. The FICBs were authorized to purchase stock in PCAs, participate in loans and loss-sharing plans with other FICBs and PCAs, own and lease equipment to borrowers, offer financial services, and issue nonvoting stock to nonfarm borrowers. This action permitted these institutions to participate more aggressively in the great farm credit expansion of the 1970s, which—counting all lenders—brought the total nonreal estate farm debt to nearly $90 billion by the early 1980s (Figure 14-2).

Meanwhile, in the late 1970s and early 1980s, inflation was running at high levels, and the Fed had shifted to a "tight money" policy to shrink the money supply and stop inflation. This action drove interest rates to record levels, strengthened the dollar in international exchange, and after some lag began to curb the rate of inflation. At the same time, the FCS had moved away from fixed interest rates to variable rates based on average-cost pricing, which substantially moderated the fluctuations in farm mortgage interest rates relative to the fluctuations of market interest rates in the late 1970s and early 1980s. By 1982, the interest earned by the FCS reached a record high of $10.8 billion, compared with $7.6 billion in 1980, and net earnings in 1981 and 1982 were at an all-time high (Table 14-1).

The mandate given to the FCS was to provide credit to agriculture at as low a cost as possible. To do this, the FCS raised capital by issuing a large volume of long-term,

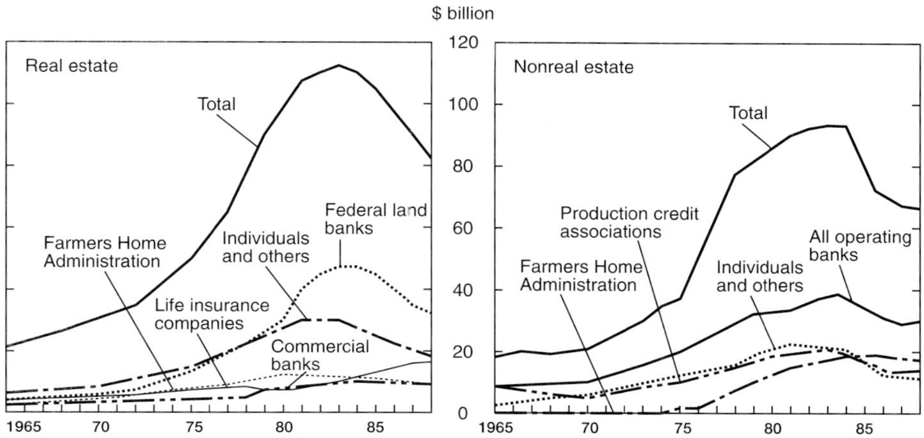

FIGURE 14-2
Trends among holders of the U.S. farm debt. "Individuals and others" include Commodity Credit Corporation real estate loans. "Production credit associations" include Federal Intermediate Credit Bank loans to other financial institutions. *Source: 1990 Agricultural Chartbook,* USDA, Agriculture Handbook No. 689, Washington, DC, p. 33.

noncallable bonds. During the period of rising interest rates, this strategy allowed the FCS to minimize its cost of capital by locking in a fixed interest rate. Buyers were attracted to these bonds because they would be assured a stable return when interest rates fell.

Beginning in 1981, however, when the Fed's inflation-control policy began to take effect, the Fed relaxed its tight control on the money supply, and interest rates began to drop. Then, the FLBs' combination of variable interest rates on farm mortgages and fixed interest rates for its bonds began to turn into disaster for the FCS. In 1983 and 1984, net income of the FCS fell sharply to less than half its level in 1981 and 1982 (Table 14-1). In 1985, a substantial loss (subsequently $2.7 billion) was projected for the year, and in September the governor of the FCA announced that the FCS might require more than $13 billion in government assistance over the next few years to stay solvent.

Three factors are usually cited for the FCS disaster: (1) the drop in farm prices due to the strong dollar and loss of export markets, which depressed prices of most export commodities, spelling trouble for a single-industry credit institution like the FCS; (2) the loose organizational structure of the FCS, which allowed most FLBAs and PCAs considerable loan initiative; and (3) the monetary policy mentioned earlier. Consequently, the high indebtedness of many farmers, together with the high interest rates and rather aggressive lending practices of the 1970s were involved in the FCS problem and its solution. In other words, the losses of the FCS can be traced to the government-directed FCS policy decisions, and to the broader monetary decisions of the Fed which, by changing the money supply, destabilized agricultural credit markets. In the early 1980s, the broad speculative farm boom of the 1970s, which had added nearly $120

TABLE 14-1
SUMMARY OF SELECTED FINANCIAL DATA FOR THE FARM CREDIT SYSTEM, 1980–1987

Balance sheet data	1980	1981	1982	1983	1984	1985	1986	1987
				$ million				
Assets								
Accrual loans	69.244	78.916	81.123	80.753	78.052	64.412	51.183	47.264
Nonaccrual loans	263	354	745	1.269	2.108	5.145	7.066	5.234
Allowance for losses	(1.229)	(1.403)	(1.430)	(1.418)	(1.336)	(3.250)	(3.635)	(2.951)
Cash and securities	3.077	3.478	3.109	2.460	3.258	8.062	10.976	9.109
Interest receivable	2.923	3.856	4.139	3.713	3.675	3.078	2.201	1.641
Other assets	949	1.141	1.232	1.429	1.769	3.072	2.310	1.941
Total assets	75.227	86.342	88.918	88.206	87.526	80.519	70.101	62.238
Liabilities								
Interest-bearing debt	64.202	72.847	74.261	74.067	73.389	69.722	62.676	55.529
Other liabilities	2.566	3.562	3.614	2.810	2.795	2.856	1.783	1.679
Total liabilities	66.768	76.409	77.875	76.877	76.184	72.578	64.459	57.208
Capital								
Borrower stock	4.739	5.356	5.525	5.430	5.266	4.672	4.183	3.684
Surplus	2.260	9.933	5.518	5.900	6.075	3.270	1.453	1.346
Income data								
Interest earned	7.614	9.909	10.808	9.571	9.848	9.077	7.169	5.783
Interest paid	(6.218)	(8.391)	(9.103)	(8.163)	(8.394)	(7.758)	(6.388)	(5.274)
Other income	184	337	279	133	118	157	129	97
Provision for loan losses	(201)	(231)	(202)	(241)	(358)	(3.007)	(1.798)	196
Net other expenses	(511)	(610)	(676)	(760)	(841)	(1.158)	(1.025)	(819)
Net income	868	1.014	1.106	540	373	(2.689)	(1.913)	(17)

Source: Federal Farm Credit Banks Funding Corporation, Farm Credit System Annual Report to Investors, various years.

billion to the estimated value of U.S. farm real estate, ground to a halt; and by 1984, more than $200 billion of the appraised farm real estate value had been lost from the peak value of 1979 (Figure 14-1).

Reorganization of the FCS[6] The FCS was not structured or financed to withstand such extreme fluctuations. In the fall of 1985, it quickly became clear in congressional hearings that collapse of the FCS, as a federally chartered financial institution with agency status, would be a disaster not only for itself, but also for other financial enterprises with agency status. For example, its collapse would threaten the financial markets for the Federal National Mortgage Association, the Student Loan Marketing Association, the Federal Home Loan Mortgage Corporation, and the Federal Home Loan Bank Board. These agencies' bonds, which exceed those of the FCS in volume, might also lose a favorable rating, and the financial integrity of most, if not all, of the federal government's guarantees might be in question. More simply put, the FCS had to be saved from disaster because failure of the FCS could weaken if not destroy the confidence of investors in perhaps all U.S. government securities. (Was this an early warning of the much larger savings and loan disaster still to come?)

In 1985, 1986 and 1987, three separate Acts were adopted to save the FCS. The Farm Credit Amendments Act of 1985, which was part of the farm income-support policy embodied in the Food Security Act of 1985, did the following: (1) restructured the FCA as an ''arms-length'' regulator and farm loan supervisor of the FCS, (2) rechartered the FCS's Capital Corporation as a specialized bank to deal with nonperforming loans of the FCS, and (3) authorized a backstop line of credit from the U.S. Treasury if the FCS failed to meet its obligations. These amendments proved inadequate, however. In 1986, even though the net income from farming was matched by record set-aside and deficiency payments, 6 of the 12 FLBs reported that they would require direct financial help to avoid *impairment of capital.*

The Farm Credit Amendments Act of 1986 was adopted as a stopgap to buy time. The Act authorized the banks of the FCS to (1) reduce interest rates to avoid losing borrowers; (2) depart from generally accepted accounting principles to defer recognizing interest expenses associated with some of their long-term, high-coupon bonds; and (3) depart from the accepted accounting principles in certain circumstances to defer making normally required provisions for loan losses.

By the end of 1986, however, in spite of these measures, nonoperating (delinquent) loans of the FCS had risen to $1.9 billion, an all-time high. In addition, many of the system's borrowers were paying off their loans and thus liquidating their required stock investments, and refinancing elsewhere. The FCS could not lower its interest rates enough to retain many of these more credit-worthy borrowers without losing more money. The staffs of the FLBs were unable to keep up with requests from borrowers to lower or defer interest payments, or to extend the terms of loans. The leaders within the FCS had been aware of the need for restructuring well before adversity struck in the 1980s, and their landmark Project 1995 provided the guidelines for the Agricultural Credit Act of 1987.

The 1987 Act provided for refinancing and restructuring the entire FCS system. It authorized the issuance of $4 billion of 15-year bonds backed by federal government guarantee (only about half of which was used by 1992). The law provided that all interest would be paid by the U.S. Treasury during the first 5 years, split evenly between the FCS and the Treasury during the next 5 years, and paid entirely by the FCS after 10 years. The Act reaffirmed the federal government's guarantee of existing capital and provided that additional stock issued within nine months after adoption of the Act would also be fully guaranteed.

The 1987 Act provided for merging each of the 12 FLBs with the 12 FICBs in the same district, and it required the 13 BCs to merge into one consolidated bank. It also provided the basic guidelines for the mergers and consolidation, based on the existing voting rights of the banks. By 1992, mergers of FLBs and FICBs had been completed in 11 of the 12 farm credit districts. There was one National Bank for Cooperatives and two district BCs remaining. Each of the three banks had a national charter, and could operate nationwide if it had occasion to do so. The 1987 Act also provided guidelines for merging the local FLBAs with PCAs, and for consolidating the 12 farm credit districts into no fewer than six districts. It established new rules for capitalizing banks and sharing losses, tightened lending and accounting practices, and specified new rights for borrowers. By 1992, almost the entire FCS had been reorganized from what

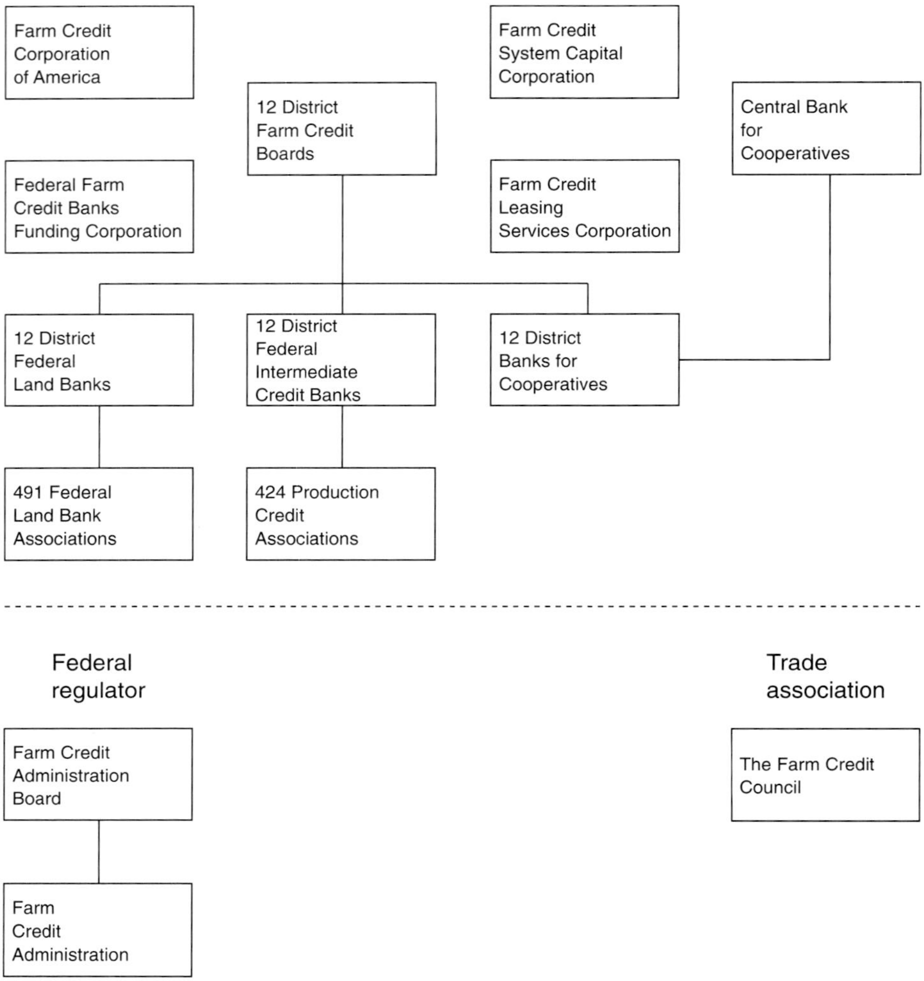

FIGURE 14-3
Farm Credit System organization, January 1, 1980. *Source: Agricultural Income and Finance, Situation and Outlook Report,* USDA, February, 1992.

it had been before (Figure 14-3), and streamlined into two types of lending institutions, Farm Credit Banks and Banks for Cooperatives (Figure 14-4).

To help implement some of these policies, the Act established a new secondary market for farm real estate loans through the Federal Agricultural Mortgage Corporation, to be called ''Farmer Mac.'' Existing lenders who had purchased stock in Farmer Mac could originate loans that could be ''pooled'' and sold in a secondary market. In addition, a secondary market called ''Aggie Mac'' was established for the FmHA loans. Aggie Mac was patterned after the Small Business Administration's secondary market, which provided market access to pension funds, trusts, and other investors. Finally, the

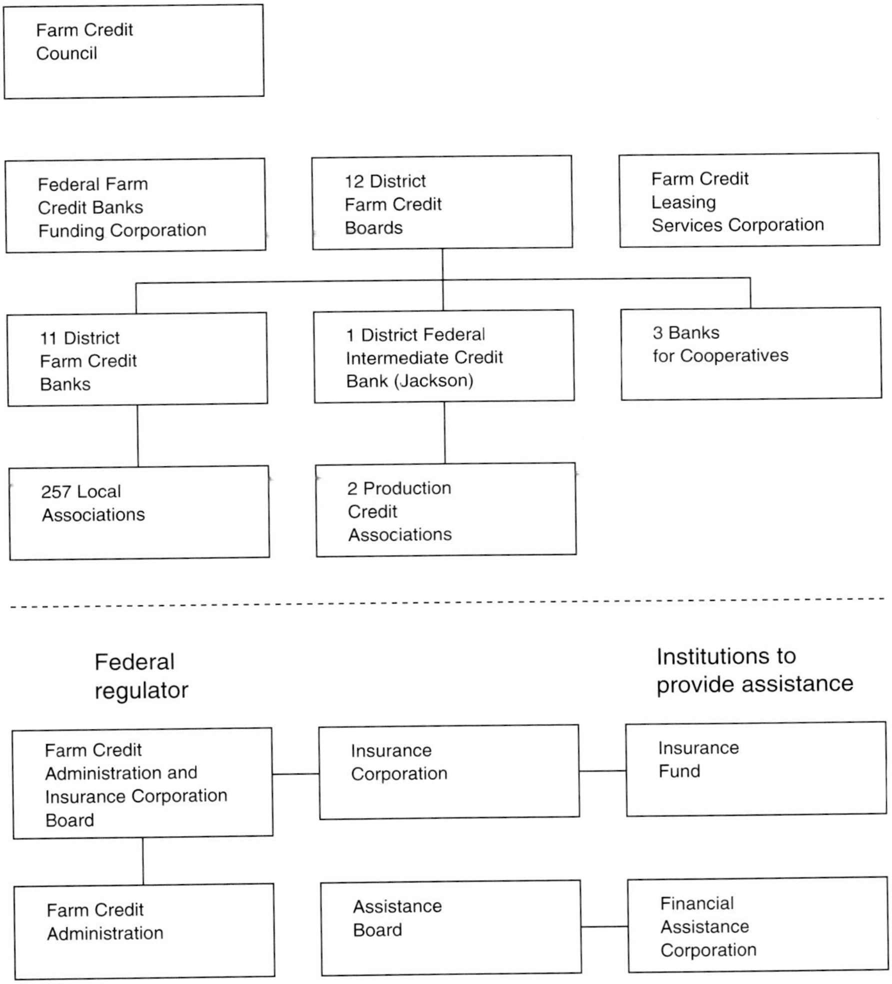

FIGURE 14-4
Farm Credit System organization, January 1, 1992. *Source: Agricultural Income and Finance, Situation and Outlook,* USDA, February, 1992.

1987 Act provided authority for establishing a new federal insurance corporation by 1993 to replace the FCS Financial Assistance Corporation (Figure 14-4).

Appraisal of FCS Reorganization In the short term of a few years, the policy of reorganizing the FCS, when combined with other policy as noted, had clearly preserved the solvency of the FCS and saved many farm families from financial disasters. Within a short time, it became clear that not all of the $4 billion authorized for new bonds would be required, and by 1992 only about half of it had been borrowed. Beginning in

1988 the net income of the FCS began to stabilize at some $600 million annually, and in the early 1990s the total farm debt stabilized at about 30 percent below its peak in the mid-1980s. Also, by 1992, farm banks in the private commercial banking system were among the strongest institutions in the system. Finally, with the weaker dollar, export subsidization, and the lowering of farm price supports, farm commodity export markets were reviving. However, the integrity of the FCS could not be guaranteed over the long term.

Indeed, the long term is less certain. The lessons from FCS financing and interest rate policy are clear, and much has been learned. However, the FCS—as a low-cost, nonprofit, cooperative lender that is largely dependent on the farm sector—is not an appropriate type of agency to withstand such an expansive lending policy as characterized the 1970s, or the types of wide swings that characterized the Fed's monetary policy during the 1970s and 1980s. Also, the granting of more local autonomy to the merged FLBAs and PCAs could lead to more overcommitment unless supervisory powers are strengthened.

A long-term policy problem is inherent in the mission and structure of the FCS. The FCS, which by its charter is prohibited from carrying large excess reserves, will continue to be more vulnerable than more diversified lenders to fluctuations and uncertainties in agricultural prices and production. Therefore, it is not surprising that the FCS will continue to require government guarantees for its bonds and back-up support in times of farm financial crisis or depression. Fortunately for the FCS, the farm sector, and rural communities, the federal government has demonstrated, both in the 1930s and in the 1980s, that it can and will protect the solvency of agricultural credit institutions, and use them to help strengthen the solvency of the farm sector and many rural communities.

The imperfectly competitive structure of the nation's private credit and financial institutions that serve the farm economy and many rural communities leaves a void that can be filled by the federally sponsored and direct lending programs. However, there are limits to these programs; and, if these limits are exceeded, the results can be damaging to the continuing development of agriculture and rural communities. It is difficult to recognize these limits and act accordingly. Economics and politics are entwined in the programs of both the FCS and the FmHA.

The Farmers Home Administration (FmHA)

The FmHA is a direct-lending government credit agency within the USDA. It is funded through the federal budget and provides loans or guarantees loans by private credit firms for qualifying farmers and others who cannot get credit from other sources. It provides credit at rates of interest based on the federal government's cost of funds and also provides publicly supported professional supervisory service for its borrowers: farmers, rural homeowners, some agribusinesses, and some rural community facilities. Its purposes are threefold: (1) to alleviate emergency hardships for the farm and rural community, (2) to provide supervised credit for capable but low-income farm families and some rural businesses and government facilities, and (3) to stimulate rural community resource development. Initially offering farm purchase and production loans, the service was expanded for low-income farm housing (1949) and for rural businesses and community facilities (1972). Periodically, the FmHA is the channel for emergency loans and other disaster assistance.

The volume of agricultural credit offered by the FmHA has varied, depending on economic conditions in agriculture, the availability of credit from the private sector, and the policy goals of various national administrations. The FmHA's mission was broadened in the 1970s, and again in the 1980s to target increased public emergency credit during the farm financial crisis of the 1980s. In the mid-1980s, the FmHA's percentage of the total farm debt reached a record of about 25 percent of the total nonreal estate and 12 percent of the real estate loans (Figure 14-2).

In the 1980s, similar to the FCS solvency crisis, loan delinquencies and potential bankruptcies reached levels high enough to warrant federal policy responses. As a result, the FmHA was granted new lending authority, authorized to make numerous short-term adjustments for borrowers in difficulty, and mandated to restructure its practices to reduce future losses of public funds. After more than half a century of specifically targeted public credit policy, the role of the FmHA has been clearly defined as a lender of last resort for a minority of financially strapped farmers and rural communities, and as an avenue for future emergency loans and disaster assistance. In addition, since the mid-1980s, a major change in policy has been a substantial shift to loan guarantees in lieu of more direct loans, in which commercial lenders provide the funds and service the loans with a financial backup provided by the federal government. Such a change spreads the risk of loaning to the targeted clientele and broadens the potential markets for the services of the FmHA.

Because of the broad mission assigned to the FmHA, its record of performance cannot be evaluated simply in terms of government costs and returns. An adequate evaluation must consider who has been helped and how effective such help has been. Also, it is necessary to consider the results in terms of broad economic and social criteria, including farm and community development and growth. This requires much detailed analysis well beyond the scope of this book.

The Rural Electrification Administration

For its contribution to the almost total electrification of the U.S. farming sector, the Rural Electrification Administration (REA), established in 1936, is one of the most important public policy instruments for development of agriculture and rural communities. It is seldom understood to be primarily a credit agency, however. At the onset of The Great Depression, central-station electricity was available to fewer than 10 percent of the nation's farms, usually prosperous farms near population centers where private power companies deemed it profitable to extend their transmission lines and services.

The new policy was designed to expand rural electrification primarily in two ways: through subsidized loans to locally organized REA farmer cooperatives for power substations and lines, and through supervisory assistance. The objective was to extend this vital service to all farm and rural communities on an equal-usage cost basis, regardless of remoteness. After World War II, with the increased use of electricity by a rapidly industrializing agriculture, the REA cooperatives paid off the loans on schedule. Direct public funding through a revolving fund of repayments has gradually shifted to guarantees of private lending.

The REA accelerated rural development and the productivity of farms by substituting electrical energy for labor and other resources. By 1968, more than 98 percent of all farms were receiving central-station electricity, with the REA serving about half of the farms in remote areas and many of the smaller more isolated communities. The REA continues to request federal financing based on the federal government's cost of funds, to replace facilities, extend service to the few unserved areas, and for a new associated mission of cooperative telephone service to more remote rural communities. The basis for continuing such policy is controversial for policy making.

Although this unique public policy has helped revolutionize traditional agriculture and has aided many rural communities, the REA has always been involved in controversy surrounding the need for it, the kind and level of public support required, and the appropriate balance between the public and private sectors. The public cost of this program is incurred primarily through the federal government's provision of low-interest-rate credit to REA cooperatives, and the government's underwriting and guaranteeing of loans made to the REA by banks and other investors. As long as credit is supplied to REA cooperatives at below the commercial market rates, it may be assumed that this controversy will continue. The answer to the basic policy questions concerning how much and how long public support is justified depends on the values attached to the services provided. These can be determined only on a case-by-case basis.

FEDERAL CROP INSURANCE POLICY

All-risk crop insurance is linked to farm credit policy because this insurance is an alternative to federal grants and emergency loans to farmers and agribusinesses in times of crop failure, and the risk reduction potentially provided by such insurance may also encourage improved credit worthiness for farmers as they seek financing from commercial lenders. The involvement of the federal government in all-risk crop insurance began with the Federal Crop Insurance Act of 1938, following a series of failures by state and private insurance companies to provide such insurance. The Federal Crop Insurance Corporation (FCIC) was given the task of developing a program that could insure farmers against all risks of crop failure on specified crops, and be self-supporting except for administrative costs, which would be paid by the federal government. Subsequently, crop insurance premiums were set at a uniform level for each county and indemnities depended on the individual farmer's yield experience. Until 1980, the FCIC sold insurance to only a small percentage of eligible farmers, and consistently lost a substantial amount of money almost every year. However, because disaster aid in the form of emergency grants and loans to farmers was difficult to administer and more costly to the federal government, the Federal Crop Insurance Act of 1980 expanded the program of the FCIC to make it the primary means of protecting farmers against crop failure.

The 1980 Act provided that the FCIC should pay 30 percent of the insurance premium that was charged to insure a farmer up to 65 percent of the normal average yield of the farm, and it provided for coverage of additional crops and counties. Subsequently, from 1981 to 1988, FCIC losses—not counting administrative costs—exceeded $2 billion, while indemnities were more than twice the aggregate of premiums paid by farmers (not counting the 30 percent of premiums paid by the government). Even so, in 1988, less than one-fourth of the eligible crop acreage was insured. In 1990, the USDA recommended liquidating the program. Senators and representatives in Congress from important agricultural states did not agree to accept this recommendation, however, and the program was continued at a lower level of funding.

In 1989, a National Commission on Agricultural Finance concluded that: (1) the existing private and public sources of credit were adequate for agriculture; (2) the federal government should not offer emergency loans and grants to farmers along with all-risk crop insurance, because many farmers would not buy crop insurance if such disaster aid would be available; and (3) possibilities for improving federal crop insurance should be studied further.[7]

Also, in 1989, a National Commission on Crop Insurance, which had been organized to make a comprehensive study of crop insurance, recommended that a new actuarial plan should be tried for such insurance.[8] The plan recommended by the Commission would base indemnities on average yields in a county instead of the actual yield received by a farmer. The Commission concluded that such a program could provide farmers with a more adequate hedge against low yields in their county or crop area, and also eliminate the major actuarial faults in all-risk crop insurance. These faults are adverse selection and moral hazard, which become important when indemnities are based on the yield experienced by the individual farmer.[9]

Adverse Selection This exists in an insurance program when eligible clients who have high probabilities of loss are more frequently attracted to the program than clients who have lower probabilities of loss. Adverse selection occurs in crop insurance when (1) the insurer uses an indemnity formula that does not accurately reflect the probabilities of loss among different counties, or (2) individual farmers within the same county have different probabilities of loss, and these differences are known by those who are eligible to buy insurance. Adverse selection may arise from either case, but the means available for compensating for it are quite different for all-risk crop insurance.

The first type of adverse selection may be eliminated by changing the county insurance formula for premiums or indemnities, or both. However, when all-risk crop insurance indemnities are based on yields reported by each individual farmer, the correct facts to consider may be illusive for the insurer, because it is very difficult if not impossible for the insurer to determine whether the insured farmers in a county come from a sample of farmers that is biased or unbiased. If it is a biased sample, then the farmers who are not in the insured group may have decided correctly that the terms of the insurance contract are not attractive to them. If so, then the contract cannot be made attractive to them without overcompensating the already insured farmers. If the insured group is an unbiased sample, then all farmers in the county will face the same yield probabilities, and adverse selection can be eliminated by changing the insurance formula to equalize it among the different counties.

However, in the second type of adverse selection, in which individual farmers in the same county have different probabilities of loss, the insurer cannot correct the adverse selection by changing the county insurance formula. To develop an actuarially sound crop insurance formula, the insurer would have to reset the insurance formula for each small group of farmers within each county. The statistical basis for doing this would be very expensive to develop and the results might not be fair or reliable. In addition, the insurer would still have the problem of dealing with the moral hazard.

Moral Hazard This exists when a policy holder can influence the outcome of the event on which insurance is based. For example, a farmer with an all-risk crop-insurance policy may increase the net farm income in some instances by (1) buying insurance on a crop, (2) reducing the variable costs to a very low level, and (3) claiming that there is a crop failure on the land covered by the insurance. Alternatively, a farmer might know prior to the deadline for buying crop insurance that crop prospects were very poor, and that it would almost certainly pay to buy insurance.

Hence, when all-risk crop insurance is based on premiums that are uniform for a county and indemnities that depend on low yields experienced or reported by the individual farmer, the FCIC is exposed to both adverse selection and moral hazard. On such a basis, the FCIC cannot fully separate the losses due to bad weather or other natural hazard from crop losses due to inaccurate indemnity formulas or poor farm management. The risks of the former can be determined statistically and are therefore an insurable risk. The latter is an uncertainty for the FCIC and, therefore, uninsurable. Farmers often know when their crop prospects are poor, and this practically assures failure of the program through adverse selection and moral hazard.

Actuarial Structures for Crop Insurance[10] There are two other alternatives for federal crop insurance, either one of which could provide more adequate protection against bad weather and other natural hazards, at reduced cost to the federal government. These have been called "area-yield crop insurance" and "weather crop insurance." They would be used to cover different situations or different hazards, or they could be used in combination. For example, weather crop insurance might be added to an area-yield crop insurance contract to cover a more localized specific risk such as hail. Or weather crop insurance could stand alone as discussed below.

Area-Yield Crop Insurance Both the premium and indemnity would be based on crop yield in an area, such as a county or a type-of-farming area in a county, or more than one county. Adverse selection and moral hazard would be eliminated because the indemnity, if any, would be based on the yield in the area, which would not be determined by the individual farmer. (A type of crop insurance that embodies the basic principles of area-yield crop insurance has been called a "Group Risk Plan," the key feature of which is to pay indemnities to policy holders only when county yields are below average).[11]

Weather Crop Insurance Both the premium and the indemnity would be based on specific risks, such as hail, low rainfall, or high temperature. For example, a spring-wheat farmer in North Dakota might buy insurance (in essence a hedge) against low rainfall or hail in May, June, and July. A farmer in the Corn Belt might hedge against summer drought, or other important risks such as too much rain in the spring or fall. A winter-wheat farmer in Kansas or Oklahoma might hedge against low precipitation from August through April.

Conclusion

Area-yield crop insurance, or the analogous Group Risk Plan mentioned above, can be developed as an actuarially sound crop insurance program to fill an important policy niche in contributing to increased farm income stability. The program could increase credit worthiness of farmers, and reduce political pressures and incentives for the federal government to provide emergency farm loans and grants, or other types of disaster relief. Reducing farm income uncertainties due to weather can contribute to making more accurate investment decisions in agriculture and help stabilize the fiscal structure of rural communities.

FISCAL POLICY AFFECTING AGRICULTURE AND RURAL COMMUNITIES

Fiscal policy includes the taxation, borrowing, and public expenditure policies of federal, state, and local governments. Issues in fiscal policy affecting agriculture and rural communities arise out of the fiscal interrelations between and among these governments. The primary policy issues relate to changes in government taxing, borrowing, and spending.

Federal Policy

The federal income tax affects agriculture in various ways. Federal tax policy is guided by the principle that taxpayers with similar incomes should be treated similarly. However, farmers receive special treatment in federal estate-tax law, which reduces the estate tax in transfers of farm property from one generation to another. Between 1981 and the early 1990s, however, these gains from the treatment of estate taxes were far outweighed by transfers in the total tax burden from federal to state and local tax sources. Reductions in federal income tax rates, the cancellation of federal revenue sharing, and the comparative declines in the federal share of public expenditures for schools and other public activities have increased the magnitude of tax problems faced by state and local governments.

State Policy

During the 1970s, the sharing of federal revenues with state and local governments helped to support the services provided by these governments. The termination of revenue sharing in the early 1980s shifted more of the total tax burden to state and local revenue systems. Under these circumstances, if states do not increase their state income or sales taxes, the fiscal burden on local governments increases, other things being equal.

Local Policy

The chief revenue source available to local governments in the United States is the property tax, which is based almost entirely on ownership of land and other real estate. It is of special interest to farmers because farming generally requires a relatively higher ratio of land and other real estate to net income than occurs in most other occupations. Therefore, the property tax paid by farmers is generally much larger relative to their net income than it is for taxpayers in most other occupations.

The relatively large magnitude of the property tax relative to net farm income has important implications for farmers' tax policy. Under stable economic conditions, the tax on property held and used for income is a burdenless tax, provided a purchaser of such property takes the tax into account as a fixed cost. The higher the existing tax, the less the buyer will (or at least should) pay for real estate or other taxable property. The higher the tax, the lower will be the amount of money required to buy it. In other words, the property tax is a rent levied by government and, if this rent is known at time of purchase, it becomes burdenless.

However, because of the relatively large amount of property to net income in farming, any increase in property taxes, either to offset reductions in other federal or state tax revenue, or to offset increases in local public expenditures, transfers proportionately greater fiscal burdens to the farm sector. In addition, in a country as large and diverse as the United States, financial conditions of farms and rural communities vary widely from state to state as well as within individual states; and the results from such transfers are far from uniform. They may be negligible in some more urban-type or wealthier farm communities, but relatively large for farmers in poor communities. Property-tax increases become a fixed cost for farmers, which cannot be shifted in the short run, so the incidence of a property-tax increase falls on the farm sector.

One general solution to the farm property-tax problem is to shift the support of schools away from property taxes to state income taxes or other state taxes.[12] This shift can provide more uniform support of schools within a state and also reduce the total tax paid by most farmers. The policy issue has been highlighted by several decisions of both state and federal courts to mandate a reorganization of local school financing.[13]

Generally the courts have found that reliance on local property taxes to fund elementary and secondary education discriminates against children who live in relatively poor taxing districts. However, whether this is unconstitutional or not depends on the provisions of the individual state's constitution that pertain to equal opportunity. In any case, the policy solution involves political action. How many states will respond to the courts' mandates, and how fully they will respond remains an open question.

Another policy solution relates solely to farm property-tax relief. State laws generally provide that all real estate shall be valued for tax purposes at a specified percentage of full and true value, as appraised by duly elected or appointed officials. Many states, however, have installed a classified property-tax system that provides for taxing certain classes of property, such as farm real estate, at a lower percentage of full and true value than property in other uses. Another form of relief, sometimes called *use-value assessment,*[14] is to capitalize the net income produced by farmland and to use this as a basis for taxing farm property. Many states have adopted laws to provide for use-value assessment of farmland, but the guidelines vary widely and the results are far from uniform. In addition, if the valuation of farm real estate fluctuates uniformly with changes in farm income, use-value assessments will destabilize local revenues, which would add to the difficulties of financing rural schools and other rural community services in periods of low farm income. Hence, the general fiscal solution for most rural communities and school districts is to shift to broader-based revenue sources at the state and federal level.

LOOKING AHEAD

Chapter 15 discusses human capital policy with some emphasis on applications of this policy to agriculture and rural communities. Publicly supported education and research is an important part of this policy, particularly as it applies to agriculture. Chapter 15 discusses the necessary conditions for economic progress with emphasis on the farm sector, and alternatives in policy for enhancing economic progress. Some emphasis is placed on the effectiveness of human capital policy in respect to its effect on agricultural productivity and growth, agricultural incomes, and income distribution.

APPENDIX: HOW THE FED MANAGES MONEY AND CREDIT

Interest rates in a competitive money (credit) market are equilibrium prices determined by the intersections of the supply of money (credit) and the various demands for money (credit). The various demands for money (credit) are influenced by a wide range of economic conditions and public policy. The supply is regulated by decisions of the Board of Governors of the Federal Reserve System (FRS), commonly called ''the Fed.'' Such decisions are usually based on recommendations of the Fed's major committees, especially the Fed's open-market committee. The 12 federal reserve banks are under direct control of certain regulations promulgated by the Fed's Board of Governors. These regulations give the Fed the power to set the policy that governs the entire monetary system.

Policy-Making Powers of the Fed

The power of the Fed arises from its policy decisions to regulate the following:

1 Open-market operations of the federal reserve banks. These operations involve the Fed in regulating the buying and selling of federal bonds or bank notes by the 12 federal reserve banks in the open (auction) market.

2 The legal reserve ratio. This ratio is the percentage of deposits that each commercial bank operating under a state or a federal charter is required by the Fed to hold as reserves to pay depositors.

3 The discount rate. This rate is the interest rate that federal reserve banks are permitted by the Fed to charge on the loans they make to qualified (sound) commercial banks.

Regulating Open-Market Operations

Among the three policy instruments that the Fed uses, by far the most important is its control of open-market operations of the 12 federal reserve banks. The Fed's policy is implemented through its power to order the banks to buy or sell federal bonds or bank notes.

When ordered to buy federal bonds in the open market, the federal reserve bank simply credits the reserves of a selling commercial bank (or depositor) by an equal amount. This transaction increases the commercial bank's loanable funds several times because the commercial bank is required to hold only a small fraction of its outstanding loans as a legal reserve for depositors, usually 15 to 20 percent of deposits. Whether the seller of the bond is an individual, a business firm, or other institution, the general effect on commercial bank deposits is much the same. The money supply is increased and, other things being equal, interest rates will fall.

When the Fed orders the federal reserve banks to sell government bonds in the open market, this reduces the commercial bank's reserves and decreases the bank's loanable funds by some multiple of the amount of bonds sold. The effect is to decrease the nation's money supply. Other things being equal, interest rates will rise.

Regulating the Legal Reserve Ratio

The Fed has limited power to raise or lower the legal reserve ratio that it requires all banks to hold as reserves against customers' deposits, usually 15 to 20 percent of deposits. Most banks do not operate right at their legal minimum, so raising or lowering the legal reserve by a few percentage points usually affects relatively few banks directly.

Most of the time, banks are not fully loaned up. However, if they are close to the legal minimum when the Fed raises the legal reserve, such banks must increase their reserves to retain their charter. They can do this by (1) selling bonds to a federal reserve bank, thereby increasing their reserve, (2) declining to renew maturing loans or to make new ones, or (3) borrowing directly from a federal reserve bank. Whichever options the banks choose, their actions reduce the money supply and thereby tend to raise interest rates.

If the Fed's policy lowers the legal reserve ratio, the loaning capacity of those banks that had been fully loaned up will increase. To the extent that they take advantage of this opportunity, the money supply will increase and interest rates will fall, other things being equal.

Regulating the Discount Rate

Because commercial banks must maintain minimum legal reserve levels, and sometimes do so by borrowing from a federal reserve bank, raising or lowering the cost (discount rate) for such interbank loans ultimately affects the money supply. Thus, when the Fed raises the discount rate, the money supply is reduced, and interest rates rise. As an alternative, the commercial bank may increase its own interest rate to discourage new borrowing, or it may decline to renew loans as they become due. Usually, only a small percentage of all banks are directly affected by the Fed's discount rate, but this power of the Fed still significantly affects the nation's money supply. In addition, the Fed has the power to raise or lower the federal funds rate, which is the interest rate that banks are permitted to charge for usually large or overnight interbank loans.

How the Fed Operates

By combining the three powers, the Fed at any time may expand the money supply and lower interest rates, which, under certain economic conditions, could also be inflationary (an easy-money policy). That is, dollars become cheaper in respect to all other goods and services. Expanding the money supply not only affects domestic interest rates, but also tends to make the dollar weak in international trade. Alternatively, if the Fed reduces the money supply (a tight-money policy), dollars become more scarce for domestic loans and more valuable in international exchange, and the dollar becomes stronger. International rates of exchange will shift so that, other things being equal, a given number of dollars will buy more of another nation's money. Or other nations will have to spend more of their own money to buy the same amount of U.S. goods. However, raising the interest rate may also make the United States a more attractive market for foreign investors. Consequently, foreign individuals, businesses, or even governments may increase their purchases of U.S. bonds, stocks, and real estate. Hence, some of the effects of the Fed's tightening of the money supply, which makes borrowing more expensive for U.S. investors, may be offset in part by more foreign investment in the U.S. economy.

ENDNOTES

1 John D. Hicks, *The Populist Revolt: A History of the Farmers' Alliance and the People's Party,* University of Nebraska Press, Lincoln, 1968 (first published by the University of Minnesota Press, Minneapolis, 1931).

2 John Bowman, "Trends in Midwestern Farm Values, 1860–1914," unpublished Ph.D. dissertation, Yale University, 1964; Allan G. Bogue, *From Prairie to Corn Belt,* The University of Chicago Press, Chicago, 1963, as discussed by Douglass C. North, *Growth and Welfare in the American Past: A New Economic History,* Prentice-Hall, Englewood Cliffs, NJ, 1966, pp. 145–148.

3 Readers interested in this topic may enjoy John Kenneth Galbraith, *The Great Crash, 1929,* Hamish Hamilton, London, 1955, or Gordon Thomas and Max Morgan-Witts, *The Day the Bubble Burst,* Doubleday and Company, Garden City, NY, 1979.

4 Milton Friedman and Anna Jacobson Schwartz, *The Great Contraction, 1929–1933,* a study by the National Bureau of Economic Research, New York, published by Princeton University Press, Princeton, NJ, 1965. This is a reprint of Chapter 7 of the authors' *A Monetary History of the United States, 1867–1960,* Princeton University Press for the National Bureau of Economic Research, Princeton, NJ, 1965.

5 For more extensive discussion and analysis up to the mid-1950s, see R. J. Saulnier, Harold G. Halcrow, and Neil H. Jacoby, *Federal Lending and Loan Insurance,* a study by the National Bureau of Economic Research, New York, published by Princeton University Press, Princeton, NJ, 1958, especially pp. 151–199. For further discussion and experience up to the

mid-1970s, see Halcrow, *Food Policy for America,* McGraw-Hill, New York, 1977, pp. 398–414.

6 This section draws on David Freshwater, "The Political Economy of Farm Credit Reform: The Agricultural Credit Act of 1987," and Neil E. Harl, "Commentary," in Carol S. Kramer (ed.), *The Political Economy of U.S. Agriculture: Challenges for the 1990s,* National Center for Food and Agricultural Policy, Annual Policy Review, 1989, *Resources for the Future,* Washington, DC, pp. 105–143. See also Robert N. Collender, "Changes in Farm Credit System Structure," *Agricultural Income and Finance: Situation and Outlook Report,* USDA, ERS, February 1992, AFO 44, pp. 37–44.

7 *Report of the National Commission on Agricultural Finance,* Harold B. Steele, Chairman, USDA, Washington, DC, February 22, 1989.

8 *Report of the National Commission on Crop Insurance,* USDA, Washington, DC, 1989. See also G. A. Barnaby, Jr. and Jerry Skees, "Public Policy for Catastrophic Yield Risk: An Alternative Crop Insurance Program," *Choices: The Magazine of Food, Farm and Resource Issues,* Second Quarter 1990, pp. 7–9.

9 Mario J. Miranda, "Area-Yield Crop Insurance Reconsidered," *American Journal of Agricultural Economics,* 73(2), May 1991, pp. 233–242.

10 For further discussion, see the following by Harold G. Halcrow: "The Theory of Crop Insurance," Ph.D. dissertation, The University of Chicago, 1948; "Actuarial Structures for Crop Insurance," *Journal of Farm Economics,* 31(3), August 1949, pp. 418–443; *Agricultural Policy of the United States,* Prentice-Hall, Englewood Cliffs, NJ, 1953, pp. 407–420; *Food Policy for America,* McGraw-Hill, New York, 1977, pp. 425–429; "A New Proposal for Federal Crop Insurance," *Illinois Agricultural Economics,* Vol. 18, July 1978, pp. 20–29 (a study invited by the Chairman, U.S. Senate Committee on Agriculture, Nutrition and Forestry); *Agricultural Policy Analysis,* McGraw-Hill, New York, 1984, pp. 242–247.

11 Jerry R. Skees, *Background Research and Educational Material for a Pilot Test of the Group Risk Plan for Soybeans.* Presented to the Board of the Federal Crop Insurance Corporation, September 17, 1992, Federal Extension Service and the University of Kentucky, Lexington, KY, 1992.

12 For further discussion and a test of the effects of displacing the property tax for schools with state income taxes or sales taxes, see H. G. Halcrow et al. *School Tax Options Affecting Illinois Agriculture,* University of Illinois, Agricultural Experiment Station Bulletin 744, April 1973.

13 *Serrano v. Priest,* cited in *The United States Law Week,* Vol. 40, September 14, 1971, p. 2128; *Van Dusartz v. Hatfield,* cited in *The United States Law Week,* Vol. 40, October 26, 1971, p. 2228; *Rodriguez v. San Antonio Independent School District,* 337 F. Supp. 280 (D.C. Tex. 1971); *Milliken v. Green, Michigan Supreme Court, The United States Law Week,* Vol. 41, January 9, 1973, p. 2348; *San Antonio Independent School District v. Rodriguez,* cited in *The United States Law Week,* Vol. 41, March 20, 1973, p. 4407; *Edgewood Independent School Dist. v. Kirby, West's South Western Reporter,* 777 S.W. 2d, 391 (Tex. 1989); *Helena Elementary Dist. v. State, West's Pacific Reporter,* 769 P. 2d 684 (Mont. 1985); *Rose and Blandford v. The Council for Better Education,* No. 88–SC–804–TG, Supreme Court of Kentucky, 1989 Ky. LEXIS 55; *Kukor v. Grover, West's North Western Reporter,* 436 N.W. 2d 568 (Wis. 1989).

14 For further discussion, see David L. Chicoine and John T. Scott, Jr., "Farmland Use–Value Assessment: Illinois Approach," *Journal of the American Society and Farm Managers and Rural Appraisers,* 46(2), 1982, pp. 16–22; Chicoine and Scott, "Agricultural Use–Valuation Using Farm Level Data," *Property Tax Journal,* 2(1), 1983, pp. 1–12.

TERMS AND CONCEPTS

Populist Revolt
Federal Reserve System
Monetary policy
Farm Credit System (FCS)
Farm credit acts
Federal land banks (FLBs)
Federal intermediate credit banks (FICBs)
Banks for cooperatives (BCs)
Farmers Home Administration (FmHA)
Rural Electrification Administration (REA)
Federal Crop Insurance Corporation (FCIC)
Crop insurance theory
Adverse selection
Moral hazard
Crop insurance alternatives
Fiscal policy affecting agriculture and rural communities

QUESTIONS AND STUDY SUGGESTIONS

1 What is a weak dollar? A strong dollar? What is the effect of each on: inflation? deflation? U.S. agricultural exports? U.S. agricultural imports? Have U.S. farm policy makers generally favored a weak dollar, or a strong dollar? Why?

2 What types of credit have been supplied by the federal land banks? The federal intermediate credit banks? The banks for cooperatives?

3 How has the role of the Farm Credit System (FCS) changed since 1916? What prompted the major changes? What is the current justification for having the FCS? Do you believe that this justification is adequate? Why, or why not?

4 How does the policy role of the FCS differ from the role of the Farmers Home Administration (FmHA)? What is the current justification of public support for the FmHA? Do you believe this justification is adequate? Why, or why not?

5 What is the Rural Electrification Administration (REA) and what does it do? What has been the role of the REA in development of agriculture and rural communities? Does this role justify continued public funding of the REA? Why, or why not?

6 How did monetary policy and federal farm credit policy affect agriculture in the 1970s and 1980s? Was the great fluctuation in farm real estate values shown in Figure 14-1 related more closely to monetary policy or to agricultural credit policy, or to both? Explain.

7 What factors contributed to the farm financial crisis of the 1980s? Was any one factor more important than another? Why or why not? How might the crisis have been avoided or prevented?

8 The Agricultural Credit Act of 1987 appears to have solved most of the short-term problems of the FCS. Has the Act solved the long-term problems as well? Why or why not? What long-term problems do you believe have not been solved? Can they be solved? Why, or why not?

9 Summarize the history of the federal crop insurance program. What accounts for the failures of the program? What are the main alternatives for all-risk crop insurance?

10 Since 1980, what has been the effect of changes in federal fiscal policy on farmers and rural communities in the state in which you live? What are some of the major policy alternatives for federal fiscal policy that might help farmers and rural communities?

CHAPTER 15

HUMAN CAPITAL AND RURAL DEVELOPMENT POLICY

Human capital involves those attributes of people that contribute to their productivity and economic well-being. The modernization of agriculture is characterized by investments in human capital—education, skills, and health—of people in farming, agribusiness, and economic sectors related to agriculture. Results of such investments are seen in rising rates of productivity of human resources, and in declining labor requirements for given levels of agricultural output. Some rural communities are persistently poor due to the low stock of human capital of their residents and lack of resources for investments in human capital. In this chapter our policy interests focus on public actions that affect human capital throughout the food and agricultural system. We apply the economic concepts of human capital development to farm families and other rural people because of some critical problems of human productivity and income among this population.

IN THIS CHAPTER

By studying Chapter 15, you will learn how to answer the following questions:

1 How do supply and demand influence income and productivity of human capital in rural areas?
2 What are the implications for farm labor productivity and farm income distribution of (a) the U.S. educational policy, (b) federally funded economic development programs, (c) the declining rate of growth in off-farm income of farm families, and (d) policies influencing the hired farm work force?
3 What are the major economic problems of rural areas that are linked to the development of human capital?

4 What are some alternative policies for improving the well-being of farmers and other rural people?

HUMAN CAPITAL SUPPLY AND DEMAND

The concept of *human capital* views people as unique products of investments in their talents and abilities. Starting with innate ability, humans develop their capacity to make a contribution to production by a process of investment in which current resources are sacrificed for future returns through formal education, on-the-job training, migration, health care, nutrition, and other activities that improve their knowledge, skills, and abilities. This investment, combined with investments in other resources, yields returns over the lifetime of the individual and succeeding generations.

General Theory Applied to Agriculture

In a freely competitive labor market, which is characteristic of much of agriculture, supply and demand tends to determine the basic wage or return to human capital. The supply of human capital for farming depends on opportunity cost, the opportunities for employment in nonfarm occupations, or the pecuniary and nonpecuniary rewards those same persons could earn in other available alternatives. Although an individual or a family can supplement farm income, a farmer typically must give up the opportunity of a full-time career in another occupation. The demand for human capital in farming depends on the marginal value of the product attributable to the input of the farmer, or the marginal value product (MVP) of what that capital will produce.

In the short run, with a given upward sloping supply curve for agricultural commodities, an increase (decrease) in the prices of agricultural commodities will increase (decrease) the demand for human capital. Unless the supply of human capital is perfectly elastic, this shift will result in higher (lower) wages for farm workers and an increase (decrease) in net farm income. Over the long run, however, when the supply curve for human capital to an economic sector in pure competition is highly elastic, if not perfectly elastic, changes in demand for human capital in farming will have little, if any, effect on that sector's price or wage, which is fixed largely by its opportunity cost. Hence, when the returns to human capital in farming are below the opportunity cost, the farm population declines.

The U.S. farm population peaked at slightly more than 32 million shortly after the beginning of the twentieth century; sharp declines did not occur until after 1940 (Table 15-1). Then, improved opportunities in nonfarm employment resulted in higher opportunity costs for farm labor. This development, accompanied by advancing technology, displaced farm labor or reduced the demand for it, thus stimulating much sharper reductions in the farm population. By 1990, it had dropped to about 4.6 million, which was less than one-fifth of its 1940 level (based on a comparable definition of a farm). More modest declines are expected in the future, given the relatively small size of today's farm population. The pace of this decline reflects the levels of marginal returns to human capital in farming compared with returns in other occupations. This change follows a general theory regarding human capital that has important implications for food and agricultural policy.

TABLE 15-1
FARM POPULATION OF THE UNITED STATES

Year	U.S. farm population	Percent of total U.S. population
1910	32,077,000	34.9
1920	31,974,000	30.2
1930	30,529,000	24.9
1940	30,547,000	23.2
1950	23,048,000	15.3
1960	15,635,000	8.7
1970	9,712,000	4.8
1980[a]	6,051,000	2.7
1990[a]	4,591,000	1.8

Note: The size of farm families is also declining at a steady pace, with fewer children and potentially fewer heirs to take over the business. This fewer number of heirs could mean less pressure to divide up the family farm.

[a] Data based on current definition of a farm ($1,000 or more in annual agricultural production sales). Under the previous definition, the 1980 farm population would be 7.2 million because more rural places were counted as farms.

Sources: U.S. Bureau of the Census; and *Economic Report of the President,* U.S. Government Printing Office, Washington, DC, 1992.

Policy Implications of Human Capital Theory

According to a theory first applied to agriculture by Professor Theodore W. Schultz at the University of Chicago, two conditions are necessary for economic progress in agriculture: (1) an investment in the education of people in agriculture, especially young people, sufficient to enable them to compete on an equal basis with others in nonfarm employment; and (2) opportunities for employment outside of agriculture, or at least outside the farm sector, sufficient to raise the opportunity cost above the MVP of what that person is earning in farming.[1]

As long as this opportunity cost is above the MVP in farming, entry into the farm sector will be discouraged, and out-migration will occur until the MVP in farming rises to equal the opportunity cost. The productivity and MVP of the farm family in farming affects the level of its economic gain and the magnitude of labor adjustment continuing to face that family. Consequently, although various farm programs may raise farm income in the short run, the only way to raise farm incomes in the long run is to raise family members' opportunity cost to enter or remain in farming.

Important qualifications and limitations influence the application of this theory. Investments in human capital usually take several years for the full benefits to be realized. Returns for any individual are far from certain. Labor markets are imperfect. Farm people often hesitate to leave the relative security of farming to invest in many years of education or other human capital improvement because of their sunk costs and rural background. Finally, education typically combines investment and personal consumption. Yet the theory is important in developing concepts that apply especially to income in the farm sector.

Advances in labor-saving technology decrease the demand for farm labor and, if opportunity costs do not rise because of new opportunities outside the farm sector, an excess supply of farm labor occurs at prevailing wage rates. Real returns to both family farm labor and hired farm labor will generally fall. Farmers who successfully apply the new technologies may continue to save and invest as their total productivity increases. Other farm families facing adverse economic conditions or having less success with new technologies will tend to have lower incomes unless increasing demand raises the prices of farm outputs relative to the prices of farm inputs.

When we attempt to apply this theory, however, we should understand that farm labor is not a homogeneous form of capital. High technology substitutes for many types of traditional family farm and wage labor. Generally there is increasing demand for high level organizing and managing skills, which are required for efficient use of advanced technology. Unless the supply is very elastic, marginal returns to this type of human capital will rise. The demand for skilled farm managers increases as the demand for unskilled farm workers declines. What are some of the implications for human capital investment and educational policy?

IMPLICATIONS FOR EDUCATIONAL POLICY

Two general concepts have guided U.S. educational policy as it relates to human capital formation in agriculture. The first, an extension of a general principle inherent in public elementary and secondary education, is that every child should have the opportunity to complete the level of education that the child's tastes and abilities warrant. This historic principle is rooted in some of the earliest public policy decisions made by European settlers in North America. The second concept is that publicly supported science shall be used to increase the efficiency and productivity of industry and agriculture. This precept is of more recent vintage.

The Beginning of Rural Public Education

The public policy decisions made only a decade after the founding of the nation set a course of free education for all youths, even those in the most remote reaches of the developing territories. The Common Schools Section of the Northwest Ordinance of 1787 not only set the guiding principles for rural public education in the Midwest, but also became the model for 31 of the 50 states. Prophetically, it set aside section 16 of every township, where a "common school"—now called an elementary school—was to be established for the children of the surrounding community.

Never before in the history of any nation had such a commitment been made so dramatically to public education. What began as the one-room school for farm children later became the educational model for a highly urbanized, industrialized country. Four percent of the total public domain, which was the bulk of the wealth of the nation apart from its people, was eventually committed to this initial public educational policy. The policy technique was the simple act of granting public land to establish a public school system.

Problems Relevant to Agricultural Research and Education

What were the problems that prompted such a unique set of public educational policies? Policy makers justifiably might have devoted exclusive attention to urgent issues like defense, conflicts among the recently independent colonies, and land settlement. Yet innovative and lasting educational policies were fashioned out of the crisis periods of the revolutionary and civil wars. Why? Because certain persistent concerns were widely felt across the new nation. These concerns included supporting a private proprietorship, family farm system; enabling citizens to participate in government processes; increasing agricultural productivity; and preparing citizens for military leadership.

Supporting a Private Proprietorship, Family Farm System The United States had been settled primarily by tillers of the soil and owners of small shops who were rebelling against societies in which only the elite were worthy of educational opportunities and most workers on the land were pawns for the landed gentry. Even though the early boarding schools and private universities of the growing eastern cities were thriving, they were distant in both spirit and location from the numerous small family farms. In striving for a just or equitable society, policy decisions were reached to offer educational opportunity to each individual through publicly supported schools.

Enabling Citizens to Participate in Government Processes If the new experiment in a participatory democracy was to succeed, then citizens, whether pioneers on the frontier or professionals in the cities, needed access to educational opportunities. Thus citizens were concerned about schooling in the territories as well as in the eastern colonies, in rural as well as urban areas, for youths working on farms as well as for those living in boarding schools, and for subsistence families as well as the wealthy.

Increasing Agricultural Productivity for a Growing Population and for Export Earnings During the nation's formative years, farming was the dominant occupation. With the expanding richness of the land resources, agriculture seemed destined to continue as an important industry. A soaring population demanded more food and fiber, and a similar demand originated in Europe, which resulted in expanded exports.

Preparing Citizens for Military Leadership Outside Officer Academies In reaction to the control that European military academies exercised over civilian governmental affairs, means were sought to prepare a cadre of reserve officers outside the military schools for national emergencies.

Evolution of Public Higher Education

In the early 1800s, an idea evolved that public higher education should shift from its traditional liberal arts foundation and encompass more practical knowledge and skills to help promote economic development and growth. The French philosopher Alexis de Tocqueville, after extensive travels in the United States in the early 1800s, may have expressed this idea best. "It is evident," he wrote, "that, in democratic communities the interests of individuals, as well as the security of the commonwealth demands that the education of the greater number should be scientific, commercial, and industrial, rather than literary."[2] This theme was taken up by others, including many of the rural societies organized largely to promote agriculture. These societies published magazines, newspapers, journals, almanacs, and books. They also promoted fairs, meetings, socials, debates, and other activities through which they advanced the idea of practical education for agriculture and other industry.

In 1851, Jonathan Baldwin Turner, a teacher of rhetoric, Greek, and Latin at Illinois College in Jacksonville, proposed a plan to establish "a University for the Industrial Classes in each of the States."[3] He spoke at great length about this idea to almost any person or group that would listen. In 1852, in a letter published in the *Prairie Farmer*—a magazine of great influence—Turner proposed that if

> farmers and their friends exert themselves they can speedily secure for this State and for each State in the Union, an appropriation of public land adequate to create and endow in the most liberal manner, a general system of popular Industrial Education, more glorious in its design and more beneficent in its results than the world has ever seen before.[4]

Such ideas gained political support in Illinois. In 1853, the Illinois legislature sent to Congress a set of resolutions declaring that a system of industrial universities, liberally endowed in each state, would develop the people and "tend to intellectualize the rising generation."[5] It took nine years, however, for policy makers to agree on a policy response. First, the Congress could not agree, and then the President opposed. Finally, policy agreement between President Abraham Lincoln and Congress resulted in the Morrill Act of 1862, named after Justin Smith Morrill of Vermont, who introduced the bill.

The 1862 Act, often called the Land Grant College Act, was the founding legislation for the national system of land-grant colleges and universities. The policy technique was to appropriate 30,000 acres of public land for each senator and representative in Congress, on the basis of the 1860 census. Each state or territory was charged with establishing a college or university according to the principles embodied in the legislation. The teaching of agricultural, mechanical, and domestic sciences was the major thrust and opportunity for education in these areas and was to be available to people in all walks of life. It was also decreed in the early history of the land-grant university that all male students would take Reserve Officer Training Corps (ROTC) programs for future military leadership as needed.

Development of these colleges and universities came slowly. As a basis for support, the grant of land proved inadequate. State legislatures were slow to appropriate more money. Farm and rural people generally did not give the idea strong political support, and development also lagged from lack of scientific information to support teaching programs.

A full quarter of a century elapsed before another historic act followed. Then, in 1887, William Hatch of Missouri introduced a bill to establish a center where agricultural research would be conducted in each state in connection with a land-grant college or university. The resulting Hatch Act of 1887 directed the centers, which were referred to as agricultural experiment stations, to aid in acquiring and diffusing useful and practical information on subjects connected with agriculture and promote scientific investigations and experiments on the principles and applications of agricultural science. These experiment stations increased the efficiency and progress of agriculture.

Concerns also emerged that the research and classroom teaching in the land-grant universities were not reaching the farmers. Years of controversy swirled around the issues of whether the federal government should intrude into any local educational activity and, if so, whether a program should involve the high schools or a separate adult educational service. Finally, 27 years after the Hatch Act of 1887 and 52 years after the Morrill Act of 1862, the Smith-Lever Act of 1914 was adopted as a compromise. This Act authorized the appropriation of federal funds on a matching basis with the states to distribute and encourage application of

> useful and practical information on subjects relating to agriculture and home economics as mutually agreed by the secretary of agriculture and the state agricultural college or colleges receiving the benefits of the Act.

Although the three missions of the land-grant system—teaching, research, and extension—were launched at different times, together they constituted an evolutionary concept of human capital development and its application. In addition, in 1917, as part of a compromise in the Smith-Lever Act of 1914, the Smith-Hughes Act was adopted to provide matching federal funds for local rural high schools to offer instruction and supervised practice in agriculture and home economics. This Act contributed greatly to the development of rural human resources and substantially changed the face of secondary rural education throughout the nation.

Many additional acts were adopted over the next several decades.[6] In 1955, legislation consolidated all the previous laws as an amendment to the Hatch Act,[7] declaring a policy

> to continue the agricultural research at the State agricultural experiment stations . . . to promote the efficient production, marketing, distribution, and utilization of products of the farm as essential to the health and welfare of our peoples and to promote a sound and prosperous agriculture and rural life as indispensable to the maintenance of maximum employment and national prosperity and security . . . to assure agriculture a position in research equal to that of industry, which will aid in maintaining an equitable balance between agriculture and other segments of our economy. It shall be the object and duty of the State agricultural experiment stations . . . to conduct original and other researches . . . , including researches basic to the problems of agriculture in its broadest aspects . . . , the development and improvement of the rural home and rural life and the maximum contribution of agriculture to the welfare of the consumer, as may be deemed advisable, having due regard to the varying conditions and needs of the respective states.[8]

Hence the mission of the land-grant system has expanded to encompass farming, agricultural marketing, natural resources, the rural community, the home, and an emerging urbanized, international society. This emphasis on broadly based industrial education has been described as a unique system in which "the essential features of method, of curriculum, of organization, or purpose . . . can be identified as American, distinct from those of any other people.''[9] The two main features, one emphasizing science and industrial education and the other extending the opportunity to far greater numbers of students, are strongly developmental and egalitarian, at least as compared with the elitist concepts of higher education.

Widespread opportunities have been extended to many more young people, with great effects on productivity, technological advance, and human welfare. Yet from the founding of the United States to the mid-1900s, some minority groups, notably blacks and Native Americans, were disadvantaged in respect to the availability of educational opportunity. Let us consider this situation for blacks.

Historically Black Land-Grant Colleges

For many years, a dual system of land-grant colleges existed legally in parts of the United States. In the Southern and Border states, blacks were not allowed to attend the colleges established under the Morrill Act of 1862 and a separate system evolved for educating them.

From the end of the Civil War to the late 1880s, several Southern state legislatures established colleges for blacks. Despite the urging of the federal government, only Mississippi gave the designation ''land-grant'' to a black college.''[10] Thus, in 1890, a second Morrill Act was passed by Congress and signed into law. The Act contained the following clause:

> No money shall be paid . . . for the support or maintenance of a college where a distinction of race or color is made in the admission of students, but the establishment and maintenance of such [separate] colleges for white and colored students shall be held to be a compliance.[11]

To comply with the Morrill Act of 1890, some Southern states designated existing black colleges as land-grant institutions, while others later established separate land-grant colleges for blacks.

The historically black land-grant colleges, commonly referred to as the 1890 institutions, have been severely underfunded by both the federal and state governments.[12] Much of the early federal land-grant legislation provided financial support mainly to the colleges established under the first Morrill Act. For example, initially the 1890 institutions did not receive funds for extension services under the Smith-Lever Act of 1914. Funds were not made available to these colleges until 1972. Furthermore, hard money for research was not forthcoming until 1967. The inadequate funding hampered the growth and development of the 1890 institutions and their ability to provide services to their mostly black, rural, disadvantaged clientele.

Finally, federal legislation has begun to address the inequities in funding of the 1890 institutions. For example, the Food and Agriculture Act of 1977 authorized the appropriation of funds (primarily grants) to enhance the extension and research programs at these colleges. The Agriculture and Food Act of 1981 provided for increased funding through appropriations rather than grants and permitted the funds to be used for long-range planning and construction of buildings.

Until 1954, when the U.S. Supreme Court ruled that the separate but equal doctrine did not provide equal opportunity and was unconstitutional, the 1890 institutions were the only land-grant colleges in the South where blacks could obtain an education in agricultural, mechanic, and domestic sciences and where black farmers could avail themselves of extension services. Further, the human capital of many black rural residents was enhanced by matriculating at these colleges. With the dismantling of de jure segregation, the 1890 institutions, like the colleges established under the Morrill Act of 1862, have served both black and white clientele. However, the 1890 institutions and other historically black colleges and universities continue to graduate a disproportionately high number of blacks, and they allocate a relatively large percentage of their agricultural research budget to the problems of the economically and socially disadvantaged, particularly small-scale farmers and the rural poor. Relatively greater increases in funding and other educational support will be required therefore to achieve true equality in educational opportunity. This conclusion raises a broad policy issue concerning the appropriate level of public support.

Level of Public Support

Questions persist concerning the appropriate level of public support for research and education at all levels. Research and education directed toward food, agriculture and the rural economy have made remarkable progress in recent decades, but have not been supported as strongly at either the state, local, or federal levels, as many scientists, educational leaders, and consumers believe to be appropriate. Also, in spite of greatly expanded student scholarships, grants, and loan programs, many people from middle and lower income families have been denied a university education because of rising tuition and restricted admissions. However, the mission and performance of the land-grant college complex continue to be evaluated, and its effects on various societal values are repeatedly scrutinized. The returns from this type of human capital investment are very high and are also positively correlated with wealth, ability, available opportunities, and other factors. Therefore, although the returns rank high in terms of productivity, the opportunities are not equally available to all.

The gains in productivity are substantial, but because the benefits are unevenly distributed, those people who are not involved may be worse off than before. This situation occurred in the United States[13] and is generally true around the world, especially in developing countries.[14] Under economic development, the structural transformation of traditional agriculture is not neutral with regard to income distribution because incomes become increasingly concentrated. Although applied most recently to late-developing countries,[15] this generalization apparently has applied to the U.S. economy for several decades.[16] Because of the adverse effects on some people and communities, the federal government has been involved in economic development programs that help low-income people and depressed areas of the economy. But the policy question remains: What is an equitable, desirable, or adequate level of support?

The postwar period of the 1950s and 1960s focused on improving education through federal financial support, including the National Defense Education Act of 1958, Vocational Training Act of 1963, Higher Education Act of 1963, Library Act of 1964, and Elementary and Secondary School Act of 1965. The federal contribution to public schools grew from about $150 million in 1955 to $9.1 billion in 1980. At the beginning of the 1980s, the federal contribution amounted to 9 percent of all expenditures for public elementary and secondary education. This percentage then declined substantially throughout the 1980s.

The federal programs have important implications for human capital policy and income distribution, especially for food, agriculture and the rural economy. Most of the federal educational initiatives emphasized support for equalizing educational opportunities, a principle that is of particular significance to poorer rural schools. However, it may be fair to say that, at least up to the early 1990s, direct federal funding fell below the earlier stated goals. Generally, federal policy has continued to encourage and emphasize rural development as an instrument for increasing human capital in the farm and rural sectors.

HUMAN CAPITAL POLICY APPLIED TO RURAL AREAS

As the nation's farm economy has become more commercial and integrated into urban markets, some rural communities have faced increasingly difficult adjustment problems (see Box on p. 283). Employment opportunities have declined, and rural institutions and services are often less than adequate. Rural development programs and policies have been implemented to enhance rural infrastructure, develop rural businesses, and improve human resources. By and large, public policy has failed to provide rural residents with the quality of education and access to public services equal to that offered urban residents. The history of rural development programs and policies that contribute to human capital formation can be traced back to the 1930s. How has rural development policy evolved and what are its goals?

Early Background for Rural Development

With the unprecedented economic deprivation of the Great Depression, the New Deal under President Franklin D. Roosevelt launched many new agricultural programs to improve the economic well-being of farm families and other rural residents. In addition, employment programs such as those under the Works Progress Administration (WPA) were established, but a relatively small amount of additional federal aid was directed specifically toward education. Income security programs established during the New Deal era such as Aid to Families with Dependent Children (AFDC) and Social Security

How is *rural* defined?

The terms *rural* and *nonmetropolitan* (*nonmetro*) are commonly used interchangeably to refer to areas outside a Metropolitan Statistical Area (MSA). An area qualifies for MSA recognition if it includes a city of at least 50,000 with a total metropolitan population of at least 100,000 (75,000 in New England). In addition to the county containing the main city or urbanized area, a MSA includes other counties where residents have strong commuting ties to the central county. Areas designated by the federal government as a MSA are commonly referred to as *urban* or *metropolitan* (*metro*).

provided a "safety net" for the poor and contributed to the development of human capital. In 1950, a presidential commission established by President Harry Truman criticized the use of children as farm laborers and pointed out that there were no federal laws to prevent this practice.[17] Later in the 1950s, however, instead of focusing specifically on the needs of children and educational policy, the Eisenhower Administration recognized rural development as a relatively high policy priority.[19] The need for rural development was linked in part to the declining farm population. From the early 1930s to the end of the 1950s, the policy objectives and public commitments were not consistently or adequately funded. The money appropriated was either for administrative staffing only or was too uncertain for long-term professional commitments. The goal of substantially improving the human capital of rural residents, especially those with low incomes, continued as a hope rather than a mission accomplished.

New Policy Directions

Under President John F. Kennedy, the Area Redevelopment Act of 1961 established an administrative framework for supporting rural development.[19] Following President Kennedy's death on November 22, 1963, President Lyndon B. Johnson searched for a major domestic theme for the next administration. On January 8, 1964, in the State of the Union address, the president announced that "this Administration today, here and now, declares unconditional war on poverty in America." Subsequently, programs were initiated to improve education, housing, health care, and job training. The Economic Opportunity Act of 1964 created the Office of Economic Opportunity and gave it the task of coordinating or administering most of the programs included in the War on Poverty.[20] These programs proved to be invaluable to some of the rural poor. They improved their quality of life and aided human capital formation. However, some economists and sociologists believe that they tended to result in a "culture of poverty," which keeps them from taking advantage of opportunities to escape poverty. An underlying assumption of the culture of poverty theory is that the poor do have opportunities to escape poverty. According to the theory, they fail to take advantage of the opportunities because they lack sufficient motivation to escape poverty.

In the Manpower Act of 1965, the Economic Development Administration absorbed the programs of the previously established Area Development Administration. Next, economic development regions were established under the Public Works and Economic Development Act of 1965. Within each region, districts were formed with an economic administrative center to organize development programs. Redevelopment areas in a district were eligible for a 10 percent bonus on grants for public works projects. Initial action in creating a district had to be taken by a state or agency. Each state was responsible for technical assistance on a project that could include: (1) studies to identify area needs or to solve industrial or economic problems, (2) grants-in-aid amounting to 75 percent of the costs of planning and administering local economic development programs, and (3) management and operational guidance for private firms. In addition, low-interest, long-term loans were offered to help private businesses establish plants in redevelopment areas. Loans could cover up to 65 percent of the total project cost and could run for 25 years at interest rates commensurate with federal borrowing costs.

Under some of the work-training programs, funds were appropriated to hire the chronically unemployed to provide long-term jobs at decent wages for workers to improve rural towns and depressed rural areas.

By the mid-1960s, economic expansion had generated more employment opportunities, and unemployment had fallen below 4 percent, the level then used by the Department of Labor to define full employment. Manpower concerns turned to the special labor market problems of those broadly designated as disadvantaged. New legislation included the Civil Rights Act, the Public Works and Economic Development Act, the Economic Opportunity Act, Model Cities legislation, the Vocational Education Act and amendments, new provisions for work incentives in the Social Security Act, and amendments to the Manpower Development and Training Act of 1962. The legislation provided new directions and some increased funding for programs for the disadvantaged. Emphasis was placed on helping youth, the undereducated, and unskilled members of various minority groups. Further, the War on Poverty began to address rural poverty explicitly. In 1967, President Johnson created a National Advisory Commission on Rural Poverty, which concluded that the plight of the rural poor was a national disgrace.[21]

The Rural Development Act of 1972 combined a number of related programs into the first generally comprehensive program of rural development. The Act reaffirmed some of the earlier goals of enhancing depressed rural areas and the human capital of its citizenry. However, grant and loan funds were not provided on the scale originally authorized. The cooperative extension services in each state continued to be responsible for providing educational leadership for rural development and to suggest policy alternatives and priorities. In addition, four regional centers for rural development were organized by the land-grant universities and the USDA to coordinate research and educational programs on rural development. These centers have continued to serve as supporters, catalysts, and avenues of communication for research and educational programs, conferences, publications, and newsletters on rural development issues.

During the 1970s, revitalization of rural America was a dominant theme.[22] The yearly rate of population growth in rural areas surpassed that of urban areas; industries such as agriculture, manufacturing, and mining, which were important to the rural economy, appeared to be sound; and the quality of life was improving. Rural outmigration ended and people flocked back to rural communities to take advantage of the job opportunities and the quality of life. These conditions were short lived, however.

The decade of the 1980s began with passage of the Rural Policy Act. This Act established broad policy goals, but did not authorize additional funds to meet the goals. Although it required the USDA to submit annual reports on rural development strategy to Congress, when the department did not meet the deadline or failed to submit a report, there was no Congressional action. There was little interest, from either the administration or Congress, in the 1980s, in providing assistance to the rural economy, other than to the farm sector.[23] Furthermore, New Federalism, which is characterized by reliance on market forces and shift of some responsibilities from the federal to state and local governments, resulted in a decline in expenditures for rural development programs as well as in some of the ''safety net'' programs.

Analysts often characterize the 1980s as a decade of *economic stress* in rural America.[24] During the 1980s rural unemployment rates increased more rapidly than the rates

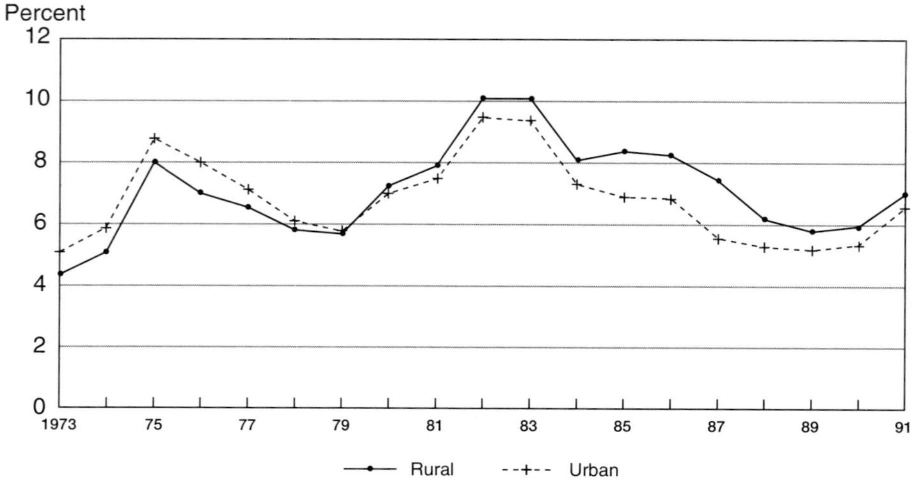

FIGURE 15-1
Rural and urban unemployment rates, 1973 to 1991. *Sources:* David L. Brown and Kenneth L. Deavers, "Rural Change and the Rural Economic Policy Agenda for the 1980s," in David L. Brown et al., editors, *Rural Economic Development in the 1980s: Prospects for the Future,* Economic Research Service, U.S. Department of Agriculture, Rural Development Research Report No. 69, September 1988; *Rural Conditions and Trends,* Economic Research Service, U.S. Department of Agriculture, Fall 1992, Vol. 3, No. 3. Note: The urban (metro) area delineation was updated in 1985 and the subsequent data are not directly comparable with the data in earlier years.

in urban areas, peaked at a higher level, and remained above them (Figure 15-1). Job growth in rural manufacturing counties increased slowly, agricultural counties experienced virtually no growth, and mining and energy counties experienced declines in total employment. Fueled by reduced entry and net out-migration, rural population growth declined. Moreover, many who left rural America were the best educated. Structural changes in the rural economy, combined with demographic changes, contributed to deteriorating incomes for rural residents, and the income gap between rural and urban families widened. Further, while poverty rates rose dramatically for both rural and urban residents in the early 1980s, they peaked at a higher level in rural areas, and remained relatively higher at the end of the decade (Table 15-2). The continuing economic stagnation of rural areas, especially during the recovery from the recessions of the early 1980s, contributed to legislative activity which culminated in Title XIII (Rural Development) of the 1990 Food, Agriculture, Conservation, and Trade Act.

The 1990 Act consolidated rural development programs, in particular the Farmers Home Administration's Community and Business Programs, into a new USDA agency—the Rural Development Administration. In addition, it provided for rural business assistance in the form of grants, loans, and technical assistance and for improvements in infrastructure (i.e., water and waste facilities, telecommunications). The human resources provisions of the Act related to enhancing learning and expanding rural opportunities by increasing access to telecommunications services, computer networks and other technology.

TABLE 15-2
POVERTY RATES BY RESIDENCE, 1970–91[a]

Year	Total	Farm	Nonfarm	Nonmetro	Metro
			Percent		
1970	12.6	21.1	12.2	16.9	10.2
1971	12.5	20.9	12.1	17.2	10.4
1972	11.9	14.9	11.7	15.3	10.3
1973	11.1	13.4	11.0	14.0	9.7
1974	11.2	16.2	10.9	14.2	9.7
1975	12.3	16.4	12.1	15.4	10.8
1976	11.8	15.9	11.6	14.0	10.7
1977	11.6	17.1	11.3	13.9	10.4
1978	11.4	12.2	11.3	13.5	10.4
1979	11.6	13.2	11.6	13.7	10.7
1980	13.0	17.5	12.9	15.4	11.9
1981	14.0	23.0	13.8	17.0	12.6
1982	15.0	22.1	14.8	17.8	13.7
1983	15.2	23.7	15.0	18.3	13.8
1984[b]	—	—	—	—	—
1985	14.0	20.3	13.8	18.3	12.7
1986	13.6	19.6	13.4	18.1	12.3
1987	13.5	12.6	13.5	16.9	12.5
1988	13.0	12.3	13.0	16.0	12.2
1989	12.8	11.1	12.8	15.7	12.0
1990	13.5	11.2	13.6	16.3	12.7
1991	14.2	12.5	14.2	16.1	13.7

[a] Data are for all persons.
[b] Data for 1984 are not available.
Sources: U.S. Department of Commerce, Bureau of the Census, Current Population Reports. *Characteristics of Population Below Poverty Level* and *Poverty in the United States,* various issues.

Improving the Human Capital and Economic Conditions of the Rural Poor

In some important ways for public policy, the rural poor differ from the urban poor. If public policy is to adequately address rural poverty, then the causes of rural poverty must be understood and the differences recognized. The causes of rural poverty are complex and multifaceted. Human capital deficiencies are a major factor. Historically, rural areas have underinvested in education, in part because of the difficulty of local areas to capture the benefits of human capital investments relative to physical investments in private property or public services such as highways. On three key educational variables—average educational level of adults, percentage of adult population that completed four years or more of college, and the high school dropout rate for youth—rural areas lag behind urban areas.[25] Adult illiteracy and marginal literacy are pervasive in many rural communities.

The nature of jobs also contributes to relatively high poverty rates. For example, many rural jobs are characterized as "poor jobs" because the wages are low or the work is seasonal or intermittent. Researchers at the Massachusetts Institute of Technology estimated that 42 percent of rural workers in 1987 earned wages too low to lift a family of four out of poverty with full-time, year-round work.[26] Other factors contributing to poverty include racial discrimination and urban biases in programs designed to aid the disadvantaged.

Public policies designed to aid the poor in general could differentially affect the rural poor if their characteristics are substantially different from those of the urban poor. In 1987, for example, 65 percent of poor rural families had at least one member who worked, compared with 54 percent of poor urban families. These data indicate that expansion of the earned income tax credit could benefit a sizable proportion of rural poor families. Welfare reform, while important to many of the rural poor, may not benefit most of them because, unlike the urban poor, a relatively high proportion live in two-parent households. Other salient characteristics of the rural poor and poverty rates by demographic characteristics are presented in Table 15-3. Furthermore, while nearly three-fourths of the rural poor are white, poverty falls disproportionately on certain racial and ethnic minorities, particularly Blacks, Hispanics, and Native Americans.

Persistent rural poverty is heavily concentrated in particular areas of the United States, notably Appalachia, the Mississippi Delta, the Ozark–Ouachita Plateau, and Native American reservations. This geographic concentration has led to the formation of regional commissions to address rural poverty. For example, the Lower Mississippi Delta Development Commission was established under Public Law 100–460 in October 1988 to "study and make recommendations regarding economic needs, problems, and opportunities in the Lower Mississippi Delta region, and to develop a ten-year regional economic development plan."[27]

Through several decades of evolutionary policy, the goals of rural development have not changed from the original ideas of a balanced development of rural and urban society, but the means of implementing the policy have changed.[28] This change in emphasis shifts the programs from large federal grants for construction of facilities (except for transportation) to programs that promote community initiative and leadership. The community college movement exemplifies this change as explained in the next section. The movement developed and expanded because many four-year colleges were inaccessible to large numbers of people because of cost, admission standards, location, and other factors. Community colleges have provided educational opportunities to many rural residents.

IMPLICATIONS OF THE COMMUNITY COLLEGE MOVEMENT

The junior college or community college movement is another type of educational program that is distinctly American. Although it has some roots going back to the early days of education in this country, its major economic and social effects have occurred primarily since the end of World War II. In many respects the State of California pioneered the movement in terms of financial resources and by integrating two-year junior colleges into the administrative structure of the Greater University of California.

TABLE 15-3
SELECTED CHARACTERISTICS OF THE POOR BY RESIDENCE, 1990

Item	U.S.	Urban	Rural
	Percent	Percent	Percent
Poverty rate for:			
Total population	13.5	12.7	16.3
Related children in households headed by female, no spouse present	53.4	52.5	56.8
Whites	10.7	9.9	13.5
Blacks	31.9	30.1	40.8
Aged[1]	12.2	10.8	16.1
Disabled	35.9	34.8	39.0
Percentage of poor who are:			
Related children in households headed by female, no spouse present	21.9	23.7	17.0
Whites	66.5	64.1	72.9
Blacks	29.3	31.4	23.6
Aged[1]	10.9	9.8	14.0
Disabled	8.8	8.6	9.4

[1]At least 65 years old.
Source: Rural Sociological Society Task Force on Persistent Rural Poverty, *Persistent Poverty in Rural America,* Westview Press, Boulder, Colorado, 1993.

The state also encouraged more community colleges independent of that system. Community colleges are now authorized under state law in nearly all states, usually with their own governing boards. Financing is largely based on revenues from property taxes levied on specially created districts either with or without additional state revenues.

Most community colleges offer a wide array of subjects under two distinct types of curricula: (1) university preparatory and (2) vocational. Under the university preparatory curriculum, students are enrolled in subjects similar to those given in the first two years of study in a four-year undergraduate university program. Under the vocational curriculum, subjects may vary much more widely and may include specialized courses for teaching skills used in a specific occupation. Tuition is lower in community colleges, the criteria for enrollment are more flexible than in most universities, and enrollments are not so restricted. Community colleges have grown rapidly therefore, with enrollments in many states equaling or exceeding those of four-year undergraduate universities. By increasing human capital investment, community college programs will have an important influence on expanding the opportunity costs of both agricultural and nonagricultural rural residents in many occupations. This outcome can significantly reduce the skewness in farm income distribution by opening up broader opportunities for many young people and some older people as well.

The long-term effect of the community college movement is still to be determined because it depends on employment levels in the general economy and on the continuing growth of education and training. According to most studies, as people have migrated out of the farm sector, their former earnings in agriculture are strongly correlated with their subsequent nonfarm earnings. Merely migrating from one occupation to another usually is only a limited solution to their income problems. It is the effect of human capital investment itself that is important in upgrading the productivity of human capital. This is the unique contribution of the growing community college movement as well as other educational policy. This conclusion is supported by data showing that part-time, off-farm employment of current farm operators is declining in importance as the farm population continues to decline in response to rising opportunity costs.

How has the rate of increase in off-farm income slowed, and what are the implications for policy?

DECLINING GROWTH OF OFF-FARM INCOME[29]

For many years after the beginning of the technological revolution in U.S. agriculture, off-farm income generally increased, both in absolute terms and as a proportion of total farm-family income. Since the early 1960s, however, the rate of increase in constant dollars has slowed substantially, reaching essentially no growth in the 1980s (Table 15-4). Multiple job holding appears to be driven by the employment opportunities offered in the nonfarm economy and may continue to be important at the margins of lower skill levels and part-time jobs, where some farm labor operates. The decrease in the rate of growth of off-farm income implies that the benefits from this source of income may be nearing exhaustion.

As long as new farm production technology is forthcoming, farm size will increase and excess farm labor will continue to be generated. Farm labor and management must

TABLE 15-4
CHANGE IN U.S. FARM AND FAMILY INCOME, NUMBER OF FARMS, AND FARM POPULATION RELATED TO LEVELS OF EMPLOYMENT, 1960 TO 1985

	Percent Change in constant dollars[a]			Percent Changes[a]		Percent
Period	Net farm income per farm[b]	Off farm income per farm[b]	Total family income per farm[b]	Number of farms	Total farm population	Unemployed U.S. average
1960–1965	4.2	10.5	7.2	−3.3	−4.6	5.4
1965–1970	.9	4.6	2.9	−2.6	−4.7	3.8
1970–1975	7.8	2.7	4.9	−3.1	−1.8	6.0
1975–1980[b]	−7.8	.9	−3.3	−.7	−.8	6.7
1980–1985[b]	2.5	.1	1.1	−1.3	−2.4	8.2

[a] Compound annual rates of change for period indicated, constant 1982 dollars derived using GNP deflator.
[b] Data for all 1980 income rates are the average of years 1979 to 1981. The average is used because of the unusual drop in farm income in 1980.
Sources: U.S. Department of Agriculture. *Economic Indicators of the Farm Sector, National Financial Summary,* Economic Research Service, U.S. Department of Agriculture, various years; *Economic Report of the President,* U.S. Government Printing Office, Washington, DC, various years.

constantly adjust to the squeeze between decreased prices of farm commodities, resulting from rapid technological advance and the slow growth in demand for farm commodities. If the opportunity costs for farm labor and management are low, either because of unemployment or their own lack of training for higher paying jobs, then excess labor will be kept on the farm. Thus the economic welfare of both consumers and farmers depends on the reliability and stability of employment opportunities for off-farm employment, and adequate investment in their human capital to take advantage of these opportunities.

These opportunities in turn are determined by macroeconomic conditions and policies in the United States and world economies, state and local economic conditions, the structure of future U.S. industrial growth, and investments in the human resources of rural America, especially on the farm. If opportunities for alternative employment are readily accessible within commuting distance of the farm, then off-farm employment can serve as a strategic variable in the adjustment process. Without such opportunities, however, further out-migration and reduced entry will be necessary to solve the farm income problem.

The push for farm labor adjustment comes from the farm-labor supply side through the employment-reducing effects of advancing farm technology. The pull from the demand side comes from increasing employment opportunities in the nonfarm economy, which increase the opportunity costs for farm labor. The adjustment is enhanced through investments in human capital that farm people—especially young people—can make to be productive in nonfarm as well as farm employment. Fulfillment of the two conditions necessary for economic progress in agriculture—human capital investment and increasing nonfarm employment or increasing opportunity cost—will be the critical elements in policy for raising farm incomes and readjusting agriculture.

This conclusion also applies to the hired farm work force, which merits a separate discussion. The economic and social problems of this labor force are unique in many ways, and are more severe, if not more intractable, than problems of the general farm population. Why is this so, and what are the inferences for policy?

HUMAN CAPITAL: THE HIRED FARM WORK FORCE

Wages of most hired farm workers are determined by the interaction of supply and demand in relatively free labor markets. Agriculture was not covered under the first federal minimum wage law, which was passed in 1938, and since then the efforts to achieve minimum wage standards to cover farming have failed in Congress. With important exceptions in California and Hawaii, attempts to organize farm workers in labor unions have generally failed.

The California Agricultural Labor Relations Act of 1975 established an Agricultural Labor Relations Board and extended to agricultural employees the same privileges and protection generally extended to other organized workers under the National Labor Relations Act. As California employs more farm labor than any other state, its 1975 law is significant. In Hawaii, the International Longshoremen's and Warehousemen's Union obtained for sugar workers some of the highest agricultural wages in the world, along with other privileges. In most of the other states, hired farm workers remain unorganized, which means that the supply of farm labor is unrestricted by labor organizations. Farm wages, living conditions, and other perquisites thus depend on supply and demand.

Composition of the Hired Farm Work Force

A great variation has existed within the hired farm work force. Among many of the regular or year-around workers, earnings and working conditions have compared favorably with those of the majority of farm owner–operators. However, the conditions of the migratory work force are generally well below those of the farm owner–operators. Many migrant workers live in extreme poverty. Not only have they suffered from the usual disadvantage of an unorganized or nonunion labor market, many have been exploited by unscrupulous crew bosses.

Structural changes in agriculture have led to a reduction in the number of hired farm workers, from 4.3 million in 1950 to 2.5 million in 1987. The hired workers are a heterogeneous group varying in skills and productivity, demographic characteristics, and the amount of time they are employed (Table 15-5). The typical hired farm worker

TABLE 15-5
DEMOGRAPHIC CHARACTERISTICS OF HIRED FARM WORKERS BY DAYS OF FARM WORK, 1987

		Days of farm work		
Characteristics	**All hired farmworkers**	**Casual Fewer than 25**	**Seasonal 25 to 149**	**Regular and year-round 150 and over**
		Thousands		
All hired farmworkers	2,463	856	816	790
Racial/ethnic group				
White	1,917	719	627	571
Hispanic	338	68	112	158
Black and other	208	69	77	60
Sex				
Male	1,961	648	637	676
Female	501	208	179	114
Age (years)[a]				
14–17	531	270	200	60
18–24	621	201	228	191
25–34	563	167	157	239
35–44	316	99	109	109
45–54	216	49	63	104
55–64	148	50	35	64
65 and older	67	20	23	24

[a] Data are not reported for persons younger than 14 years of age.

Source: Victor J. Oliveira and E. Jane Cox, *The Agricultural Work Force of 1987: A Statistical Profile,* Agricultural Economic Report No. 609, Economic Research Service, U.S. Department of Agriculture, May 1989.

is under 34 years of age, male, and white. Fewer than one-third are employed year round or at least 150 days or more. A little more than one-third are seasonally employed, working 25 to 149 days a year. About one-third are casual or itinerant workers who are employed for fewer than 25 days a year. The numbers in each category have been declining rather steadily over many years.

The incomes and living conditions are as varied as the composition of the work force. On most of the medium and large family farms, some hired workers must have high-level mechanical or managerial skills. This raises their opportunity costs, provided the workers are mobile and can compete for off-farm employment. Many jobs on large farms are routine, however, requiring few skills except the ability and willingness to work hard for long hours and sometimes in adverse weather. The bargaining power of such workers is low because they are unorganized and have low opportunity costs. Also, the supply of workers is very elastic and is replenished by immigrants from other countries, particularly Mexico. Hired farm workers with few skills and low opportunity costs often face low income and poor living conditions.

Effects of U.S. Policy on Incomes of Hired Farm Workers

National policy has tended to augment the supply and hence depress the wages of some farm workers. Under regulations of the Immigration and Naturalization Service, the legal requirements for immigrants to be admitted to the United States as *farm workers* have been more liberal at times than for workers claiming to have skills in other occupations. However, the restrictions were tightened in the 1980s. The Immigration Reform and Control Act of 1986 provided that, if the U.S. Department of Labor and the U.S. Department of Agriculture find farm labor shortages, workers can be allowed to enter the United States. Not unique to labor law applying to agriculture, this provision attests to the political strength of the Western Growers Association and other groups representing large farmers. It tends to assure a continuing supply of workers who have low opportunity costs and who are available for farm employment.

In 1970, Congress passed a law (only sporadically enforced) specifying minimum housing standards for workers on large farms, but all attempts in Congress to establish minimum wages for farm workers have failed. There have been many protests—such as strikes, boycotts, and other actions—by farm-worker organizations over low wages, poor working conditions, inadequate housing, and the inability of itinerant workers to obtain good medical care or good schooling for their children. Most protests have had limited influence in improving the conditions, but workers in California have gained some bargaining power.

The California Labor Relations Act[30]

The California Labor Relations Act of 1975 was the first law in the United States to deal specifically with farm wage labor. This law declared it to be the policy of the state government of California to encourage and protect the right of agricultural employees to full freedom of association, self-organization, and designation of representatives of their own choosing to negotiate the terms and conditions of their employment. Labor organizations were to be free from interference, restraint, or coercion by employers or their agents.

The Act also provided for an Agricultural Labor Relations Board and regulations for hearings, rights of witnesses, publicity, free speech, and bargaining. It was a significant step in the long policy-making struggle to organize farm workers, and its influence is far-reaching, especially in California, where many of the largest of all U.S. farms are located.

The long history of labor organization and bargaining clearly shows that success at the bargaining table depends on controlling the supply of labor, but immigration policy and other laws make it difficult to do so. Because of the large number of potential farm workers, the long-term success of farm labor organizations depends on increasing the opportunity costs of the potential hired farm work force. This can come only through investing in human capital and increasing nonfarm employment opportunities relative to the supply of labor.

GENERAL INFERENCES FOR POLICY

Human capital is a crucial input for food and agricultural production and marketing. For a given population, the supply of human capital for agriculture depends on the opportunities for employment outside of agriculture. Supply varies inversely with these opportunities, other things being equal. The demand depends on the MVP produced by human capital, which is related to its productivity. Combined with other technological advances, human capital investment resulting from public policy has resulted in the output per unit of labor rising rapidly and consistently over several decades. However, despite the significant increase in productivity, income from farming has tended to be depressed and more skewed than in most other industries. This phenomenon can be partly attributed to skewed investment in human capital, which in turn skews the MVP of employees and reduces the demand for poorly trained workers. It also reduces their opportunity costs and tends to perpetuate poverty.

Once investments in human capital are made, they may continue to yield returns to individuals, and to society. All measurements tend to highlight the generally high rates of return from such investments. The general policy conclusion is clear: In an age of high technology especially, increasing investment in human capital enhances the general welfare and is necessary for economic progress in agriculture. A current policy issue is whether food and agricultural policy can or should be the primary means for enhancing human capital in rural areas or whether the primary means should be a broader based education and training policy.

National public investments in human capital, and in research and education generally, must compete with other demands on state and federal governments. With continuing budget deficits, these conflicts may sharpen. Within the total scope of food and agricultural policy, there is competition for funds between human capital investment and other programs such as farm price and income supports, low-income food aid, farm credit programs, conservation needs, trade assistance, and food safety. Policy makers must seriously assess the relative benefits from expenditures in each of these types of programs and the many conflicts among them that must be resolved. In final summary regarding rural poverty, our conclusion is the nation's policy effort could more directly attack the problem of rural poverty. To more effectively attack this problem in the long term, much more attention must be given to early learning and to other investments in human capital at early ages, with more explicit targeting of children from poor families.

LOOKING AHEAD

In the next chapter, we will consider future long-term issues in food and agricultural policy. As there is a great deal of uncertainty involved in forecasting, we will focus on factors that are likely to influence future policy.

ENDNOTES

1 See Gary S. Becker, *Human Capital: A Theoretical and Empirical Analysis, with Special Reference to Education,* published by the National Bureau of Economic Research, Columbia University Press, New York, 1964.

2 Alexis de Tocqueville, *Democracy in America,* as quoted by Earle D. Ross, *Democracy's College: The Land-Grant Movement in the Formative Stage,* Iowa State College Press, Ames, 1942, p. 11.

3 Mary Turner Carriel, *The Life of Jonathan Baldwin Turner,* University of Illinois Press, Urbana, 1961, as quoted by Richard Gordon Moores, *Fields of Rich Toil: The Development of the University of Illinois College of Agriculture,* University of Illinois Press, Urbana, 1970, p. 5.

4 *Prairie Farmer,* March 1852, p. 114, as quoted by Moores, *ibid.,* p. 6.

5 Allen Nivens, *The State Universities and Democracy,* University of Illinois Press, Urbana, 1962, pp. 15–16.

6 For a summary of laws quoted in this section, see *United States Code, Annotated, Title 7, Agriculture,* U.S. Government Printing Office, Washington, DC, Chapters 13 and 14, Sec. 301–390k. For further discussion, see Harold G. Halcrow, *Food Policy for America,* McGraw-Hill, New York, 1977, pp. 451–473.

7 *Public Law 352,* 84th Congress, 1st Session, chap. 790, 1955.

8 *U.S. Code Congressional and Administrative News,* 84th Congress, 1st Session, House Report No. 1298, p. 2977.

9 Paul Monroe, "Historic Foundations of American Education," *Essays in Comparative Education,* as quoted by Earle D. Ross, *Democracy's College: The Land-Grant Movement in the Formative Stage,* Iowa State College Press, Ames, 1942, p. 1.

10 B. D. Mayberry, ed., *Development of Research at Historically Black Land-Grant Institutions,* School of Applied Sciences, Tuskegee Institute, Tuskegee, AL, 1976.

11 William Payne, "Forgotten . . . but Not Gone: The Negro Land-Grant Colleges," *Civil Rights Digest,* Spring 1970, pp. 12–17.

12 Payne, *Ibid.* Thomas T. Williams and Handy Williamson, Jr., "Teaching, Research, and Extension Programs at Predominantly Black Land-Grant Institutions," *Southern Journal of Agricultural Economics* 17, July 1985, pp. 31–41.

13 Jim Hightower, *Hard Tomatoes, Hard Times: The Failure of the Land Grant College Complex,* Preliminary Report of the Task Force of the Land Grant College Complex, Agribusiness Accountability Project, Washington, DC, 1972.

14 Irma Adelman and Cynthia Taft Morris, *Economic Growth and Social Equity in Developing Countries,* Stanford University Press, Stanford, 1973.

15 Bruce F. Johnston and Peter Kilby, *Agriculture and Structural Transformation: Economic Strategies in Late-Developing Countries,* Oxford University Press, New York, 1975. For a more specific criticism of the role of the World Bank, see Joseph Collins, Frances Moore Lappe, and David Kinley, ''The Poverty Brokers: The Trickle Down Theory at Work at the World Bank,'' *The Progressive,* December 1980, pp. 40–44. See also James Bovard, ''The Alarming Truth About the World Bank,'' (condensed from *Reason*) *Reader's Digest,* June 1989, pp. 108–114.

16 *Rural People in the American Economy,* U.S. Department of Agriculture, Economic Report No. 101, U.S. Government Printing Office, Washington, DC, 1966.

17 For more general background on child labor in farming, see Harold G. Halcrow, *Food Policy for America,* McGraw-Hill, New York, 1977, pp. 477–481.

18 *Ibid.,* pp. 457ff.

19 *Ibid.,* p. 458.

20 For a popular account, see Nicholas Lemann, ''The Unfinished War,'' Part I, *The Atlantic Monthly,* December 1988, pp. 37–56; Part II, *The Atlantic Monthly,* January 1989, pp. 53–68.

21 National Advisory Commission on Rural Poverty. *The People Left Behind,* U.S. Government Printing Office, Washington, DC, 1967; Marian Clawson, ''Rural Poverty in the United States,'' *Journal of Farm Economics,* 49 (5), pp. 1227–1234 (1967).

22 See for example, *Rural Development; Research Priorities; Communities Left Behind; Alternatives for Development; Rural Industrialization; Problems and Potentials,* from three symposia under general sponsorship of the North Central Regional Center for Rural Development, Iowa State University, and other cooperating institutions, the Farm Foundation, South Dakota State University, and Purdue University, 1973 and 1974. See also reports sponsored by or prepared for the Subcommittee on Rural Development, Committee on Agriculture and Forestry, U.S. Senate: *Library Services in Rural Areas,* October 5, 1973; *Getting a Hand on Rural Development: The Colorado Approach,* October 10, 1973; *Where Will All the People Go?* October 23, 1973; *Training and Employment Programs Serving Rural America,* October 31, 1973; *Proceeding of a National Conference on Rural Development* (November 25–27, 1973), January 2, 1974; *The Transportation of People in Rural Areas,* February 27, 1974; *Rural Industrialization: Prospects, Problems, Impacts and Methods,* April 19, 1974.

23 David Freshwater, ''The Historical Context of Federal Rural Development Policy,'' *Western Wire.* Western Rural Development Center, Oregon State University, Corvallis, Oregon, 1991.

24 William P. O'Hare, *The Rise of Poverty in Rural America.* Population Trends and Public Policy Report No. 14, Population Reference Bureau, Inc., Washington, DC, July 1988; David L. Brown et al., ed., *Rural Economic Development in the 1980s: Prospects for the Future,* Rural Development Research Report No. 69, Economic Research Service, U.S. Department of Agriculture, Washington, DC, September 1988.

25 Linda L. Swanson and Margaret A. Butler, ''Human Resource Base of Rural Economies,'' in David L. Brown et al. (ed.), *Rural Economic Development in the 1980s: Prospects for the Future.* Rural Development Research Report No. 69, Economic Research Service, U.S. Department of Agriculture, Washington, DC, September 1988, pp. 159–179.

26 Scott Barancik, *The Rural Disadvantage: Growing Income Disparities Between Rural and Urban Areas,* Center on Budget and Policy Priorities, Washington, DC, April 1990, p. 10.

27 *The Delta Initiatives, Realizing the Dream . . . Fulfilling the Potential,* Final Report by the Lower Mississippi Delta Development Commission, Memphis, TN, May 1990.

28 For a broad reappraisal of rural development and some policy evaluations, see *Choices, The Magazine of Food, Farm and Resource Issues,* Second Quarter 1989, pp. 4–15.

29 R. K. Mahoney, Agricultural Economist, Agricultural Cooperative Service, U.S. Department of Agriculture, Washington, DC, has contributed to this section. See also R. K. Mahoney, "An Analysis of Off-farm Income to Illinois Farm Households," unpublished M.S. thesis, University of Illinois, 1987.

30 *Topical Law Reports, Labor Law Reporter,* California Par. 47, 200–47, 488, June 13, 1975.

TERMS AND CONCEPTS

Human capital
Supply and demand for human capital
Opportunity cost of human capital in agriculture
Two conditions necessary for economic progress in farming
Impacts of technological advance on farm income
Concepts guiding U.S. educational policy
Problems giving rise to public education
Impacts of public higher education for agriculture
Development of agricultural research and educational policy
Evolution of the land-grant university system
1890 institutions
Rural development policy
Rural poverty
Off-farm income of farm families
Hired farm work force
Effects of U.S. policy on the hired farm work force
Impacts of human capital investment

QUESTIONS AND STUDY SUGGESTIONS

1 Human resources may be visualized as a unique form of capital. Given this proposition, on what does the demand for human capital depend? On what does the supply depend?

2 What are the two conditions necessary for economic progress in agriculture? How can these two conditions be created?

3 Trace the evolution of the land-grant university system through public policies. What are the interrelations between land policy and education policy?

4 What were the primary concerns that undergirded the development of the U.S. public education and research system?

5 Why has rural development policy tended to emphasize the development of infrastructure rather than investments in human capital?

6 What are the trends in rural poverty? What are some of its causes, and how do the rural poor differ from the urban poor? What are the policy implications?

7 What are the policy alternatives for raising farm income per capita? What is the role of opportunity cost in determining the level of per capita income? Will increasing the productivity of labor tend to affect its demand, its supply, or both?

8 In general economic terms, explain why there is a continuing skewness in farm income distribution. What is required to decrease the skewness? On what grounds could such a policy be supported? On what grounds could it be criticized?

9 Why are migrant farm workers generally low-paid? In terms of supply and demand, what might be done to raise their incomes? Who might support such a policy? Who might oppose it? Why?

CHAPTER 16

LOOKING FOWARD TO THE TWENTY-FIRST CENTURY

As we look forward to the twenty-first century, some things are relatively certain, but most are uncertain. Sorting out the difference and learning how to use the best available knowledge to reduce the risks associated with the unknown is a challenge to the scholarship of policy and an opportunity for students of policy. You can help shape that future public policy for the food and agricultural sector, if you choose, by keeping informed and participating in the policy-making process. You can participate through one role or a combination of roles: citizen, leader of an interest group, professional educator or researcher, or elected or appointed policy maker.

Future food and agricultural policy will be determined by: (1) existing policy, (2) economic and social conditions, (3) the knowledge base of the participants, (4) values of the participants, and (5) relative political influence of the participants. You may contribute through the latter three, but the central purpose of your study of policy is to enhance your level of knowledge, which has been the focus of the previous chapters. Policy in the United States for the decade of the 1990s will be firmly rooted in the Food, Agriculture, Conservation, and Trade Act of 1990. This policy affects the national economy directly, and it will have effects around the world. Consequently, it is our point of departure in looking forward to the twenty-first century.

IN THIS CHAPTER

By studying Chapter 16, you can explore answers to the following questions:

1 What policies in the Food, Agriculture, Conservation, and Trade Act of 1990 are most important for future policy?
2 What are the implications of these policies for the national economy?
3 What are the long-term implications for the expanding world economy?

FOOD, AGRICULTURE, CONSERVATION, AND TRADE ACT OF 1990[1]

To understand current policy and its effects, you need to appraise: (1) the situation or the setting for policy, (2) the political process for making it, and (3) the various provisions embodied in it. This appraisal will provide a basis for analyzing the effects of policy on the national economy and on the links between it and the world economy.

Economic and Political Setting

Two forces dominated the economic setting for the 1990 Act: the recovery of the farm sector from the devastating financial crisis of the mid-1980s, and the need to reduce the federal budget deficit. Farm income recovered in the late 1980s as a result of higher commodity prices combined with large deficiency payments. The larger payments were necessary to raise total returns from market prices, which were declining initially, to the target-price levels mandated in the Food Security Act of 1985. Higher prices for most farm commodities subsequently were generated by new limits on supply and by increases in demand. Supply was limited by an almost nationwide drought in 1988 and by record program-induced set-asides of 15 to 20 percent of the nation's total cropland. The quantity of exports demanded was increased by a substantial drop in the value of the dollar in international markets, and demand was increased by continuing crop production shortfalls in some other countries and by a massive export subsidization program. In addition, the farm population continued to decline by two to three percent per year, average farm size increased, increases in farm labor productivity continued to outpace increases in most other sectors, and off-farm income provided more of farm families' incomes.

The financial soundness of the farm sector was also helped by the refinancing of the federally sponsored farm credit agencies and federal direct lending programs, which was carried out largely through the Agricultural Credit Act of 1987. Even so, agricultural markets continued to have other financial problems, which some people felt could not be solved without ongoing government participation. Farm and agribusiness organizations were also exerting strong political pressure for improving the 1985 Act, not for changing policy directions or eliminating such policy.

The other important force that dominated the economic setting was the growing imperative to reduce the federal budget deficit. The urgency of the situation interrupted the final stages of the 1990 policy-making process. During the spring and summer of 1990, a prolonged summit of Congressional and Administration leaders made substantial cuts in projected costs of all food and agricultural programs in order to meet previous budget-deficit mandates. These cuts eliminated the increases initially planned for food stamps and other food-aid programs and reduced farm program outlays planned for the five years by about 25 percent for all commodity programs. The budget pressure prompted policy makers to resist raising target prices and price-support levels, and it also allowed them to adopt "flexibility" as a program approach to reduce Treasury costs and farmer benefits.

Four factors dominated the political setting: the politically determined duration of the 1985 Act, the off-year Congressional elections, the GATT negotiations, and publicly inspired environmental concerns. Because the Food Security Act of 1985 was due to expire on September 30, 1990, the policy-making cycle was scheduled to occur in that year, a politically inopportune time because of the pending Congressional elections and the uncertainty of the continuing GATT negotiations. The Bush Administration first sent out political feelers simply to extend the 1985 Act for one or two years, but Congress had the momentum to proceed on schedule. Despite the concerns about developing policy during election campaigns, particularly for the 11 of 19 Senate agriculture committee members up for reelection, the campaigns had little noticeable effect on the process of making the 1990 Act or on its substance.

Even though the prolonged and increasingly tense GATT negotiations remained outside the core of 1990 policy development, the possible outcome of the negotiations was an unknown for policy makers. Troubles in GATT may have strengthened the position of U.S. leaders who supported continuing an aggressive export subsidy program. In fact, these troubles did trigger provisions for conditional increases in export subsidies in the 1990 Act, provided there was no general agricultural trade agreement by 1992. Finally, continuing national concerns about some of the environmental effects of farming and about the effects of technology on food quality and safety filtered through the political processes to produce several new programs in the 1990 Act.

Alternative Policy Proposals

In contrast to some earlier policy-making cycles, the 1990 period did not call forth the usual proposals from prominent political leaders for eliminating all market interventions or mandatory production controls or for substituting a more welfare-oriented program for farmers. Instead, the 1985 policy was by consensus the starting place for discussion among the Administration, both political parties in both Houses of Congress, and most interest groups. The mood was to improve the existing policy, not to replace it. The first proposal, introduced by the Congressional majority leadership (Democratic Party), was the text of the expiring 1985 Act. Party members anticipated that, during the Congressional committee decision-making process, appropriate amendments would be made to improve the Act.

The second proposal, again to improve the existing policy, was advanced by the secretary of agriculture as the *1990 Farm Bill Proposal of the Administration.*[2] Its primary thrust was more flexibility and greater dependence on market forces. It called for less reliance on land retirement with reduced multiyear set-asides and for more planting flexibility. It proposed to lower the target prices and loan rates of the 1985 Act, replace federal crop insurance with a more limited crop-disaster program, continue domestic food-aid and export subsidy programs, extend the conservation reserve with an added water quality provision and a cap at 40 million acres, shift federal farm credit assistance to guaranteed private funding, and increase agricultural research through a national science initiative.

A third proposal was advanced by a new coalition of 11 nationally prominent private environmental and consumer organizations. Released early in 1990, the proposal was widely publicized through the media. A colorful, professionally designed publication set forth a broad agenda for farm, food, and natural resource policy.[3] The agenda was aimed at reducing agricultural pollutants, strengthening the conservation provisions in the 1985 Act, establishing an Environmental Stewardship Program, promoting sustainable farming, and encouraging the development of markets for organic food.

A fourth proposal emerged from a self-styled urban coalition of 19 members of Congress. Their goal was to reduce, if not eliminate, the government's agricultural price and income programs by lowering the maximum Treasury payment per farmer and by limiting the production eligible for deficiency payments. The proposal attracted at most 38 percent of the votes of the House.

Each of these alternative policy proposals had some effect on the eventual compromises in the 1990 Act, but the final version clearly extended the major aims of the 1985 Act. Hence, the final 1990 Act continued the slow evolution of public involvement in the private allocative and distributive transactions of the food and agricultural sector of the U.S. economy.

A Continuing Evolution with the 1990 Act[4]

As we discussed in Chapter 3, the 1990 Act continued most provisions of the 1985 Act, but froze target prices, reduced loan rates more gradually, and allowed more flexibility in planting. Also it gave increased attention to food safety, water quality, rural development, forestry, sustainable agriculture, and research support. Again, social concerns, health factors, ethical considerations, and other noneconomic values emerged. The estimated cost for commodity programs was $55 billion, compared with the cost of $79 billion for the 1985 Act projected during its development, both bills for five years. However, even this lower cost was too much to include in the federal budget as the longstanding deficit-reducing budget resolution called for more than $13 billion in cuts in the estimated costs of the total bill, mostly in the farm programs.

Summary of Primary Changes in the 1990 Act

In summary, the 1990 Act was the most comprehensive in the six-decade evolution of U.S. food and agricultural price and income policy. It encompassed 25 titles, six being new for forestry, fruits and vegetables, grain quality, organic products, rural development, and global warming. No substantive program was dropped, but several new ones were added and program changes made.

A synopsis of the titles of the 1990 Act, as well as a comparison with the 1985 Act, indicates the broad comprehensiveness of the Act:

- Five-year duration of the 1990 Act, the same as the 1985 Act
- Domestic food aid (food stamps, WIC, and so forth) generally the same level
- Foreign food aid (P.L. 480) also the same funding, with administrative changes to delegate some programs to the State Department and make aid more flexible to meet emergencies
- Target prices generally frozen at previous levels, compared with annual declines specified in the 1985 Act

- Price support loans generally 85 percent of recent average market prices, compared with 75 to 85 percent for grains in the 1985 Act, but with the same discretionary annual reductions up to 20 percent
- Oilseeds with new marketing loan and a higher $5.02 per bushel level for soybeans
- Dairy price support frozen at $10.10 per cwt. for fluid milk, compared with declines in the 1985 Act
- Production control with similar discretionary annual set-asides by means of the acreage reduction program (ARP)
- Deficiency payments reduced for the final two years by a new computation
- Payments also reduced by the new, no-deficiency payments, 15 percent triple base, which mandated a new set-aside for program compliance
- More flexibility by an additional but optional, no-deficiency, 10 percent set-aside
- Grain quality improvement as the focus of new standards and mandated tests
- Export programs the same, with escalation of subsidies possible in the event of a GATT failure
- Conservation Reserve with a new 40 million-acre minimum and water quality programs
- Pesticide use as the focus of new record-keeping requirements
- Reforestation as the focus of new private and public initiatives
- Rural development as the focus of new programs and funding initiatives
- Organic foods certification as the focus of new national standards
- Global warming as the focus of a new title and a mandated study
- Research funding increased for competitive, formula, and sustainable grants for food and agriculture

Compared with existing policy, the 1990 Act instituted three changes that may have strong implications for the future. The first change was *flexibility,* which became a hallmark of acceptability for some groups in policy direction. Flexibility was a priority of the Bush Administration and it also came from diverse sources such as farmer organizations and environmental groups. Flexibility carries various values and meanings and has a wide range of possible consequences, such as economic allocative efficiency, greater freedom of choice, higher production, lower prices, less cost to the Treasury, and more economic uncertainty.

Under the 1985 Act, price-support loan rates for most program crops were permitted to decline by a formula using 75 to 85 percent of an adjusted five-year moving average of market prices, and target prices were reduced an average of two percent annually. To halt these reductions and prevent inflation from cutting deeper into the farmers' safety net, the 1990 Act raised the minimum formula base for price-support loans to 85 percent of the moving price base and froze the target-price levels. To save money and keep production flowing for exports, two planting flexibility provisions were launched. One was the triple base, 15 percent set-aside of a farm's cropland to qualify for income benefits, but without deficiency payment protection against low prices. The other was the additional option of a 10 percent set-aside, also without deficiency payments. "Triple base" refers to the acreage permitted under this provision, which was in addition to the total farm base and the base associated with the minimum set-aside in the acreage reduction program (ARP). The tradeoff would be less costly to the Treasury and would give producers more options for seeking optimum resource allocation. Income uncertainty was apt to be greater, however.

The second change included new initiatives in environmental and food quality programs. These were in response to pressure from environmental, conservation, and food interest groups, the emerging concerns of which became imperatives for the 1990 Act. This change followed in the wake of the historic erosion-control programs launched in the 1985 Act with the voluntary Conservation Reserve Program (CRP). Under this program, 34 million acres had already been withdrawn from production for 10 years. In addition, compliance with conservation provisions for erodible cropland was required for any farm program benefits. The public had clamored for more protection of the soil, the water supply, and food quality.

The trade-offs were for more protective regulations and increased Treasury incentives, including a higher minimum of 40 million acres for the CRP; strengthened sodbuster and swampbuster prohibitions, which were intended to slow down the conversion of permanent pasture and swampland to cropland; expansion of the national wetlands preserve; new incentives for 30-year public easements on farm wetlands; cost sharing for remedial practices for up to 10 million acres of farmland susceptible to polluting ground water; and mandated record keeping for all applications of hazardous chemicals.

The third change involved six new titles in the 1990 Act. These launched several program initiatives, including forestry improvement efforts for both private and public owners; more organizational and financial support for rural development; strengthening of market grain quality; new national standards for organic labeling; and new research attention to global warming.

IMPLICATIONS FOR THE NATIONAL ECONOMY

The 1990 Act continued U.S. food and agricultural policy on an evolutionary course, with no abrupt or distinct change from the past. The provisions for target prices maintained the pattern of farm income stability in nominal terms, although real farm income would not be stabilized against a strong inflationary surge or spiral.

The combination of target prices and food-aid programs reaffirmed a minimum safety net for farmers and low-income consumers, but only if budgeted expenditures keep pace with inflation. Federal budget cuts, rising input prices, and inflation could erode the purchasing power of both farmers and consumers, especially smaller full-time family farmers and consumers who rely on food aid for a major part of their support. Also, as noted in Chapter 12, budget cuts could weaken soil conservation programs. Unlike the U.S. Social Security System, which is indexed for inflation, the farm income supports and some food programs are not indexed. A volatile, delicate balance between the growth in food supply and the growth in domestic demand for food plus exports will result in more price uncertainty than existed under most of the earlier acts.

At the same time, the 1990 Act failed to address the fundamental problem of farm income security related to crop failures due to adverse growing conditions. Changing the actuarial basis for the federal crop insurance program would help resolve this problem (see Chapter 14).

More generally, the income-enhancing programs embodied in the 1985 and 1990 Acts, combined with the refinancing and restructuring of federal credit agencies in the Agricultural Credit Act of 1987, have contributed substantially to the stability of agriculture in the 1990s. However, these credit agencies are more economically vulnerable than most other more diversified lenders because they depend on a single industry, agriculture, and more specifically the farm sector.

Although the environmental program initiatives in the 1985 and 1990 Acts launched major efforts to reduce farmland soil erosion and began to focus on water quality, they approached the problems indirectly, sometimes primarily serving the goals of supply control. Much more reliable information is needed to protect underground water supplies. However, new approaches to integrated pest management, combining a professionally recommended mix of crop rotations, fertilizer and pesticide applications, farm management, and tillage practices, promise to improve effectiveness and efficiency. Also, genetic improvements in many crops are developing at a very rapid pace, which increases some options for improving food quality and protecting the environment.

Finally, reliance on the several export-enhancement and promotion programs embodied in the 1985 and 1990 Acts, which are in practice forms of export subsidies, raises questions about the economic links between national policy and the global market. The 1990 Act strengthened the U.S. policy of remaining an aggressive, subsidizing exporter in the world food and agricultural trade unless GATT developments unmistakably move toward freer trade.

However, great uncertainty remains as we approach the twenty-first century. The world environment is indeed unsettled by long-range, large-scale population growth, unpredictable multilateral negotiations, the gathering clouds of a possible economic slowdown among nations undergoing revolutionary change, the revamping of European national sovereignties, the uncertain control and supply of fossil fuel resources, the conflicts among the developed agricultural exporting nations, and the widening economic gap between developed and developing nations. Although the threat of world war has been reduced greatly, the threat of nuclear proliferation is a recurring phenomenon, and worldwide peace and stability remain a distant goal.

We may be reminded that, from World War II until the early 1980s, U.S. agriculture became increasingly dependent on exports for its demand.[5] For example, in 1950, 10 percent of the total sales from farms was exported, while an all-time high of more than 30 percent was reached in 1981. Since then, foreign markets have grown slowly and erratically for U.S. farm products available for export. Slow growth in domestic demand—around one percent of total U.S. production per year—suggests agriculture's continuing and generally increasing dependency on exports. For the next several decades, the volume of exports demanded could continue to increase, but still exhibit as much variability as in the past. The influence of the United States will likely be of crucial importance in liberalizing world trade in agricultural commodities, but whether more liberal trade is achieved depends on the consistency of U.S. policy as well as the policies of other governments.

Any excess potential production beyond the 1.5 to 2 percent associated annually with increases in domestic and export markets will need to be balanced with federal programs of either production curtailment or demand enhancement. If they are not, U.S. farm prices and incomes are likely to be depressed. However, major changes in the world economic environment are always possible. For example, U.S. export demand would be increased by substantially liberalizing trading policies, slowing production growth in other exporting countries, or achieving more rapid income growth in developing countries, which can lead to larger food purchases by importing nations.

Let us consider what conditions are necessary if policy is to deal with such problems. How can the major trading nations move toward development of a more efficient and prosperous global market?

TOWARD A GLOBAL ECONOMY

In little more than half of a century, most of the U.S. food and agricultural industry has been transformed from a horse-and-buggy economy to an efficient system dependent on advanced, dynamic technology. Judging by our current technology and the rate at which scientific knowledge is expanding, we can conclude that the agricultural economy is capable of still greater advances in productivity and growth, given favorable political and economic conditions. Visitors to state fairs, farm progress shows, and agribusinesses can marvel at the advanced technologies and economics of management displayed. Yet, these advances and their economic benefits are far from being realized over most of less-developed countries.

What are the necessary conditions for a truly global food and agricultural economy? First, political stability and security are essential. To achieve this end, international institutions must be strengthened and governance shared so that conflicts can be resolved nonviolently, and mutual trust and respect among nations can develop. Sovereign survival requires that nations strive for self-sufficiency and that they have secure food supplies until this kind of world economic environment evolves.

Second, living standards must be raised above mere subsistence levels throughout the world. To do so, societies must be given the knowledge and opportunities to balance their population growth with resources under sustainable management.

Third, trade barriers must be relaxed. The critical economic gains in food, fiber, and timber needed to improve living standards are best realized through the worldwide flow of goods and services according to comparative economic advantages. Continued international negotiations to remove the numerous trade barriers in all countries, particularly in the developed countries, will enhance this trade in food and fiber. Relaxing these impediments to freer trade will inevitably help to increase the growth of food supply per capita. The United States can continue to share in these expanding agricultural markets, provided it too is willing to open its markets to the agricultural and nonagricultural goods of other nations. Since World War II, negotiations under the GATT have been designed to liberalize trade in all sectors, including international trade in food and other agricultural commodities. For decades, however, failures to do so have distorted the world's food and agricultural economy, impoverished major sectors of it, and retarded its economic development. These conditions can be corrected in a few years if political agreements can be reached and sustained. Hence, with greater international political stability and more balanced population growth, there is hope that continuing efforts to correct them will eventually bear fruit. The United States has already provided strong leadership in these directions and can do more. Eventually all nations can do likewise.[6]

Fourth, the economic growth of the less-developed and developing nations must be enhanced if they are to be stronger commercial food importers, and also gain as agricultural exporters. Most of the developed countries have much to gain from such development and growth.

Economic Vision and Policy Implementation

If the policies for a truly global economy are to be implemented, several things will have to occur, such as: (1) increased financial support for the United Nations; (2) more flexibility in trade negotiations; (3) food aid and other help for displaced people; and (4) more efforts among nations and organizations to promote world peace. As part of this suggestion, there will be times when it may be especially important for the United States, the EC, and Japan to come to the aid of other major nations suffering from political and economic instability.[7] Yet, an international perspective must not overshadow concerns for domestic problems and the importance of domestic policy, which is presently the primary effective response of citizens.

The package of U.S. problems related to policy implementation in the twenty-first century is exacerbated by a combination of short-term cyclical problems and long-term structural problems.[8] The federal policies of the 1980s and early 1990s created a huge federal debt, which limits the opportunities to use fiscal policy to stimulate the economy, and at the same time restore or build the public infrastructure that is necessary to service a highly dynamic, technologically advanced industrial nation. One of the complicating consequences has been a significant slowing down of the rate of productivity for the economy generally.

This problem hangs over the campaigns of politicians competing for national offices and also many state and local offices. It was the dominant factor in limiting the federal financial support for the 1990 Act, and it will pervade all levels of food and agricultural policy well into the twenty-first century. During this time, if political choices are made to stabilize or reduce the federal debt, this action could be deflationary in general and diminish federal financial support for all food and agricultural programs. Furthermore, this would continue to strain the budgets of state and local governments, and it would retard the development of more adequate publicly financed infrastructure. The federal government, in contrast to state and local governments, can run substantial budget deficits indefinitely. But, if it does, and these deficits continue to increase, then the ultimate result is likely to be rampant inflation or slower economic growth.

Major issues in food and agricultural policy will be vitally affected by the choices that are made in monetary and fiscal policy. Agricultural price and income policy, agricultural credit policy, and food-aid policy are perhaps the most closely linked. In addition, substantive problems characterizing the resource, input, output, and consuming markets of the food and agricultural sector are likely to continue, and desirable policy responses will be sought. For example, public issues in conservation and environment policy, food quality, human capital policy, and economic developmental policy for agriculture and rural communities will continue unabated. Similarly, the long-term public support for scientific research and development of the food and agricultural economy will continue to confront society. In these areas, the long-term effects of policy tend to overshadow the short-term, and prudent policy choices will be made in the public policy process based on the best available knowledge of probable effects.

LOOKING AHEAD

Food and agricultural policy, carefully considered, and prudently developed and supported, will continue to play a crucial role in the twenty-first century, and the welfare of human society will continue to be affected in the most vital ways. That is why the subject of food and agricultural policy is important, and why continuing study and political activity will be important throughout your life. Improvement in our performance, in turn, depends on how well informed our citizens and policy makers are about policy alternatives and processes. Your knowledge, commitment, and participation in food and agricultural policy can help your country and others to be still more constructive forces in the new global economy of the twenty-first century.

ENDNOTES

1 Portions of this section also appeared as an authored paper by R. G. F. Spitze in *Agricultural Economics,* 7: 1992, pp. 125–139. Elsevier Science Publishers, B. V. Amsterdam.

2 U.S. Department of Agriculture, *1990 Farm Bill Proposal of the Administration,* Office of the Secretary of Agriculture, Washington, DC, 1990.

3 Environmental Coalition, *Farm Bill 1990—Agenda for the Environment and Consumers,* Island Press, Washington, DC, 1990.

4 U.S. Congress, House of Representatives, *Food, Agriculture, Conservation, and Trade Act of 1990,* House Report 101–916, U.S. Government Printing Office, Washington, DC, 1990.

5 Jerry Sharples and Tracy Hart, ''U.S. Farm Output: It's Export Dependent,'' *Choices: The Magazine of Food, Farm, and Resource Issues,* Second Quarter, 1991, pp. 22–23.

6 John A. Schnittker, Dermot Hayes, and William H. Meyers, ''GATT Negotiations: It's Important to Have a Deal on Agriculture,'' *Choices: The Magazine of Food, Farm, and Resource Issues,* Fourth Quarter, 1990, pp. 20–21.

7 Vernon W. Ruttan, ''An Emerging Vision of Global Futures,'' *Challenge, The Magazine of Economic Affairs,* 35(3), 1992, pp. 51 and 52.

8 The following discussion draws on previous discussion in this book and on a series of articles on''Choices for the 21st Century,'' *Choices, The Magazine of Food, Farm and Resource Issues,* Third Quarter, 1992.

TERMS AND CONCEPTS

Terminating policy
1990 Act
Program flexibility
Target price freeze
Congressional conference committee
Triple base

Environmental and consumer coalition
Bill mark-up
Congressional committee hearing
Organic labeling certification
Global warming
Pesticide record keeping

QUESTIONS AND STUDY SUGGESTIONS

1 In your view, what did the 1990 Act accomplish, and what did it fail to do?
2 Was the period for development of the 1990 Act politically inopportune? Why or why not?
3 How was the economic setting for the development of the 1990 Act different from that of the 1985 Act?
4 What effect did budget-deficit agreements have on the provisions of the 1990 Act? Did these constraints materially weaken the Act? Why, or why not?
5 How did the ongoing GATT negotiations relate to and affect the outcome of the 1990 policy development? Suppose that major GATT agreements on trade do not hold? What are the major implications?
6 What is the scope of provisions encompassed in the 1990 Act? What are the primary differences from previous policy?
7 What food and agricultural related problems were not directly addressed by the 1990 Act? What are some alternative policies to deal with these problems in the future?
8 What conditions are necessary for developing a more desirable global market for agricultural commodities?
9 Why have some agricultural and food policy makers resisted supporting the basic goals of free trade? Which commodity groups would be benefited by free trade? Which groups might be harmed?
10 What economic factors will have the greatest influence over future agricultural and food policy?
11 What part or parts can you play in developing future global agricultural and food policy?

GLOSSARY

The vocabulary of food and agricultural policy includes a large number of special terms and phrases, many of which have had unique meanings attached to them. This glossary has been developed to provide a convenient, alphabetically arranged list of many of these terms and phrases. Hence, it can be used to assist in identifying and understanding them. Careful study of the glossary will aid in visualizing the great breadth of food and agricultural policy, and such study may also provide additional insights for understanding and analyzing this policy.

Abbreviations as commonly used in food and agricultural policy.

AAA Refers to the Agricultural Adjustment Act of 1933, and of 1938, or the federal Agricultural Adjustment Administration of the 1930s
AAM American Agriculture Movement
AC average cost
ACP Agricultural Conservation Program
AFBF American Farm Bureau Federation, also the AFBF board, as distinguished from Farm Bureau state and county organizations.
AMS Agricultural Marketing Service
APHIS Animal and Plant Health Inspection Service
AR average revenue
ARS Agricultural Research Service
ASCS Agricultural Stabilization and Conservation Service
BC Bank for Cooperatives
BIFAD Board of International Food and Agricultural Development
BLM Bureau of Land Management
CAP Common Agricultural Policy
CBOT Chicago Board of Trade
CCC Commodity Credit Corporation

CFTC Commodity Futures Trading Commission
CGIAR Consulting Group on International Agricultural Research
CPI Consumer Price Index
CSE consumer subsidy equivalent
CSFP Commodity Supplemental Food Program
EC European Community (formerly **EEC** European Economic Community)
EPA Environment Protection Agency
ERS Economic Research Service
FAO Food and Agriculture Organization, also UNFAO (United Nations Food and Agriculture Organization)
FAPRI Food and Agricultural Policy Research Institute
FCA Farm Credit Administration
FCIC Federal Crop Insurance Corporation
FCS Farm Credit System
FDA Food and Drug Administration
FDIC Federal Deposit Insurance Corporation
FICB Federal Intermediate Credit Bank
FIFRA Federal Insecticide, Fungicide, and Rodenticide Act
FmHA Farmers Home Administration
FLB Federal Land Bank
FLBA Federal Land Bank Association
FNS Food and Nutrition Service
FOR Farmer-Owned Reserve
FRS Federal Reserve System, commonly called "The Fed," which may also refer to the FRS Board of Governors
FSIS Food Safety and Inspection Service
GATT General Agreement on Tariffs and Trade
GDP gross domestic product
GNP gross national product
IFPRI International Food Policy Research Institute
IPM Integrated Pest Management
JTPA Job Training Partnership Act
LISA low-input sustainable agriculture
MC marginal cost
MR marginal revenue
MVP marginal value product, or value of the marginal product
NFU National Farmers Union
NSLP National School Lunch Program
NTB nontariff barrier, or **NTBs** nontariff barriers
OECD Organization for Economic Cooperation and Development
PAC political action committee
PIK payment in kind
PL public law
PSE producer subsidy equivalent
RUSLE revised universal soil loss equation
SCS Soil Conservation Service
UNCTAD United Nations Conference on Trade and Development
UNFAO United Nations Food and Agriculture Organization
USDA United States Department of Agriculture, also the Department of Agriculture
USDC United States Department of Commerce

USDI United States Department of the Interior
USDL United States Department of Labor
USDS United States Department of State
USLE universal soil loss equation
VAT value added tax
WEPP water erosion prediction program
WIC Special Supplemental Food Program for Women, Infants, and Children

acre The basic unit of measurement for agricultural land in the United States and Canada, equal to an area containing 43,560 square feet or a strip of land one rod in width (16-1/2 feet, or 5-1/2 yards), and one-half mile long (2,640 feet, 880 yards, or 160 rods). A section of land one mile square contains 640 acres. The U.S. Ordinance of 1785 established these measurements for surveying unsettled land into standard townships of 36 sections, usually six miles square. An acre of land is equal to 0.4047 hectares, or one hectare equals 2.471 acres. The hectare is the basic unit of measurement for agricultural land in most countries other than the United States and Canada. See **weights, measures, and conversion factors.**

acreage allotment An individual farm's share, based on its previous acreage, of the national acreage needed to produce a targeted amount of a particular crop; currently used only for tobacco. Also see **base acreage.**

Acreage Limitation Program New in the Agriculture and Food Act of 1981, this program (when in effect) required participants who voluntarily complied in return for a payment to limit the acreage of a crop to a specified portion of their base acreage, and divert the remainder of the base to a generally nonproductive soil-conserving use. Later called **acreage reduction program (ARP).**

Acreage Reduction Program (ARP) A program relying on voluntary participation of a farmer to reduce the amount of crop planted below the farm's base acreage to qualify for price supports and target prices for that crop, if such a program is in effect for that crop.

advanced deficiency payments Payments made to crop producers when they sign up for federal commodity programs. The secretary of agriculture is required by law to make the payments when an acreage reduction program is in effect and deficiency payments are expected to be paid. Advance deficiency payments can range from 30 to 50 percent of expected payments, depending on the crop. Up to 50 percent of the advance payment may be made as generic commodity certificates. If total deficiency payments are less than the advance amount, producers must refund the excess portion.

adverse selection A term applicable to insurance contracts when insurable clients who have a high probability of collecting an indemnity are attracted to the program more often than those with a lower probability. Adverse selection is encouraged when the insured have more accurate knowledge than the insurer concerning the probability of an indemnity.

aggregate consumption The sum of all consumption of final goods in an economy.

aggregate demand The sum of all the individual demands for the output of the economy.

agribusiness Firms and industries engaged in producing and distributing agricultural inputs or in marketing, processing, or distributing food and other agricultural commodities.

agricultural adjustment A term generally referring to programs designed to regulate or control agricultural production and marketing. The Agricultural Adjustment Act of 1933 created the Agricultural Adjustment Administration in the Department of Agriculture. Since then adjustment programs have been implemented through similar agencies under various names.

agricultural association A nonprofit or cooperative association of farmers, growers, or ranchers incorporated or qualified under applicable state laws to recruit, solicit, hire, employ, furnish, or transport migrant or seasonal agricultural workers.

Agricultural Conservation Program (ACP) A program in which producers agree to carry out specified conservation practices on their farms and receive payments to help pay part of the cost.

Agricultural cooperatives Agricultural cooperatives—or farmers' cooperatives, as they may be called—are either federated or centralized types of organizations, although a few cooperatives are a mixture of both. Under a federated organization individual cooperatives join together to create a business superstructure for the purpose of more coordinated and efficient operation. The individual cooperatives own and control the federated super structure rather than vice versa. Members elect the board of directors of the federated cooperatives through the local organizations. Under a centralized organization the parent cooperative owns or controls various satellite or local cooperatives. Members vote directly for the board of directors, which controls the total organization.

Agricultural Cooperative Service (ACS) A USDA agency that provides technical and educational assistance to agricultural cooperatives to improve their organization and operation, and to guide further development. ACS also provides management analysis and conducts research, which may be published as agency educational materials or in the publication *Farmer Cooperatives.*

agricultural employment Any labor or management activity related to farming or agribusiness. Farm employment includes cultivating and tilling soil, dairying, and cultivating, growing, and harvesting agricultural, aquacultural, or horticultural commodities. Labor associated with handling, planting, drying, packing, packaging, processing, freezing, or grading prior to delivery for storage of any agricultural or horticultural commodity in its unmanufactured state is also considered agricultural employment.

agricultural inputs Components of agricultural production, such as land, labor, and capital to purchase other inputs, including machinery, fertilizer, seed, and pesticides.

Agricultural Marketing Service (AMS) A USDA agency that administers and enforces regulatory laws to make marketing more orderly and efficient. AMS's responsibilities include establishing standards for grades of cotton, tobacco, meat, dairy products, eggs, fruits, and vegetables; operating grading services; and administering federal marketing orders and agreements. The agency also operates the Federal–State Market News Service, in cooperation with government agencies in 44 states, the District of Columbia, and three territories, reporting up-to-the-minute information on prices, supply, and demand for most agricultural commodities.

Agricultural Research Service (ARS) A USDA agency that conducts basic, applied, and developmental research of regional, national, and international scope. ARS conducts research in the areas of livestock; plants; soil, water, and air quality; energy; food safety and quality; nutrition; food processing, storage, and distribution efficiency; nonfood agricultural products; and international development.

Agricultural Stabilization and Conservation Service (ASCS) The agency in the U.S. Department of Agriculture that administers the farm price and income support programs as well as some conservation and forestry cost-sharing programs. In addition to the office in Washington, DC, there are offices in each state and most counties.

alternative crops Generally any crop other than a crop under acreage allotment that a farmer may be permitted to grow on set-aside acres.

alternative farming Production methods other than energy- and chemical-intensive one-crop farming (monoculture). Alternatives include use of animal and green manure rather than chemical fertilizers, integrated pest management to reduce reliance on chemical pesticides, reduced tillage, crop rotations especially with legumes, alternative crops, or diversification of the farm enterprise.

Animal and Plant Health Inspection Service (APHIS) A USDA agency that conducts regulatory and control programs to protect animal and plant health.

appropriations bill On becoming law, this legislation gives legal authority to spend or obligate Treasury funds for specific purposes. Appropriations bills provide part or all of the funds approved in authorization legislation.

aquaculture Producing aquatic plants, fish, or animals in a controlled environment, such as ponds, streams, tanks, or cages, for all or part of their life cycle.

aquifers Subterranean water-bearing rock formations that can be tapped for wells. Aquifers supply water to about 50 percent of the U.S. population.

assessment programs Programs requiring producers to pay a fee per unit of production in order to share program costs with the government. For example, to be eligible for price supports under the 1989 flue-cured tobacco program, producers were required to contribute one cent per pound of tobacco marketed to an account to ensure the operation of the program at no net cost to taxpayers. Flue-cured tobacco buyers were also required to contribute one cent per pound. In addition, both producers and buyers had to contribute 0.12 cent per pound each for a budget deficit assessment, instead of a 1.4 percent reduction in price support. For other types of tobacco under the price support program, growers paid the full assessment. Most of the costs of such assessment programs is transferred to consumers through higher market prices.

assets Cash and other property that has a market value.

balance of payments Statement of economic transactions involving the exchange of goods, services, and capital claims between a country and its trading partners. A trade deficit occurs when funds leaving a country for imported goods and services exceed money received from other countries for exports of other goods and services.

balance of trade The difference between the value of exports and imports. The balance is positive when exports exceed imports, and negative when imports exceed exports.

balance sheet of agriculture An accounting statement showing the total value of land and other property that farmers and some agribusinesses own, the amount of debt that farmers and the identified agribusinesses owe, and the difference between the two, usually called the net worth.

base acreage The acreage of a crop on a farm used in acreage limitation programs, and to calculate farm program acreage. It is usually the actual planted acreage of the crop on the farm in the previous year, although the previous two- or three-year average could be used. It may be adjusted for disasters, crop rotations, and other factors as necessary to make it fair and equitable. See **crop acreage base.**

basic commodities In the price-support legislation of the 1930s and 1940s and again in the 1980s, six agricultural crops were designated as basic commodities—corn, cotton, peanuts, rice, tobacco, and wheat—and specific price-support programs were designated for these crops.

basic loan rate See **nonrecourse loans.**

Beef Promotion and Research Act Part of Title XVI of the Food Security Act of 1985 that provides for a check-off and promotion program for beef.

benefit–cost analysis A process used for economic appraisal of a project or program. It consists of adding up all benefits and costs of the project to society, adjusting them by using a discount rate to reflect the opportunity cost of the invested funds, and computing the value received from the project. The discount rate reflects either: (1) the assumptions of the researcher about the preferences society has for consumption today over the future, (2) the amount that could have been earned if the funds had been invested elsewhere (opportunity cost), or (3) some combination of each.

bilateral trade agreement A trade agreement between any two nations. The agreement may be either preferential, applying only to the two countries involved, or most-favored-nation, negotiated between the two countries but extending to all or most other countries that comply with the terms of the agreement between the two nations.

biotechnology Use of microorganisms, plant cells, animal cells, or parts of cells to produce new products, or to carry out life processes.

blended credit A financing plan for export sales in which government credit guarantees or government credit at lower interest rates is blended with regular commercial credit to provide lower interest rates and more favorable terms for foreign buyers.

Board for International Food and Agricultural Development (BIFAD) An organization created by the federal government to advise officials managing foreign assistance programs regarding the use of U.S. land-grant and other qualified universities to support agricultural development abroad.

bovine growth hormone (bST) Bovine Somatotropin. See **growth hormones**

buffer stocks Supplies of a product stored on farms or in commercial elevators or warehouses to moderate extreme price fluctuations by assuring a more stable supply. Buffer stocks are usually controlled by government while total stocks include both government and privately held stock.

Cairns Group An informal group of nations formed in 1986 in Cairns, Australia, seeking removal of access barriers and substantial reductions in subsidies affecting agricultural trade. The original group included Argentina, Australia, Brazil, Canada, Chile, Columbia, Hungary, Indonesia, Malaysia, New Zealand, the Philippines, Thailand, and Uruguay. Much of the group's attention was directed toward the food and agricultural policies of the United States and the European Community.

call price The price level at which farmers with grain in the farmer-owned reserve must repay the nonrecourse loan plus any accumulated interest.

capital Includes human capital, all the skills, training, abilities, etc. exercised by human beings; physical things such as land and equipment; and other productive resources.

capital gains The difference between the acquisition price and a higher current market value of a capital item such as real estate and equipment. Capital gains may be realized by actual sale, or they may be an unrealized computation by assuming the capital items were sold at current market value.

cargo preference A policy that requires a certain portion of goods or commodities exported from the United States under some governmentally assisted programs to be shipped in American ships.

Caribbean Basin Economic Recovery Act More popularly known as the Caribbean Basin Initiative (CBI), this act eliminated duties on imports of some commodities from designated Caribbean countries until September 30, 1995. The CBI also provided for import relief to U.S. industries injured or threatened by increased imports from CBI countries.

carryover The supply or volume of an agricultural commodity not used at the end of a marketing year. It may continue to be stored, or it may be used during the following marketing year. Carryover may be referred to as end-of-year stocks.

cartel An alliance or arrangement among enterprises or countries in the same field of business, such as agricultural exporters for example, aimed at securing an international monopoly. A cartel usually seeks to control production or the amount marketed to raise prices and maximize profits.

cash flow The total funds generated internally by a firm for covering costs and investment. Farming presents unique cash-flow problems when income is generated only at the end of a production cycle; for example, with annual crops and with livestock, where crop failures occur, or if input prices rise faster than product prices.

cashing out Providing food assistance in cash rather than in commodities or food stamps. Eligible residents of Puerto Rico have received cash instead of food stamps since July 1982 under the Nutrition Assistance Program.

cash-out option for generic certificates The original holder of a generic commodity certificate has the option of redeeming the certificate at its face value for cash from the Commodity Credit Corporation instead of exchanging it for commodities. To encourage exchange of certificates for surplus commodities, generic certificates cannot be redeemed for cash until five months after the issue date. Those who purchase or trade the certificates from original holders are not permitted to cash-out the certificates. Certificates issued under the Export Enhancement Program also cannot be cashed out.

Census of Agriculture A survey taken by the U.S. Bureau of Census every five years to determine the number of farms, farm acreage, crop acreage and production, livestock numbers and production, farm spending, farm facilities and equipment, farm tenure, value of farm products sold, farm size, type of farm, and so forth. Data are reported for states and counties.

cereals Generic name for certain grasses that produce edible seeds. The term is also used for certain products made from seeds. Cereals include wheat, oats, barley, rye, rice, millet, corn, and sorghum grain.

Chapter 12 Signed into law in November 1986, the Chapter 12 bankruptcy reorganization was created to accommodate the special needs of financially strapped family farmers.

check-off See **commodity promotion programs.**

Child Care Food Program Provides cash reimbursement and USDA-donated foods, or cash equivalents, to nonprofit family and group day-care homes and day-care centers, based on the number of breakfasts, lunches, suppers, and snacks served.

child nutrition programs Include the National School Lunch Program, School Breakfast Program, Special Milk Program, Child Care Food Program, and Summer Food Service Program.

coalition A combination of organizations and groups working together to influence a single piece of legislation or government decision, such as a coalition of farm organizations for example.

coarse grains See **feed grains.**

cobweb If a market is initially not in equilibrium, according to price theory the pattern of prices and quantities as the market adjusts may look like a cobweb, depending on the elasticities of supply and demand.

commodities Broadly defined, any goods exchanged in trade, but usually used to refer to widely traded raw materials and agricultural commodities such as wheat or corn.

Commodity Credit Corporation (CCC) A government owned and operated corporation authorized to borrow funds from the U.S. Treasury to operate the U.S. Department of Agriculture's price and income support programs, to manage government owned stocks of agricultural commodities and administer their disposal through domestic and export programs. Most activities have been carried out by ASCS personnel, although certain programs have been administered and implemented through the Agricultural Marketing Service, the Foreign Agricultural Service, and the Food and Nutrition Service.

Commodity Credit Corporation sales price The designated price below which the CCC may not sell its stocks of wheat, corn, sorghum, barley, oats, rye, or cotton. If loan repayments are permitted at lower than loan rate levels, the sale price is usually higher than the average loan payment rates for these crops.

commodity distribution Direct donation of food products by the federal government to needy people, schools, and institutions. The commodities donated are generally in surplus, meaning supplies exceed market food demand at the given price, or prices are below desired levels.

Commodity Futures Trading Commission (CFTC) An independent federal commission that regulates trading on the 11 U.S. futures exchanges. The CFTC also regulates the activities of numerous commodity exchange members, public brokerage houses, commodity trading advisors, and commodity pool operators.

commodity options In the commodity markets, the option to buy or sell contracts of hogs, cattle, soybeans, corn, and wheat at specified prices.

commodity programs A collective term used to include the public price and income support programs for designated agricultural commodities.

commodity promotion programs Promotional activities sanctioned by the government, usually involving a per unit check-off at first stage of marketing and often managed by a commodity trade organization.

Commodity Supplemental Food Program (CSFP) A program initiated in 1968 to provide benefits to infants, children up to age six years, women during pregnancy and up to six weeks post partum, and breast-feeding women up to 12 months post partum. Under the CSFP, USDA purchases foods, such as fruit juice, beef, egg mix, and canned corn, and distributes them to eligible women, infants, and children through state and local agencies. USDA's Food and Nutrition Service determines the quantity of food required and makes allocations.

Common Agricultural Policy (CAP) A set of regulations by which member states of the European Community (EC) seek to merge their individual agricultural programs into a unified effort to promote regional agricultural development, fair and rising standards of living for the farm population, stable agricultural markets, increased agricultural productivity, and methods of dealing with food supply security. The variable import levy, government payments to EC farmers, and export subsidies to shippers, are the three principal elements of the CAP.

common market Usually refers to the EC, but may also refer to any other group of nations organized for similar purposes.

comparative advantage The condition in which a state or country produces and exports those goods and services that it can produce relatively cheaply and imports those that other countries can produce relatively cheaply. Absolute advantage implies that a country produces all goods and services more cheaply than any other country. Whether or not one of two states is absolutely more efficient in the production of every product than the other, if each specializes in the products in which it has the greatest relative efficiency, trade will be mutually beneficial to both states. The term also applies to trade among more than two countries. Comparative advantage is the economic foundation for trade theory, and one of the economic bases for trade negotiations under the GATT.

compensatory payments See **deficiency payments.**

concessional sales Credit sales of a commodity in which the buyer is allowed more favorable payment terms than those on the open market. May apply to exports of some agricultural commodities under government export programs.

conservation compliance A provision of the Food Security Act of 1985 that required farmers with highly erodible cropland to have an approved conservation plan by 1990 and have it implemented by 1995 to remain eligible for federal farm program benefits. Renewed and extended in the 1990 Act.

Conservation Reserve Program A voluntary federal program in which highly erodible land is retired from crop production. It has been designed to reduce erosion on 40 to 45 million acres of farmland. Under the program, producers who sign contracts agree to convert highly erodible cropland to approved conservation uses for 10 years. In exchange, participating producers receive annual rental payments and cash or payments-in-kind to share up to 50 percent of the cost of establishing permanent vegetative cover.

conservation tillage Any of several farming methods that provide for seed germination, plant growth, and weed control, yet maintain effective ground cover throughout the year and disturb the soil as little as possible. The aim is to reduce soil loss and energy use while maintaining crop yields and quality. No-till is the most restrictive (soil conserving) form of conservation tillage. Other practices include ridge-till, strip-till, and mulch-till.

conservation use An approved use of land for compliance with conservation programs that protects soil from erosion by planting to grasses, legumes or to small grains that are not allowed to mature. See **cover crop.**

considered planted In calculating base acreage, land is considered planted to a program crop if it was used for conservation acreage under an acreage reduction or set-aside program, or if the producer was prevented from planting the crop because of drought, flood, or other natural disaster. It may include any acreage on the farm that the secretary of agriculture determines is necessary to be included in establishing a fair and equitable crop acreage base.

Consultative Group on International Agricultural Research (CGIAR) A group sponsored by the World Bank, the United Nations (UN) Development Program, and the Food and Agriculture Organization of the UN to coordinate and fund a majority of the international agricultural research centers, such as the International Rice Research Institute in the Philippines and the International Maize and Wheat Improvement Center in Mexico.

consumer price index (CPI) A monthly measure of changes in the prices of goods and services consumed by families and individuals, published by the U.S. Department of Labor's Bureau of Labor Statistics. It is the most widely accepted measure of changes in consumer prices.

consumer subsidy equivalent An economic concept measuring the value to consumers of government food programs. It may be an estimate of the amount of subsidy that consumers would need to receive to maintain the same economic well-being if all consumer food subsidy programs were discontinued.

converted wetlands Wetlands that have been drained to produce agricultural commodities.

cooperative A form of business owned by the customers. There are many types of farmer-owned cooperatives that provide supplies and services or buy and sell agricultural commodities. The Capper-Volstead Act of 1922 exempted agricultural cooperatives from restrictions of the antitrust law provided they did not operate to restrain trade, and it established government support and assistance to agricultural cooperatives as a national policy.

corporate farm A farm business that is legally incorporated under state law. The stock may be held by a farm family, closely held and not available for public purchase, or listed on a public stock exchange. The term may be used incorrectly when referring to large farm operations when they are sole proprietorships or partnerships.

cost-of-production In farming, the average amount, in dollars-per-unit, to grow or raise a farm commodity including all purchased inputs and sometimes including allowances for management and use of land or other fixed assets owned by the farm operator. The cost may be expressed in units of a bushel, pound, ton, or per acre, depending on the commodity involved. Cost-of-production for agricultural commodities may vary widely depending on estimates of opportunity costs and other factors.

cost sharing In certain farm conservation programs, the federal government will share the cost of certain farm practices with the farm owner or operator.

cover crop A close-growing crop, such as a grass or legume, grown primarily to protect and improve soil between periods of regular crops, or grown in orchards and vineyards. May be voluntarily planted by a farmer, or planted to comply with a government conservation program. See **conservation use.**

crop acreage base Usually the crop acreage base is the average number of acres planted and considered planted to a program crop for harvest in the previous five years. Where the program crop was not grown on the farm in each of the previous five years, the law provides for procedures to develop a crop acreage base for that farm.

crop failure A cropland use classification term. Mainly acreage on which crops were not harvested because of weather, insects, and diseases, but also includes some land not harvested due to lack of labor, low market prices, or other factors. The term excludes acreage planted to cover and soil improvement crops not intended for harvest. May have significance for crop insurance or other farm-aid policy.

crop insurance See **federal crop insurance.**

crop loans See **emergency disaster loans.**

crop year The year in which a crop is harvested and marketed. Also see **marketing year.** A crop year may be part of one calendar year or it may overlap from one year to the next.

cross compliance The requirement that if a farmer wishes to participate in a price support program and qualify for price support and loans in that program, then that farmer must also meet the program provisions for other major program crops grown on that farm. Also see **off-setting compliance.**

dairy price support program The program established in the Agricultural Act of 1949. From 1949 until 1981, milk prices were supported by law at 75 to 90 percent of parity. Since 1981, several laws have set lower minimum support prices for milk. Prices are supported by government purchases of manufactured dairy products to maintain the minimum price for milk established by law.

debt/asset ratio A measure used to determine the financial soundness of a farming operation. Farmers whose debts are equal to 70 percent or more of their assets have been considered to be in some financial difficulty.

decoupling A proposed public price and income policy approach in which income support is separated from any support involving current prices or production.

deficiency payment A direct government payment to eligible producers when the average price received by farmers falls below the target price. The payment rate is usually determined by comparing the target price with the average market price, usually for the first five months of the marketing year. If the market price is below the target level, then eligible producers may apply to the ASCS for deficiency payments equal to the lesser of the differences between the target value and either the average market price or the nonrecourse price-support loan. Sometimes called a **compensatory payment.**

Delaney Clause The 1958 Food Additive Amendment to the Federal Food, Drug, and Cosmetic Act that prohibits the use of direct food additives that have been shown through appropriate tests to cause cancer in humans or laboratory animals. The clause implies a ''zero cancer risk'' standard for processed foods.

demand curve As an illustration of demand theory, this curve relates the quantity demanded to the market price. On a graph with price on the vertical axis and quantity demanded on the horizontal axis, the demand curve connects all the points showing a price and the quantity demanded at that price by an individual or group at a specified location and period of time. The demand curve is downward sloping because typically as the price of a good decreases, the amount of the good purchased increases. The market demand curve for a good is the horizontal sum of all individual demand curves for that good.

demand schedule The relationship between various prices of a good and the quantity demanded at each of the prices.

designated nonbasic commodities Commodities other than basic commodities for which the secretary of agriculture is authorized and directed to provide price support. These commodities have included tung nuts, soybeans, honey, milk, sugarbeets, sugar cane, and sometimes a few others.

developed countries A classification developed in the United Nations and the World Bank to refer to the most highly developed industrial countries, i.e., Australia, Canada, countries in Western Europe, Japan, South Africa, United States, and New Zealand. Others may be added as a designated threshold of industrialization and per capita income is reached.

developing countries All countries not defined above as developed, and often countries whose economies are largely dependent on agriculture and primary resources, and do not have a strong industrial base. The term is often used synonymously with less developed countries and underdeveloped countries.

dietary guidelines Guidelines developed by the USDA and the Department of Health and Human Services, emphasizing variety, balance, and moderation in the total diet without making recommendations regarding specific foods to include or to eliminate.

direct payments Payments in the form of cash or commodity certificates made directly to producers including deficiency payments, annual paid land diversion payments, or conservation reserve payments.

disaster payments Federal government payments to farmers participating in certain government programs when the farmers are prevented from planting their crops or when their crop yields are abnormally low. Payments may be mandated in areas where federal crop insurance is not available, where crop insurance is inadequate, or where such payments are necessary to avert an economic emergency.

discount rate The discount rate is the interest that the Federal Reserve charges on loans to banks to cover reserve shortfalls. If the Fed raises the discount rate, the cost to a bank of having a shortfall goes up. As a result, if the rate is raised, banks become more conservative in their lending and the money supply is reduced.

disposable income Earnings remaining after deductions of all necessary expenses and direct taxes on those earnings which is available to be spent, invested, or saved as the receiver chooses.

diversion payment Payments made to farm owners and operators for diverting land from certain crops into conservation uses.

diverted acres Under acreage reduction programs, those acres that are taken out of production and diverted to some conserving use.

double crop In some areas a second crop can be planted after the major crop is harvested. For example, soybeans may be grown after a winter wheat crop is harvested. Government farm programs may be designed to permit this normal cropping practice in certain parts of the country.

dumping Sale of products on the world market below the average cost of production to dispose of surpluses, or to gain access to a market.

economic costs Economic costs are payments that a firm must make, or incomes it must provide, to attract and keep resources away from alternative opportunities or lines of production and comply with public policies. These payments or incomes may be either explicit or implicit.

economic emergency disaster payments See **disaster payments.**

Economic Research Service (ERS) A USDA agency responsible for economic data and analyses and social science information needed to develop, administer, and evaluate food and agricultural policy and programs, as well as some other rural policies and programs.

economies (diseconomies) of scale The reduction in per-unit cost of production associated with an increase in optimal size of operation usually resulting from specialization, division of labor, and technology. Diseconomies involve an increase in per-unit cost of production associated with nonoptimal growth in size resulting from inadequate management or other factors.

elastic demand A market in which a percentage change in price will bring about a greater proportional change in the amount purchased with resulting total receipts being directly related to the latter and inversely related to price changes.

elasticity An economic concept that relates the percentage change in one variable, such as price, to the percentage change in another variable, such as quantity.

elements of policy The elements of policy are a perceived problem, examined alternatives, and a chosen course of action in view of the decision makers' goals, values, and evaluation of consequences.

eligible producer A producer who has become eligible for certain government benefits by complying with farm program requirements.

embargo A government-ordered prohibition of trade with another country restricting all trade or only that of selected goods and services.

Emergency Feed Assistance Program A program that has provided for the sale of CCC grain at 75 percent of the basic county loan rate to livestock producers whose feed harvest has suffered because of drought or excess moisture.

emergency disaster loans Loans made to farm producers under emergency credit programs, usually for conditions resulting from drought, floods, or other natural disasters.

equity The net worth of an individual farmer or business firm, the net value of property after all debts are deducted.

erodibility index A value that combines a soil's inherent erodibility with its susceptibility to damage by erosion. Soil type, amount of rainfall and runoff, and slope length and steepness determine a soil's inherent erodibility. The susceptibility of a soil to erosion damage is inversely related to the depth of the top soil, and to a soil's natural rate of formation. Cropland fields containing primarily soil with an erodibility index of eight or greater may be enrolled in the Conservation Reserve Program. Cropland with an erodibility index of eight or more is also subject to the conservation compliance provisions of the 1985 and 1990 Acts.

erosion The process in which water or wind moves soil from one location to another. Types of erosion are: (1) sheet and rill—a general washing away of a thin uniform sheet of soil, or removal of soil in many small channels or incisions, caused by rainfall or irrigation runoff; (2) gully—channels or incisions cut by concentrated water runoff after heavy rains; (3) ephemeral—a water-worn, short-lived, or seasonal incision, wider, deeper and longer than a rill, but shallower and smaller than a gully; and (4) wind—the carrying away of dust and sediments by wind in areas of high prevailing winds or low annual rainfall.

Eurodollar market The name given to the operations outside U.S. boundaries in which foreign banks and foreign branches of U.S. banks make loans and security transactions denominated in U.S. dollars rather than foreign currency.

European Community (EC) An organization established by the Treaty of Rome in 1957 and also known as the European Economic Community and the Common Market. Originally composed of six European nations, it has expanded to 12. The EC attempts to unify and integrate member economies by establishing a customs union and common economic policies, including the Common Agricultural Policy (CAP).

exchange rate The number of units of one currency that can be exchanged for one unit of another currency at a given time. A decline in the value of the U.S. dollar, for example, drops the "price" of U.S. farm products in terms of the currency of many importers and enhances exports.

export The goods and services that a country sells or otherwise supplies to other countries.

Export Credit Guarantee Programs (GSM-102 and GSM-103) U.S. agricultural export credit programs begun in 1982, which guaranteed repayments of private, short-term and long-term credit up to 10 years.

Export Enhancement Program (EEP) Under EEP, exporters are awarded generic certificates that are redeemable for CCC-owned commodities, enabling them to sell certain commodities to specified countries at prices below the U.S. market to meet competition from other subsidized exporters.

export sales reporting The USDA policy has required that export sales in one day involving more than 100,000 metric tons of major grains and oilseeds shall be reported to USDA within 24 hours of the sale. For other commodities, weekly reports have been required.

export subsidy A government grant, made to a private enterprise, for the purpose of facilitating exports. In Europe, it is often termed "restitution." The EC program as well as some U.S. programs have been major targets of protest by some other exporting countries. See **Cairns Group.**

Extension Service USDA's educational agency and one of the three partners in the Cooperative Extension System with state and local governments. All three share in financing, planning, and conducting educational programs.

externality (economic) An economic effect of any activity by an individual or business on others that is not reflected in the market for that activity. The economic effect can be detrimental or beneficial. For example, water pollution from the run-off of pesticides or fertilizer, or farmland erosion and sedimentation of streams and reservoirs, can be detrimental or a cost to the public, but not a part of the farmer's costs of production. However, some crops have externalities that are beneficial for the environment. For example, the growing of a legume such as alfalfa or sweet clover may be beneficial to honey bees, and improve the off-farm environment.

extra-long staple cotton Cottons having a staple length of 1-3/8 inches or more, according to the classification used by the International Cotton Advisory Committee. This cotton is also characterized by fineness and high fiber strength, contributing to finer and stronger yarns needed by thread and higher value fabrics. American types include American Pima and Sea Island cotton.

factor of production A resource that can be used in the production of economically useful commodities. Examples include labor, farm equipment, and land.

family farm A farm in which a family provides most of the labor, and management decisions. The land may be owned, partly owned, or rented. Most family farms would have annual sales between $40,000 and $300,000.

farm Starting in 1978, the Bureau of the Census defined a farm as any place that has or would have had $1,000 in gross sales of farm products.

farm acreage base See **base acreage.**

Farm Costs and Returns Survey A nationwide survey that includes, for a given year, data on costs of production, capital expenditures, debts and assets, earnings, production practices, and other farm business and operator characteristics. These data are used to update cost-of-production estimates. They are also used to measure annual farm income and to assess changes in the financial conditions of farms. The survey is managed by USDA's Economic Research Service (ERS) and conducted by the National Agricultural Statistics Service (NASS) and its 44 state offices. The survey covers every state except Hawaii and Alaska. Also, data from the survey may be used in making food and agricultural policy.

Farm Credit Administration (FCA) The government agency responsible for examination and regulation of the Farm Credit System.

Farm Credit System (FCS) The system is made up of cooperatively owned financial institutions in districts covering the United States and Puerto Rico that finance farm and farm-related mortgages and operating loans, some agribusinesses, and agricultural cooperatives. Institutions within each district specialize in farmland loans and operating credit, or lending to farmer-owned supply, marketing, and processing cooperatives. FCS institutions rely on the bond market as a source of funds. The bonds are usually guaranteed by the federal government and in some emergency situations part or all of the interest has been paid by the U.S. Treasury.

farm marketing quota See **marketing quota.**

farm program acreage The acreage of a crop on the farm used to calculate deficiency payments.

farm program payment yield The yield used for a farm in calculating deficiency payments and disaster payments. In no case may the farm program payment yields be less than the yields actually proven by the producer on the farm.

Farmer-owned reserve (FOR) A program, established under authority of the Food and Agricultural Act of 1977 and extended with modifications through subsequent Acts, to provide protection against wheat and feed grain production shortfalls and to provide a buffer against unusually sharp price movements. Under the program, farmers who place their grain in storage receive an extended nonrecourse loan from the CCC for three to five years. The loans are nonrecourse in that farmers can forfeit the commodity held as collateral to the government in full settlement of the loan without penalty and without paying accumulated interest. In addition, farmers may receive annual storage payments from the government. The farmer may not take this grain out of storage without penalty until the market price reaches a specific "release price." At that point, the farmer may decide to remove this grain from the reserve but is not required to do so. In an emergency, the secretary of agriculture can require repayment of the loans.

Farmers Home Administration (FmHA) The agency in the U.S. Department of Agriculture that is authorized to make direct loans to farmers who cannot obtain credit from other sources. It may also guarantee repayment of some such loans made by banks. The agency also makes loans for rural housing and community facilities as authorized by law.

federal crop insurance A voluntary program administered by the Federal Crop Insurance Corporation (FCIC), available to farmers since 1938, in which farmers can purchase partially subsidized insurance against crop disasters. The program was revised under legislation in 1980 and 1990.

Federal Insecticide, Fungicide, and Rodenticide Act (FIFRA) The Act that regulates sale and labeling of many agricultural pesticides.

Federal Land Bank See **Farm Credit System.**

Federal Marketing Orders and Agreements See **Marketing Orders and Agreements.**

Federal Reserve Board See **monetary policy.**

feed grains Those grains most commonly used for livestock or poultry feed. Corn, grain sorghum, oats, and barley are the main feed grains produced in the United States. Sometimes they are called coarse grains.

financial intermediaries Financial intermediaries act to transfer savings from savers to investors. Examples include banks, savings and loan firms, mutual funds, and insurance companies.

Findley provision A provision in the Agriculture and Food Act of 1981 and carried over into the Food Security Act of 1985 and the 1990 Act. It gave the secretary of agriculture discretion to reduce crop loan rates below the basic loan rate by up to 20 percent if the average market price in the previous marketing year was not more than 110 percent of the loan rate for that year, or if the reduction was necessary to maintain a competitive market position. Findley payments are additional deficiency payments required as a result of applying the Findley provision.

fixed costs Those short-run costs, e.g., land or management, that do not change in the aggregate when the level of output changes. No costs are fixed in the long-run, or all costs are variable.

flexibility A price and income policy concept to allow farmers greater flexibility in planting crops and still allow them to participate in federal farm price and income programs.

Food and Nutrition Service The USDA agency responsible for administering federal domestic food-aid programs.

food grains The cereal grains most commonly used for human food. Wheat, oats, rice, and rye are the main food grains produced in the United States.

Food Stamp Program A program that provides food coupons to low-income households. The coupons are redeemed for food at retail food stores.

forward contracting An agreement between a buyer and seller for future delivery of a commodity specifying quantity, quality, place of delivery, price, and date. Exchange of title normally occurs at the time of delivery.

forward pricing A concept advanced in the 1940s to set prices for future farm production and often discussed in connection with food and agricultural policy.

free market A theoretical concept in which prices are set by private forces without direct government influence on the terms of trade.

free trade A theoretical concept to describe international trade in which terms are set by private market forces without direct governmental barriers such as tariffs, quotas, and embargoes.

futures contract A contract to buy (or sell) a set amount of a commodity for delivery at a future time and place. The two largest futures trading markets in the United States are the Chicago Board of Trade and the Chicago Mercantile Exchange.

General Accounting Office (GAO) An agency of Congress that investigates the operations of various programs and the expenditure of appropriated funds. Most investigations are conducted at the request of Congressional committees or individual members of Congress.

General Agreement on Tariffs and Trade (GATT) Multilateral agreement signed by some 100 countries to establish rules and guidelines for regulating world trade among members and a forum for countries to discuss and resolve trade disputes. Launched in 1947, an underlying principle of the GATT is that trade should be restricted only through the use of uniformly applied tariffs, and that domestic and export subsidies should not be used to create unfair advantages for producers.

generic certificates Certificates issued instead of cash for certain farm program payments. They can be used to pay off commodity loans or buy government owned commodities, or they can be sold for cash.

genetic engineering A highly technical process of developing new plants and animals by combining genes into new forms and combinations. The process is also called biotechnology.

Gramm–Rudman–Hollings An Act adopted in 1985 that specified rules for moving toward a balanced budget through systematic cuts in federal expenditures.

ground water Water beneath the earth's surface between saturated soil and rock that supplies wells and springs.

growth hormones Bovine Somatotropin or bST is naturally produced by cows, but additional injections of the hormone have resulted in 10 to 25 percent additional milk production per cow. Porcine Somatotropin or pST is a growth hormone for swine that increases muscle weight with less fat and also boosts feed efficiency. The policy issues that arise with growth hormones are the effects of its use on supply and prices of the affected product and the safety of the product for consumption.

guaranteed loan An arrangement by which the government guarantees repayment of a loan made by a private lender. The Farmers Home Administration may guarantee some loans made to farmers by private banks. The government may guarantee loans made to exporters by private banks to promote more agricultural exports.

highly erodible cropland Land that is in cropland use and is classified by the Soil Conservation Service as class IV, VI, VII, or VIII under the SCS land capability classification system, and is considered to be particularly vulnerable to weather conditions.

homestead protection A provision in food and agricultural legislation that permits a foreclosed farmer–borrower to remain in the principal residence on the farm even though the land is repossessed.

import The goods and services that a country buys or otherwise obtains from other countries.

import barriers Quotas, tariffs, and embargoes used by a country to restrict the quantity or value of a good that may enter that country. Import barriers other than tariffs have been called nontariff barriers or NTBs.

import quota The maximum quantity or value of a commodity allowed to enter a country during a specified time period. A quota may apply to amounts of a commodity from specific countries.

indemnity programs Payments made under certain programs to producers who sustain losses as a result of pesticides, nuclear radiation fallout, residues, or toxic substances.

indifference curve As part of demand theory, an indifference curve connects all of the combinations of goods that give an individual equal utility. An individual offered a choice among any of the combinations on a curve would be indifferent as to which combination is chosen.

inelastic demand A market situation in which a change in price will bring about a smaller proportional change in the amount purchased. For example, when demand is very inelastic consumers tend to buy and consume about the same amount of a product regardless of the change in price, with the result that less total expenditures are made for a large output than for a small one. An elastic demand has the opposite characteristics. Generally, those goods that have no close substitutes are the most inelastic in demand. Most agricultural commodities have inelastic demands, and the demand for all food is very inelastic.

inelastic supply Supply is inelastic if the change in production is relatively less than the change in price. Time is the most important factor in determining the elasticity of supply; the shorter the time period, the more inelastic supply will be.

innovators Those people or firms who are the first to develop or try new methods or practices.

input An item used in the production of food or other agricultural product such as seed, fertilizer, chemicals, feed, machinery, fuel, labor, and land.

Integrated Pest Management (IPM) The control of pests using an array of crop production strategies, combined with careful monitoring of inset pests or weed populations, and other methods.

integration The combining of various steps in the production and marketing of a commodity under the management or control of a single firm. Also see **vertical integration.**

interest buy-down A plan to assist people and firms in financial difficulty with large debts in which the lender may reduce the interest rate and the government pay part of the interest cost so total interest payments are reduced. May be combined with a program to get lenders to scale down the size of the debt. Both interest buy-down and farm debt adjustment were used to help U.S. farmers in the 1930s, and again in a more limited program in the 1980s.

irradiation The use of ionizing radiation on food to kill insects and microorganisms that cause spoilage or human health problems.

level playing field The concept that all producers have an equal opportunity to compete in all markets or have equal treatment under government policies. Commonly used currently in food and agricultural policy, it sometimes refers to U.S. farmers having trade advantages equal to farmers in the EC and Japan.

liquidity The ability to supply funds or raise money by selling assets to pay off debts.

loan deficiency payment When applied to agriculture, it is the difference between the loan repayment rate and the rate at which the loan was obtained when a marketing loan is in effect. This term may also apply to a situation whenever the secretary of agriculture lowers the formula-determined loan rate, as has been permitted for wheat and feed grains, for example. In such cases, payments must be made to producers to provide the same return as if the loan rate had not been reduced.

loan forfeiture The forfeiting of commodities placed under loan instead of repaying the loan in cash.

loan rate See **nonrecourse loan rate.**

lobbyists Individuals and organizations that represent private interests or certain public agencies and attempt to influence decisions by members of Congress and other government officials.

long run That period of time in which all costs are variable.

low-input sustainable agriculture (LISA) The concept of reducing the amount of cash inputs, particularly fertilizers and chemicals, and through cropping practices and rotations, to maintain agricultural productivity while protecting a desired level of environmental quality.

macroeconomic policies Policies that affect the general economic environment in which the total economy, or a certain sector such as agriculture, operates. Monetary and fiscal policies are important macroeconomic policies.

mandatory program A program that requires all farmers to participate if they wish to produce and market a designated commodity without prohibitive penalty. The U.S. tobacco and peanut programs are examples of mandatory programs.

marginal cost The marginal cost (MC) is the change in total cost for a one-unit change in output. $MC = \Delta TC/\Delta Q$. Note: The marginal cost is based on profit-maximizing changes in all the variable inputs.

marginal physical product The marginal physical product of an input (MPP) is the change in output for a one-unit change in that input. $MPP = \Delta TPP/\Delta Qinp$.

marginal revenue Marginal revenue (MR) is the change in revenue for one-unit increase in output. For a small firm with no control of its market price, MR is the same as the market price. For a large firm with monopoly or oligopoly power, MR is **not** the same as the market price. As increased output increases sales, the price the monopolist or oligopolist receives will fall, other things being equal.

market-oriented farm policy See **free market.**

market price The amount received or paid for a unit of a commodity. Most market prices for U.S. agricultural commodities are quoted in pounds, bushels, dozens, quarts, gallons, hundred pounds, or tons. Most other countries use grams, kilograms, liters, tons, etc. See **weights, measures and conversion factors.**

marketing agreements and orders Federal government regulatory programs that permit agricultural producers collectively to promote orderly marketing of a crop or commodity. Agreements are voluntary, usually based on discussions and deliberations between representatives of the secretary of agriculture and the duly constituted representatives of a commodity organization. Orders are requested by producers and issued by the secretary of agriculture. To go into effect, an agreement or order must be approved by producers in a referendum, usually by a two-thirds majority. The marketing order permits producers to join together to regulate marketing of the commodity by means of regulatory restrictions binding on all handlers of the commodity in the area covered by order. Regulations for fruits and vegetables may involve packing standards, grades, size, price, and limitations on quantities shipped or marketed. Agreements may be issued in conjunction with orders or may be issued without orders. If issued without orders, there are fewer restrictions on marketing procedures and these only apply to those producers or handlers who agree voluntarily to enter into the agreement with the secretary of agriculture.

marketing board A major form of government involvement used by many countries to control the marketing of a commodity. These quasigovernmental boards generally handle all exports for that commodity.

marketing certificates Certificates that have been issued by the USDA as part of a price support program, which may be redeemed for cash or commodities, depending on the specific provisions of the program.

marketing loan A loan that may be repaid at a level below the rate at which the nonrecourse loan was made. Under some Acts, marketing loans have been mandatory for upland cotton and rice, and discretionary for wheat, feed grains, and soybeans.

marketing quotas A marketing quota sets the maximum amount of a commodity that a farmer can sell without incurring a penalty. Used most consistently in policy for tobacco, this kind of mandatory quota usually requires some substantial majority approval vote by producers.

marketing spread The difference between the price the producer received for a commodity and the price paid by the consumer for an equivalent amount of the same commodity.

marketing year The period of 12 months beginning at the start of harvest of a crop and extending to the same time in the following year. The marketing year for wheat, oats, and barley begins on June 1 and extends through May 30 the following year. For corn, grain sorghum, and soybeans the marketing year is from September 1 through August 31.

milk assessments Deductions from dairy farmers' milk checks to finance a special federal government supply-reduction program. See **Milk Production Termination Program.**

milk diversion program A program adopted in 1983 by which milk producers could receive direct payments by agreeing to reduce production in their herds between January 1, 1984 and March 31, 1985.

milk marketing orders The federal or state programs that cover most of the fluid milk marketed in the United States by regulating the supply of fluid milk and setting the minimum prices that buyers must pay. The federal government may also support the prices of certain manufactured dairy products, such as butter and cheese, through purchase and storage programs.

Milk Production Termination Program (or Whole-herd Buyout Program) A program established by the Food Security Act of 1985 under which the secretary of agriculture paid some dairy farmers to reduce herd size for a period of five years.

monetary policy Policies carried out by the Federal Reserve Board to influence the supply of money and credit and the rate of interest.

moral hazard This exists in an insurance contract when the insured can influence the outcome of an indemnity or its probability.

moratorium A postponement in payment of interest and principal on debts owed by farmers, particularly those with financial problems. The proposals for a moratorium on farm debts are usually aimed at those debts farmers owe to the U.S. government through loans from the Farmers Home Administration.

multilateral agreement An international compact in which three or more countries participate.

national farm program acreage The number of harvested acres of a crop that is estimated to be needed to meet the policy targets for domestic and export use and to accomplish any desired increase or decrease in carryover.

National School Lunch Program A program that provides financial and commodity assistance for lunches served in public and private elementary and secondary schools and in public and private licensed nonprofit residential child-care institutions. Participating schools and institutions serve lunches that meet standards set by USDA.

negotiable marketing certificates Marketing certificates that can be exchanged for cash or commodities. See **marketing certificates.**

net worth See **equity.**

nominal rate of interest The actual rate of interest paid without any adjustment for inflation.

no net-cost programs Price support programs in which producers are assessed to finance the cost of the program. The U.S. tobacco program was designed to be a "no net-cost" program to the federal government.

nonrecourse loans Price support loans to farmers from the government to enable them to hold their crops for later sale. The loans are nonrecourse because if a farmer cannot profitably sell the commodity and repay the loan with interest when it matures, usually during a marketing year, the commodity for which the loan was advanced can be delivered to the government in full settlement of the loan. Alternatively, the farmer can redeem the stored commodities by paying off the loan with interest.

nontariff trade barriers (NTBs) All types of restraint on imports or exports other than tariffs.

normal crop acreage The acreage on a farm normally devoted to a group of designated crops. May be used in setting allotments or quotas for the farm.

normal yield See **program yield.**

offsetting compliance The requirement that if a farmer wishes to participate in a program for one farm then that farmer must also meet the program provisions for other farms under the same ownership or control. See **cross compliance.**

oilseed crops Those crops from which oil is extracted from the seed: primarily soybeans, peanuts, cottonseed, and flaxseed. Sunflower, safflower, castor beans, and sesame are also produced for oil and are considered oilseed crops.

open economy A term applying to a country that allows free international trade.

open market operations In the U.S., a giant market for government securities (the debt of the U.S. government) exists. When the Federal Reserve buys or sells government securities in that market it is conducting open market operations. If the Federal Reserve banks are ordered by the Fed to buy securities, they may be used as reserves, which increases the money supply. If the Fed orders the banks to sell securities, this reduces the money supply.

organic farming Farming methods that use only organic fertilizers and avoid using inorganic agricultural chemicals and herbicides.

output Total yield or quantity produced.

paid diversion program See **diversion payment.**

parity A relationship that defines a level of purchasing power for a unit of farm commodity equal to an earlier base period, and sometimes is proposed as a fair price.

parity index The index of prices paid by farmers for items used in production, interest, taxes, and wage rates.

parity price The current price for a unit of farm commodity that would give it the same relative purchasing power for goods and services that it had in 1910–1914, adjusted for the changes in its price over the past 10 years, compared with changes in the prices received by farmers for all farm products during that same period.

parity ratio The ratio of the index of prices paid by farmers to index of prices received.

Payment-in-Kind (PIK) A program that has provided payment to farmers in the form of commodities for reducing acreage of certain crops and placing that acreage in soil conserving uses. The term may also apply to export enhancement programs or other programs in which payments are made in the form of commodities.

payment limitation A limit set by law on the amount of money any individual farmer may receive in farm program payments each year under specified programs.

permanent legislation The laws on which many agricultural programs are based. For the major commodities, the permanent legislation is the Agricultural Adjustment Act of 1938 and the Agricultural Act of 1949. These laws have been amended frequently for a given number of years but would be in effect again if current amendments are not extended or amendments not enacted.

permitted acreage The acreage of a program crop that a participating farmer is eligible to plant after reducing acreage by the percentage required in that year. The permitted acreage is usually the crop acreage base less any land idled by acreage reduction and paid land diversion programs.

PIK and roll A procedure by which producers may attempt to profit from situations in which certificate exchange values (posted county prices) are below nonrecourse loan rates. With this procedure, a producer may place the eligible commodity under the nonrecourse loan at the loan rate, and use generic certificates—when available—to exchange for Commodity Credit Corporation (CCC) commodities. If the posted county price is below the nonrecourse loan rate, then the producer may be able to acquire the quantity placed under loan for less than the proceeds of the nonrecourse loan, in addition to saving interest and storage charges.

PIK certificates See **Payment-in-Kind** and **generic certificates.**

pipeline stocks The minimum quantity of any commodity required to carry on the normal processing and marketing operations.

policy A decision and planned course of action taken by individual, interest group, or government. Public policy refers to actions taken by a representative or participatory government.

prevented planting disaster payment See **disaster payments.**

price-support program A program of commodity purchases or nonrecourse loans undertaken by the CCC to raise the average price received by farmers to a desired price.

primary demand Final consumer or retail demand. Derived demand is primary demand minus processing and marketing charges. For example, the demand for corn, an important livestock input, is derived from the retail demand for meat and other livestock products.

producer assessments See **assessment programs.**

producer subsidy equivalents The level of subsidy that would be necessary to compensate producers (in terms of income) for the removal of government programs affecting a particular commodity.

production control programs Any government program intended to limit production. At various times these programs have been called acreage reduction, reduced acreage, set-aside, diverted acreage, acreage allotments, marketing quotas, PIK, and soil bank.

Production Credit Association See **Farm Credit System.**

program benefits The various forms of financial assistance available to those farmers who sign up and agree to comply with the requirements of government farm programs. These benefits may include eligibility for CCC loans, deficiency payments, diversion payments if offered, and disaster payments under certain circumstances.

program commodities Certain agricultural commodities for which federal support programs are available. In the United States, these have been wheat, corn, grain, sorghum, barley, oats, rye, cotton, rice, milk, peanuts, soybeans, sugar, honey, wool, mohair, and tobacco.

program yield The yield for a crop on a given farm used to calculate deficiency payments. Program yields are based on history of past yields, and records of crop sales provided to the local ASCS office by the individual producer. Program yields have been frozen in some Acts to the average yield for a previous base period, dropping some high and low years.

progressive A policy or program that works to the disadvantage of higher income people.

protectionsim Usually a reaction by an industry, company, or national government to foreign competition. The most common type of protectionism is expressed in food and agricultural policy as an import tariff or quota imposed to reduce imports and shield domestic producers.

proven yield Yields substantiated by records of crop sales or other documentation acceptable to local ASCS offices.

Public Law 480 (P.L. 480) Common name for the Agricultural Trade Development and Assistance Act of 1954, which seeks to expand foreign markets for U.S. agricultural products, combat hunger, and encourage economic development in developing countries. It has been amended and extended several times since it was first passed.

rangeland Land, usually in the West, that is used for grazing of animals rather than for growing crops.

real rate of interest The rate of interest earned after adjusting for the average rate of inflation.

recourse loan A loan in which commodities could be used as collateral but which would have to be repaid in cash rather than delivering the commodities, as permitted under a nonrecourse loan. See **nonrecourse loans.**

reduced yield disaster payments See **disaster payments.**

referendum In relation to food and agricultural policy, a referendum involves a vote by producers of a specific commodity for a proposed program that will obligate all producers to participate in the program if a specified percentage of producers vote for it.

regressive A policy or program that has a greater effect on, or works to the disadvantage of, lower income people.

release price The price at which farmers who have grain stored in the reserve may sell it without incurring penalities. The release price may be stated as a specific nationwide price or as a percentage level substantially above the support level or loan. For example, farmers who placed 1982 wheat in the reserve received a reserve premium of 45 cents a bushel, and for 1982 corn, they received a premium of 35 cents a bushel.

reserve requirements The Federal Reserve has control over how much of the deposits in the banking system must be held in reserve rather than lent out. If the Fed raises reserve requirements, banks that are "loaned up," or nearly so, will have to reduce their loans and increase their reserves. This action reduces the money supply.

Section 32 A section of the Agricultural Act Amendment of 1935 that authorizes use of customs receipts funds to encourage increased consumption of agricultural commodities. It is funded by a continuing appropriation of 30 percent of the import duties imposed on selected agricultural and nonagricultural products.

sequestration The required reduction of government expenditures under the Gramm–Rudman–Hollings Act if the federal budget does not meet deficit reduction goals.

set-aside A program under which a farmer must divert a proportion of the farm's total cropland to soil conserving uses to be eligible for program benefits.

short run The period of time in which certain costs, called fixed costs, cannot be eliminated even by shutting down the firm. In the long run, all costs are variable.

slippage in demand In reference to food-aid programs, this occurs when low-income households substitute a portion of their cash subsidy or food subsidy for their previous (prior to receiving food aid) cash expenditure on food.

slippage in supply In reference to programs to control agricultural supply, this is the difference between the percentage reduction in harvested acreage and the percentage reduction in crop output. This difference may be measured in terms of acreage, yields, or production.

sod buster The 1985 and 1990 Acts each had sodbuster provisions that restricted plowing up highly erodible land for crop production and required approved soil conservation plans on highly erodible land in crop production in order to be eligible for government farm program benefits.

soil bank A program operated in the 1950s to achieve both soil conservation and production control objectives. Under the program farmers signed contracts for varying periods of time to place part of their acreage into conserving uses.

Special Supplemental Food Program for Women, Infants, and Children (WIC) A food program that provides either vouchers redeemable for specified foods at retail food stores or a food package to pregnant women, post partum mothers, infants, and children up to five years old who are at nutritional risk due to inadequate income and nutrition.

stocks and flows A person uses a **stock** of resources—that is human capital such as skills, talents, savings—to earn income, a **flow.** In the same way, a country uses its **stock** of resources—the number of productive workers, the amount of capital equipment, the acres of crop land—to earn income, which is called Gross Domestic Product, Gross National Product, or National Income. The number of productive workers is a **stock;** we can count the number of workers in existence on any given day without having to worry about the time dimension. Some imports such as gasoline, for example, are used up, and would be counted as a **flow.** Also, income is a **flow.** It makes a little sense to say ''My income is $800,'' without saying ''per year,'' ''per month,'' or ''per week.''

storage payment The payment a farmer receives from the government when a commodity is placed in the farmer-owned reserve (FOR). The commodity must be held in storage facilities approved by the ASCS, either on the farm or in commercial storage facilities.

structure of agriculture The make-up of the food and agricultural sector usually described in terms of numbers and sizes of farms, types and sizes of agribusiness firms and industries, and other features that determine the control of agricultural assets and management decisions.

subsidy A government payment. Subsidies are paid to many people and firms under many different programs.

supplementary payments A payment similar to deficiency payments made to producers of wool, mohair, and extra-long staple cotton.

supply curve The curve that shows the relationship between the price of a good and the quantity a firm operating in a competitive market is willing to produce. In the short run under pure competition it is the marginal cost curve above the minimum point of the average variable cost curve.

supply management A term used to describe a policy in which government programs are used to influence and control the supply of a commodity to maintain a desired price.

swampbuster bills Legislation that places restrictions on the draining of natural wetlands for crop production if producers are to be eligible for program benefits.

T by 2000 The conservation program of reducing soil erosion adopted in some states to meet what is known as the soil loss tolerance level, or the T value, by the year 2000. The T value for most cropland usually falls between 3 and 5 tons of soil loss per acre per year.

target price A price for certain crops established by law. If the average market price does not equal the target price, qualifying farmers may receive a deficiency or compensatory payment to make up part or all the difference. Generally deficiency (or compensatory target price) payments are made if average market prices do not equal the target price for the first five months of the marketing year. See **deficiency payments.**

Targeted Export Assistance (TEA) A program that reimburses private business to promote exports of specific U.S. food and agricultural commodities in specified markets. Eligible participants, such as exporting firms, may receive generic certificates in payment for promotional activities approved by the secretary of agriculture.

tariffs Taxes imposed on commodity imports by a government. A tariff may be either a fixed charge per unit of product imported or a fixed percentage of value (ad valorem tariff).

tax-loss farming Farming operations carried out with the main objective to produce a loss to be claimed for tax reporting purposes. Such losses from farming operations may be used to reduce taxes owned from other businesses or nonfarm income sources.

technology In agriculture, the methods, techniques, and systems used in producing and marketing food and other agricultural commodities.

tenancy The renting or leasing of land by a farm operator rather than owning it.

tenant protection Provisions in some state and federal legislation designed so that both tenants and landowners are entitled to farm program benefits.

tenure The relationship or type of control that a farmer has on the land that is rented. Farms are classed as owner operated, part-owner operated, or tenant operated.

tiering A method of directing benefits of federal price support programs toward smaller or medium sized farms. Under tiering, deficiency payments per unit of production would be higher for small farms and lower for quantities beyond a given amount. For example, federal payments to sugar-beet growers have been tiered.

trigger price The point at which the market price reaches or exceeds the release price of grain in the reserve for the specified length of time, so farmers can sell their reserve grain without penalty.

triple base plan A plan to reduce government farm program costs by making deficiency payments on only part of the permitted acreage that a farmer could plant under that program, in which case there would be a base acreage, a program acreage, and a payment acreage.

two-price plan A plan that involves supporting that part of production used in the domestic market at one price and selling the remainder for export at whatever it will bring. A two-price plan will increase aggregate farm income if demand is more inelastic in the domestic market than in the foreign market, other things being equal.

U.S. trade representative The cabinet-level head of the Office of the U.S. Trade Representative, the principal trade policy agency of the U.S. Government.

value-added tax Taxes collected at each designated stage of production or marketing.

variable costs Costs that vary in aggregate with a change in output of a firm. In the short run there are both fixed and variable costs. In the long run all costs are variable.

vertical integration A situation in which two or more firms at different stages of production and processing or marketing combine under a single ownership or management.

voluntary program A program that allows farmers to choose whether or not to comply without loss of market access.

weights, measures, and conversion factors

Standard bushel weights:
wheat or soybeans = 60 lbs.
corn, sorghum, or rye = 56 lbs.
barley (grain) = 48 lbs., malt = 34 lbs.
oats = 32 lbs.

To convert bushels of wheat to metric tons of 2,240 lbs. each, divide 2,240 by the above bushel weights. For example, 2,240 ÷ 60 = 37⅓ bushels of wheat in one metric ton. Then, given the number of bushels of wheat, divide this figure by 37⅓ to get the number of metric tons. For example, if the U.S. wheat crop is approximately 2.5 billion bushels, this is equivalent to approximately 67 million tons (2.5 billion bushels ÷ 37⅓ = 66.9 million tons).

To convert metric tons to bushels, multiply tons by the number of bushels in each ton. For example, if the wheat crop in India is reported to be 50 million metric tons, this is equivalent to approximately 1.9 billion bushels (50 million tons × 37⅓ = 1.92 billion bushels).

To convert acres to hectares, multiple acres by 0.405.

To convert hectares to acres, multiply hectares by 2.471.

The conversion table that follows shows conversions from United States units to metric units and vice versa for weight, volume, length, area, pressure, yield or rate, and temperature.

To convert from United States units to metric units, find the appropriate unit in column 2 and multiple the measurement by the number in the column to the right of it.

To convert from metric units to United States units, find the appropriate unit in column 1 and multiply the measurement by the number in the column to left of it.

zero–92 (0–92) A program adopted to encourage farmers to reduce the number of acres they plant. Officially known as the optional acreage diversion program, it has allowed producers growing wheat and feed grains to forgo planting on any of their farms, but still receive 92 percent of their deficiency payment just as if they had planted their crops. Variations of this have also involved planting of only 50 percent of base acreage.

CONVERSION BETWEEN UNITED STATES UNITS AND METRIC UNITS

To convert column 1 to column 2 multiply by:	(1) Metric units	(2) U.S. Units	To convert column (2) to column (1), multiply by:
Weight			
1.102	ton (metric)	ton (U.S.)	0.907
2.205	kilogram, kg	pound, lb	0.454
0.035	gram, g	ounce (avdp), oz	28.349
Volume			
0.264	liter, L or l	gallon (U.S.), gal	3.785
1.057	liter, L or l	quart (fluid), qt	0.946
2.838	hectoliter, hl	bushel, bu	0.352
0.034	milliliter, ml	ounce (fluid), oz	29.573
Length			
0.621	kilometer, km	mile, mi	1.609
1.094	meter, m	yard, yd	0.914
3.281	meter, m	foot, ft	0.305
0.394	centimeter, cm	inch, in	2.540
Area			
2.471	hectare, ha	acre, A	0.405
247.1	kilometer^2, km^2	acre, A	0.004
0.386	kilometer^2, km^2	mile^2, mi^2	2.590
10.765	meters^2, m^2	feet^2, ft^2	0.093
0.155	centimeters^2, cm^2	inches^2, in^2	6.452
Pressure			
0.987	bar	atmosphere, atm	1.013
14.504	bar	lb/in^2, psi	0.069
14.223	kg/cm^2	lb/in^2, psi	0.070
14.70	atmosphere, atm	lb/in^2, psi	0.068
0.968	kg/cm^2	atmosphere, atm	1.033
Yield or rate			
0.446	1 ton (metric/hectare)	ton (U.S./acre)	2.242
0.892	kg/ha	lb/A	1.121
1.149	hl/ha	bu/A	0.870
0.107	l/ha	gal/A	9.346
8.347	kg/l	lb/gal	0.120
Temperature			
1.80° C + 32	Celsius,°C	Fahrenheit, °F	0.55 (F − 32)

Source: From R. Gordon Halfacre and John A. Barden, *Horticulture,* McGraw-Hill, New York, 1979, app. table 2, p. 695.

BIBLIOGRAPHY

Aaron, Henry J.: "Politics and the Professors Revisited," Richard T. Ely Lecture. *American Economic Review,* 79:1–15 (1989).

Allen, Joyce E, and Doris Epson Newton: "Existing Food Policies and Their Relationship to Hunger and Nutrition," *American Journal of Agricultural Economics,* 68: 1247–1252 (1986).

Allen, Joyce E. and Alton Thompson: "Rural Poverty among Racial and Ethnic Minorities," *American Journal of Agricultural Economics,* 72:1161–1168 (1990).

Archibald, Sandra: "Regulating Chemicals: A Public Policy Challenge," *Choices: The Magazine of Food, Farm, and Resource Issues,* First Quarter, 1990, pp. 20–21, 24–25.

Ayer, Harry, ed.: *Choices: The Magazine of Food, Farm, and Resource Issues,* First Quarter, 1993, pp. 2–43.

Barry, R. D., Luigi Angelo, Peter Buzzanell, and Fred May: *Sugar Background for the 1990 Farm Legislation,* AGES No. 9006, USDA, Economic Research Service, Washington, DC, 1990.

Basiotis, Peter and others: "Nutrient Availability, Food Costs, and Food Stamp Program Participation for Low-Income Households," *American Journal of Agricultural Economics,* 65: 685–693 (1983).

Becker, G. S.: *Provisions of the 1990 Farm Bill.* CRS 90–553 ENR. Congressional Research Service, Library of Congress, Washington, DC, 1990.

Benedict, M. R.: *Farm Policies of the United States, 1790–1950,* The Twentieth Century Fund, New York, 1953.

Bosselman, Fred P. and David Callies: *The Quiet Revolution in Land-Use Control,* Government Printing Office, Washington, DC, 1972.

Bottom, J. Carroll: "Policy Formation and the Economist" (Fellows Address), *American Journal of Agricultural Economics,* 57: 764–768 (1975).

Breimyer, H. F.: *Individual Freedom and the Economic Organization of Agriculture,* University of Illinois Press, Urbana, 1965.

Browne, W. P.: *Private Interests, Public Policy and American Agriculture,* University of Kansas Press, Lawrence, 1988.

Buchanan, James M., and Gordon Tullock: *The Calculus of Consent, Logical Foundations of Constitutional Democracy,* University of Michigan Press, Ann Arbor, 1962.

Calef, Wesley: *Private Grazing and Public Lands, Studies of the Local Management of the Taylor Grazing Act,* The University of Chicago Press, Chicago, 1960.

Caswell, Julie A., ed.: *Economics of Food Safety,* Elsevier Science Publishing Company, New York, 1991.

Chavas, Jean-Paul: "Future Directions for Domestic Food Policy," *American Journal of Agricultural Economics,* 66:225–231 (1984).

Chavas, Jean-Paul and Keith O. Keplinger: "Impact of Domestic Food Programs on Nutrient Intake of Low-Income Persons in the United States," *Southern Journal of Agricultural Economics,* 15: 155–163 (July 1983).

Christy, Ralph D. and Enrique E. Figueroa: "The Impacts of Structural Change and Public Policy on the Rural Disadvantaged," *American Journal of Agricultural Economics,* 72:1169–1178 (1990).

Clancy, Katherine L., ed.: *Consumer Demands in the Marketplace: Public Policy in Relation to Food Safety, Quality and Human Health,* Resources for the Future, Washington, DC, 1988.

Clawson, Marion: *The Land System of the United States,* University of Nebraska Press, Lincoln, 1968.

Cochrane, Willard W.: *Farm Prices: Myth and Reality,* University of Minnesota Press, Minneapolis, 1958.

Cochrane, Willard W.: "Some Nonconformist Thoughts on Welfare Economics and Commodity Stabilization Policy," *American Journal of Agricultural Economics,* 62:509–511 (1980).

Cochrane, Willard W. and Mary E. Ryan: *American Farm Policy,* 1948–1973, University of Minnesota Press, Minneapolis, 1976.

Commons, John R.: *Legal Foundations of Capitalism,* University of Wisconsin Press, Madison, 1957.

Davis, Carlton G. and Benjamin Senauer: "Needed Directions in Domestic Food Assistance Policies and Programs," *American Journal of Agricultural Economics* 68: 1253–1257 (1986).

Day, Richard H., ed.: *Economic Analysis and Agricultural Policy,* Iowa State University Press, Ames, 1982.

Dewey, John: *Logic, the Theory of Inquiry,* Henry Holt & Co., New York, 1938.

Dicks, Michael R. and David Harvey: "Issues Behind Mandatory Seafood Inspection," *National Food Review,* October–December 1989, NFR–12, pp. 30–33.

Eleveld, Bartelt and Harold G. Halcrow: "How Much Conservation Is Optimum for Society?" *Soil Conservation Policies, Institutions, and Incentives,* Harold G. Halcrow, Earl O. Heady, and Melvin L. Cotner (eds.), Soil Conservation Society of America, Ankeny, IA, 1982, pp. 233–250.

Faeth, P., R. Repetto, K. Kroll, Q. Dai, and G. Helmers: *Paying the Farm Bill: U.S. Agricultural Policy and the Transition to Sustainable Agriculture.* World Resources Institute, Washington, DC, 1991.

Fisher, Eric O'N. and Harry de Gorter: "The International Effects of U.S. Farm Subsidies." *American Journal of Agricultural Economics,* 74: 258–267 (1992).

Food and Agricultural Policy Research Institute (FAPRI): *The Commodity Supply Management Program,* FAPRI Staff Report No. 2-87. University of Missouri, Columbia, February 1986.

Foss, Phillip O.: *Politics and Grass, The Administration of Grazing on the Public Domain,* University of Washington Press, Seattle, 1960.

Fox, K. A. and D. G. Johnson: *Readings in the Economics of Agriculture,* Richard D. Irwin, Homewood, IL, 1969.

Fuell, Lawrence D., David C. Miller, and Merritt Chesley: *Dictionary of International Agricultural Trade,* USDA, Foreign Agricultural Service, Agricultural Handbook, No. 411, Washington, DC, Revised June 1988.

Gardner, Bruce L.: *The Economics of Agricultural Policies,* McGraw-Hill, New York, 1990.

Green, Robert and Harry Baumes: ''Supply Control Programs for Agriculture,'' *Agricultural Food Policy Review: U.S. Agricultural Policies in a Changing World,* USDA, Economic Research Service, Agricultural Economic Report No. 620, U.S. Government Printing Office, Washington, DC, 1989, chap. 10.

Guither, Harold D.: *Tough Choices—Writing the Food Security Act of 1985,* Occasional Papers. American Enterprise Institute, Washington, DC, 1985.

Guither, Harold D. and Harold G. Halcrow: *The American Farm Crisis: An Annotated Bibliography with Analytical Introductions. Resources on Contemporary Issues,* Pierian Press, Ann Arbor, MI, 1988.

Guither, Harold D., Robert G. F. Spitze, and Darrel L. Good: *The 1990s Farm Policy Handbook,* University of Illinois, Urbana-Champaign, Department of Agricultural Economics in Cooperation with Illinois Fertilizer and Chemical Association, and Grain and Feed Association of Illinois, Urbana, 1990.

Gureff, Jim and Sharon Bylenga: ''Harmonizing Food Safety and Other Health-Related Regulations for Agricultural Trade,'' *National Food Review* 12:37–39 (July–September 1989).

Halberg, M. C.: *Policy for American Agriculture, Choices and Consequences,* Iowa State University Press, Ames, 1992.

Halcrow, Harold G., *Agricultural Policy Analysis,* McGraw-Hill, New York, 1984.

Halcrow, Harold G.: *Agricultural Policy of the United States.* Prentice-Hall, New York, 1953.

Halcrow, Harold G.: *Food Policy for America,* McGraw-Hill, New York, 1977.

Halcrow, Harold G., Earl O. Heady, and Melvin L. Cotner (eds.): *Soil Conservation Policies, Institutions and Incentives,* published for North Central Research Committee III: Natural Resource Use and Environmental Policy by the Soil Conservation Society of America, Ankeny, IA, 1982.

Hardin, Charles M.: *The Politics of Agriculture, Soil Conservation and the Struggle for Power in Rural America,* The Free Press, Glencoe, IL, 1952.

Hathaway, D. E.: *Government and Agriculture,* Macmillan, New York, 1963.

Heady, Earl O.: *A Primer on Food, Agriculture and Public Policy,* Random House, New York, 1967.

Heady, Earl O.: ''The Adequacy of Agricultural Land: A Demand-Supply Perspective,'' *The Cropland Crisis: Myth or Reality?* Pierre R. Crosson (ed.), published for Resources for the Future by the Johns Hopkins University Press, Baltimore and London, 1982, pp. 23–56.

Helmberger, P. G.: *Economic Analysis of Farm Programs,* McGraw-Hill, New York, 1991.

Hibbard, Benjamin H.: *A History of Public Land Policies,* The Macmillan Press Company, New York, 1924.

Hillman, Jimmy S.: *Technical Barriers to Agricultural Trade.* Westview Press, Boulder, CO, 1990.

Houck, James P.: *Elements of Agricultural Trade Policies,* Macmillan, New York, 1986.

House, V. W.: *Shaping Public Policy,* Westridge Publishing Co., Bozeman, MT, 1981.

Iowa State University Center for Agricultural Economics Adjustment: *Goals and Values in Agricultural Policy,* Iowa State University Press, Ames, 1961.

Jesness, O. B., ed.: *Readings on Agricultural Policy,* Blakiston Co., Philadelphia, 1949.

John, Dewitt, Sandra S. Batie, and Kim Norris: *A Brighter Future for Rural America? Strategies for Communities and States.* National Governors' Association, Washington, DC, 1988.

Johnson, Scott L., Richard M. Adams, and Gregory M. Perry: ''The On-Farm Costs of Reducing Ground-water Pollution,'' *American Journal of Agricultural Economics,* 73(4), pp. 1053–1073 (1991).

Just, Richard, Darrel L. Hueth, and Andrew Schmitz: *Applied Welfare Economics and Public Policy,* Prentice-Hall, Englewood Cliffs, NJ, 1982.

Knutson, R. D., J. B. Penn, and W. T. Boehm: *Agricultural and Food Policy,* Prentice-Hall, Inc., Englewood Cliffs, NJ, Second Edition, 1990.

Kramer, Carol S.: ''Food Safety and Public Policy: What Can Economists Contribute?'' in Robert G. F. Spitze (ed.), *Agricultural and Food Policy: Issues and Alternatives for the 1990s,* University of Illinois Press, Urbana, 1990.

Kramer, Carol S.: ''Food Safety: The Consumer Side of the Environmental Issue,'' *Southern Journal of Agricultural Economics,* 22: 33–40 (1990).

Kramer, Carol S., ed.: *The Political Economy of U.S. Agriculture,* Resources for the Future, Washington, DC, 1989.

Kramer, Carol S. and Eileen O. van Ravenswaay: ''Proposition 65 and the Economics of Food Safety,'' *American Journal of Agricultural Economics,* 71:1293–1299 (1989).

Larson, Bruce A. and Mary K. Knudson: ''Public Regulation of Agricultural Biotechnology Field Tests: Economic Implications of Alternative Approaches,'' *American Journal of Agricultural Economics,* 73(4), pp. 1074–1082 (1991).

Leman, Christopher: ''Political Dilemmas in Evaluating and Budgeting Soil Conservation Programs: The RCA Process,'' *Soil Conservation Policies, Institutions, and Incentives,* op. cit., pp. 47–88.

Levedahl, J. William and Masao Matsumoto: ''U.S. Domestic Food Assistance Programs: Federal Costs and Public Benefits,'' *Agricultural Food Policy Review: U.S. Agricultural Policies in a Changing World.* Agricultural Economic Report No. 620. USDA, Economic Research Service, Washington, DC, November 1989, chapter 12.

Lipton, Kathryn L., *Agriculture, Trade, and the GATT, A Glossary of Terms,* USDA, ERS, Agriculture Information Bulletin No. 625, June, 1991.

McCalla, Alex F. and Timothy E. Josling: *Agricultural Policies and World Markets,* Macmillan Publishing Company, New York, 1985 (McGraw-Hill, New York, 1991).

McConnell, Campbell R. and Staley L. Brue, *Economics,* McGraw-Hill, New York, Twelfth Edition, 1993.

McCune, W.: *Who's Behind Our Farm Policy,* Frederick A. Praeger, New York, 1956.

Mellor, John W.: ''Agricultural Development in the Third World: The Food, Poverty, Aid, Trade Nexus,'' *Choices: The Magazine of Food, Farm, and Resource Issues,* First Quarter 1989, pp. 40–48.

Morgan, Robert J.: *Governing Soil Conservation: Thirty Years of the New Decentralization,* The Johns Hopkins Press, published for Resources for the Future, Inc., Baltimore, 1965.

National Public Policy Education Committee: *Speaking of Trade: Its Effect on Agriculture,* University of Minnesota Agricultural Extension Service Special Report No. 72. University of Minnesota, Minneapolis, 1978.

O'Rourke, A. D.: *The Changing Dimensions of U.S. Agricultural Policy,* Prentice-Hall, Englewood Cliffs, NJ, 1978.

Paarlberg, D.: *American Farm Policy,* John Wiley and Sons, New York, 1964.

Parnell, Jack C.: "Food Safety: Some Changes Are Needed," *Choices: The Magazine of Food, Farm, and Resource Issues,* Third Quarter 1990, pp. 4–7.

Parsons, K. H.: "The Logical Foundations of Economic Research," *Journal of Farm Economics* 31: 656–686 (1949).

Perspectives on Prime Lands, Background papers for Seminar on Retention of Prime Land, U.S. Department of Agriculture, Committee on Land Use, Washington, 1975.

Phipps, Tim T., Kristen Allen, and Julie A. Caswell: "The Political Economics of California's Proposition 65," *American Journal of Agricultural Economics,* 71:1286–1292 (1989).

Phipps, Tim T., Pierre E. Crosson, and Kent A. Price, eds., *Agriculture and the Environment,* National Center for Food and Agricultural Policy, Resources for the Future, Washington, DC, 1986.

Rasmussen, Wayne D.: "History of Soil Conservation, Institutions, and Incentives," *Soil Conservation Policies, Institutions, and Incentives,* op. cit., pp. 3–18.

Raup, Philip M.: "Competition for Land and the Future of American Agriculture," *The Future of American Agriculture as a Strategic Resource,* Sandra S. Batie and Robert G. Healy (eds.), The Conservation Foundation, Washington, DC, 1980.

Rawls, John: *A Theory of Justice,* The Belknap Press of Harvard University Press, Cambridge, MA, 1971.

Robbins, Roy M.: *Our Landed Heritage,* The Princeton University Press, Princeton, NJ, 1942.

Roberts, Tanya: "Irradiation's Promise: Fewer Foodborne Illnesses?" *National Food Review,* Spring 1986, NFR–33, pp. 7–10.

Robinson, K. L.: *Farm and Food Policies and Their Consequences,* Prentice-Hall, Englewood Cliffs, NJ, 1989.

Roll, Eric: *A History of Economic Thought,* Prentice-Hall, New York, 1946.

Rowley, W. D.: *M. L. Wilson and the Campaign for the Domestic Allotment,* University of Nebraska, Lincoln, 1970.

Sadoulet, Elizabeth and Alain de Janvry: "Agricultural Trade Liberalization and the Low Income Countries: A General Equilibrium-Multimarket Approach," *American Journal of Agricultural Economics* 74: 268–280 (1992).

Salter, Leonard A., Jr.: *A Critical Review of Research in Land Economics,* The University of Minnesota Press, Minneapolis, 1948.

Sawhill, Isabel V.: "Poverty in the U.S.: Why Is It So Persistent?" *Journal of Economic Literature,* 26:1073–1119 (1988).

Senauer, Ben and Nathan Young: "The Impact of Food Stamps on Food Expenditures: Rejection of the Traditional Model," *American Journal of Agricultural Economics,* 68:37–43 (February 1986).

Shane, Matthew: "World Agricultural Markets at Crossroads," in *Agricultural–Food Policy Review: U.S. Agricultural Policies in a Changing World.* Agricultural Economics Report No. 620. USDA, Economic Research Service, Washington, DC, November 1989, Chap. 15.

Shepherd, G. S.: *Agricultural Price and Income Policy,* Iowa State College Press, Ames, IA, 1952.

Smith, Mark E., Eileen O. van Ravenswaay, and Stanley R. Thompson: "Sales Loss Determination in Food Contamination Incidents: An Application to Milk Bans in Hawaii," *American Journal of Agricultural Economics,* 70:513–520 (1988).

Spitze, R.G.F. ed.: *Agricultural and Food Policy: Issues and Alternatives for the 1990s,* University of Illinois, Department of Agricultural Economics, Urbana, 1990.

Spitze, R. G. F.: ''Agricultural and Food Policy Issues and the Public Decision-making Environment,'' *Agricultural and Food Price and Income Policy: Alternative Directions for the United States and Implications for Research.* University of Illinois, Urbana, College of Agriculture Special Publication 43, 1976, pp. 5–16.

Spitze, R. G. F.: ''Economic Redirections in Recent U.S. Agricultural Policy,'' *Journal of Agricultural Economics,* 19:327–338 (1968).

Spitze, R. G. F.: ''Policy Direction and Economic Interpretations of the U.S. Agricultural Act of 1970,'' *Journal of Agricultural Economics,* 22: 99–108 (1972).

Spitze, R. G. F.: ''Problems of Adequacy in Rural Human Resource Development—Concepts and Accomplishments,'' in E. O. Heady (ed.), *Benefits and Burdens of Rural Development: Some Public Policy Viewpoints.* Iowa State University Press, Ames, 1970, Chapter 10, pp. 197–218.

Spitze, R. G. F.: ''The Political Economy of the 1990 Agricultural and Food Policy,'' *Choices: The Magazine of Food, Farm, and Resource Issues,* Second Quarter, 1991, pp. 18–21.

Spitze, R. G. F. and M. A. Martin, (eds.): *Analysis of Food and Agricultural Policies for the Eighties,* North Central Regional Research Bulletin No. 271, University of Illinois Agriculture Experiment Station, Urbana, 1980.

Spitze, R. G. F. and R. A. Mahoney: ''Evolving Dimensions of Rural Employment of Illinois Farm Families,'' in M. C. Hallberg, J. S. Findeis, and D. Lass (eds.), *Multiple Job Holding Among Farm Families in North America,* Iowa State University Press, Ames, 1991, Chapter 9.

Talbot, R. B. and D. F. Hadwiger: *The Policy Process in American Agriculture,* Chandler Publishing Co., San Francisco, 1968.

Taylor, C. Robert, John B. Penson, Jr., Edward G. Smith, and Ronald D. Knutson: ''Economic Impacts of Chemical Use Reduction on the South,'' *Southern Journal of Agricultural Economics,* 23:15–23 (1991).

Taylor, H. C. and A. D. Taylor: *The Story of Agricultural Economics in the United States, 1840–1932,* The State College Press, Ames, 1952.

Tweeten, Luther G.: *Farm Policy Analysis.* Westview Press, Boulder, CO, 1989.

U.S. Congress: *Agriculture, Rural Development, and the Use of Land,* Hearings, Subcommittee on Rural Development of the Committee on Agriculture and Forestry, 1974.

U.S. Department of Agriculture: *Economic Effects of Mandatory Production Controls,* by C. E. Young et al. U.S. Government Printing Office, Washington, DC, March 1989, 36 pp.

U.S. Department of Agriculture: *Federal Marketing Orders for Fruits, Vegetables, Nuts, and Specialty Crops,* by N. J. Powers. U.S. Government Printing Office, Washington, DC, March 1990.

U.S. Department of Agriculture: *Provisions of the Food, Agriculture, Conservation, and Trade Act of 1990.* Economic Research Service, Agriculture Information Bulletin No. 624, U.S. Government Printing Office, Washington, DC, 1991.

U.S. Department of Agriculture: *The Second RCA Appraisal,* Washington, DC, 1989.

van Ravenswaay, Eileen: ''The Food Industry Responds to Consumers' Pesticide Fears,'' *National Food Review,* 12(3), pp. 17–20 (1989).

Wallace, T. D.: ''Measures of Social Costs of Agricultural Programs,'' *Journal of Farm Economics,* 44:581–597 (1962).

Wharton, Clifton: ''Reflections on Poverty,'' *American Journal of Agricultural Economics,* 72: 1131–1138 (1990).

Wilcox, W. W.: *The Farmer in the Second World War,* Iowa State College Press, Ames, IA, 1947.

Young, Douglas L., David J. Walker, and Paul L. Kanjo: "Cost Effectiveness and Equity Aspects of Soil Conservation Programs in a Highly Erodible Region," *American Journal of Agricultural Economics,* 73: 1053–1062 (1991).

NAME INDEX

(Does not include citations from the Bibliography.)
Page numbers followed by ***n.*** indicate footnotes.

SUBJECT INDEX

Page numbers followed by *n.* indicate footnotes.